The Golden Age

This volume is dedicated to
the memory of
Sidney Pollard (1925–1998),
a scholar of breadth and depth

The Golden Age

Essays in British Social and Economic History, 1850–1870

Edited by
Ian Inkster

with
Colin Griffin, Jeff Hill
and Judith Rowbotham

Ashgate
Aldershot • Brookfield USA • Singapore • Sydney

© 2000 Ian Inkster, Colin Griffith, Jeff Hill, Judith Rowbotham

All rights reserved. No part of this publication may be reproduced, stored in a retrieval system, or transmitted in any form or by any means, electronic, mechanical, photocopied, recorded, or otherwise without the prior permission of the publisher.

The editors have asserted their right under the Copyright, Designs and Patents Act, 1988, to be identified as the editors of this work.

Published by

Ashgate Publishing Ltd
Gower House, Croft Road,
Aldershot, Hampshire GU11 3HR
England

Ashgate Publishing Company
131 Main Street
Burlington, VT 05401–5600
USA

Ashgate website: http://www.ashgate.com

ISBN 0 7546 0114 5

British Library Cataloguing-in-Publication Data
The golden age: essays in British social and economic history, 1850–1870
   1. Great Britain — Economic conditions — 19th century 4. Great Britain — Social conditions — 19th century
   I. Inkster, Ian
   941'.081

US Library of Congress Cataloging-in-Publication Data
The golden age: essays in British social and economic history, 1850–1870 / edited by Ian Inkster ... [et al.; contributors, Ian Inkster ... et al.].
   p. cm.
Selected essays from a series of regular fortnightly seminars of the Nottingham Trent History Workshop at the Nottingham Trent University to research Golden Age patenting during the academic years of 1997–1999.
Includes bibliographical references.
   1. Great Britain — Social conditions — 19th century. 2. Great Britain — Economic conditions — 19th century. 3. Industries — Great Britain — History — 19th century.
   4. Great Britain — History — Victoria, 1837–1901. I. Inkster, Ian. II. Nottingham Trent History Workshop. III. Nottingham Trent University.
306'.0941—dc21   00-059428

This volume is printed on acid-free paper in Great Britain by MPG Books Ltd. Bodmin

Contents

List of Tables and Charts	*vii*
Notes on Contributors	*ix*
Events of the Golden Age	*xiii*
Preface and Acknowledgements	*xix*

1. Introduction: A Lustrous Age? — 1
   Ian Inkster
2. 'Nor all that Glisters ...': The Not So Golden Age — 9
   Harold Perkin

Part I Industry

Introduction to Part I: Industry — 29
Colin Griffin

3. Coalmining in Mid-Victorian Britain: A Golden Age Revisited? — 32
   Colin Griffin
4. A Golden Age of Agriculture? — 46
   Stephen Caunce
5. The Cotton Industry in the 1850s and 1860s: Decades of Contrast — 61
   Geoff Timmins
6. The Golden Age of Electricity — 75
   Gillian Cookson

Part II Technology

Introduction to Part II: Technology — 89
Ian Inkster

7. Michael Faraday and Lighthouses — 92
   Frank A.J.L. James
8. Lies, Damned Lies and Declinism: Lyon Playfair, the Paris 1867 Exhibition and Contested Rhetorics of Scientific Education and Industrial Performance — 105
   Graeme Gooday
9. Machinofacture and Technical Change: The Patent Evidence — 121
   Ian Inkster

Part III Social Institutions

Introduction to Part III: Social Institutions — 143
Jeff Hill

10 'Why Should Working Men Visit the Exhibition?': Workers and the Great Exhibition and the Ethos of Industrialism — 146
Su Barton

11 Estimating a Public Sphere: Intellectual and Technical Associations at the Time of the Great Exhibition — 164
Vicky Brown and Ian Inkster

12 'Golden Age' and 'Better Days': Narratives of Industrialism in the Cotton Trade of North-East Lancashire, 1860s to 1920s — 175
Jeff Hill

13 Popular Culture and the 'Golden Age': The Church of England and Hiring Fairs in the East Riding of Yorkshire c. 1850–1875 — 184
Gary Moses

14 In Defence of Respectability: Financial Crime, the 'High Art' Criminal and the Language of the Courtroom 1850–1880 — 199
Sarah Wilson

Part IV Gender

Introduction to Part IV: Gender — 219
Judith Rowbotham

15 'Physically a splendid race' or 'hardened and brutalised by unsuitable toil'?: Unravelling the Position of Women Workers in Rural England during the Golden Age of Agriculture — 225
Nicola Verdon

16 The Respectability Imperative: A Golden Rule in Cases of Sexual Assault? — 237
Kim Stevenson

17 Keep the 'Whoam' Fires Burning: Domestic Yearnings in Lancashire Dialect Poetry — 249
Catriona M. Parratt

18 'All our Past Proclaims our Future': Popular Biography and Masculine Identity during the Golden Age, 1850–1870 — 262
Judith Rowbotham

Index — 276

List of Tables and Charts

Tables

3.1	Output of the Principal Coal-Producing Countries, 1850–70	33
3.2	Death Rate in Mines from Different Causes, 1851–71	40
4.1	Comparative Corn and Wheat Acreage, 1871–1901, Selected English Counties	50
4.2	Paid Male Farmworkers, Selected Counties	51
4.3	Employment in Agriculture, 1871 Census Regions	57
5.1	Volume Retained Raw Cotton Imports, 1850–59	63
5.2	Numbers Employed in the Lancashire Cotton Industry, 1851 and 1871	66
5.3	Spinning and Weaving Capacity in the UK Cotton Industry, 1850–70	67
5.4	Export of Cotton Goods from Britain 1850–70	72
9.1	Regional Distribution of Patenting, 1855–70	122
9.2	Patentees per 1,000 Population, 1855–70	124
9.3	Urban Ranking of British Patents, 1700–1881	124
9.4	Numbers of Patents per City or Town and Ranking Per Capita, 1855–70	126
9.5	Patentees in Yorkshire, Lancashire and Warwickshire	127
9.6	References to Patent Specifications in Manchester, 1857–60	129
9.7	Occupations of Patentees in Britain, 1855–70	129
9.8	Occupations of Patentees, 1855–70: Regional Distributions	130
9.9	Partnerships, 1855–70: (a) Number of Patents; (b) Partnerships Involving Engineers	132
9.10	Patents Involving a Patent Agent, 1855–70	133
9.11	Foreign Patents in Britain, 1855–70: (a) Direct and by Communication to a Patent Agent; (b) As Percentage of Total Foreign Patents; (c) Percentage of Foreign Patents Direct and Communicated	134
9.12	Occupations of Patentees in Holborn, London, 1855–70	134

9.13	UK Patent System, 1853–77: Applications and Elite Patents	136
9.14	The Belgian Patent System 1860, 1865, 1870: Patents Granted	137
11.1	Membership of Institutions, 1851 per 1,000 of Population	165
11.2	Size of Associations and Female Membership	167
11.3	Urbanism and Female Membership	167
11.4	Steam Intellect by Size of Membership and Library Books per Member	168
11.5	Extension of the Census: 'Steam Intellect' in Liverpool	169
15.1	Female Labourers on Flitcham Hall Farm, 1851	233

Charts

9.1	Direct Patentees in Britain, 1855–70	122
9.2	Lancashire Patentees per 1,000 Population	125
9.3	Yorkshire Patentees per 1,000 Population	125
9.4	Occupations of Patentees in Manchester 1855–70	130
11.1	Regional Distribution of Institutional Membership, 1851	173
11.2	Male and Female Membership of Institutions in Per Capita Terms, 1851	173
11.3	Distribution of Institutions by Cost	174
11.4	Female Membership and Books per Member by Size of Steam Intellect Institution, 1851	174

Notes on Contributors

Su Barton is an independent writer and researcher who formerly taught at North Warwickshire and Hinckley College. She has recently completed her PhD at De Montfort University, Leicester, on working-class tourism in the nineteenth and twentieth centuries, and has written (with Rhianydd Murray) *Twisted Yarns: the Story of the Hosiery Industry in Hinckley*.

Vicky Brown is a full-time research assistant and doctoral student in History at the Nottingham Trent University. Her main interests are in urban culture and gender in nineteenth-century England, and in theories of urban association.

Stephen Caunce is Senior Lecturer in History in the Department of Historical and Critical Studies at the University of Central Lancashire in Preston. Recent work includes 'Not Sprung From Princes: the Nature of Middling Society in Eighteenth-century West Yorkshire', in D. Nicholls (ed.), *The Making of The British Middle-class? Studies in Regional and Cultural History Since 1750* (1999); 'Communities in Economic Development: Lessons from the Past', in G. Haughton (ed.), *Community Economic Development* (1999); and articles in *Agricultural History Review*, *Business History* and the *Journal of Regional and Local Studies*.

Gillian Cookson is County Editor for the Victoria County History of Durham, based in the History Department of the University of Durham. Recent work includes *A Victorian Scientist and Engineer: Fleeming Jenkin and the Birth of Electrical Engineering*, with C.A. Hempstead (2000).

Graeme Gooday is Lecturer in social history of science at the University of Leeds. Interests centre on the historical sociology of experimental physics and electrical engineering, and his recent work has concerned Victorian experimental science and instrumentation, precision measurement, laboratories and the work and environs of Michael Faraday, William Ayrton and T.H. Huxley.

Colin Griffin is Reader in Economic and Social History at Nottingham Trent University. Recent work includes editing and contributing an essay for the twentieth-century section of John Beckett et al., *The Centenary History of Nottingham* (1997) and 'Not Just a Case of Baths, Canteens and Rehabilitation Centres: The Second World War and the Recreational Provision of the Miners' Welfare Commission in Coalmining Communities', in *Millions Like Us? British Culture in the Second World War* (1999).

Jeff Hill is Associate Head of the Department of International Studies at Nottingham Trent University. His research interests are in popular politics and sport in twentieth-century Britain, on which he has published various articles, and recently *Nelson: Politics, Economy, Community* (1997) and with Nick Hayes *Millions Like Us? British Culture in the Second World War* (1999).

Ian Inkster is Research Professor of International History at Nottingham Trent University and Visiting Professor of European History at the Nanhua University, Peoples' Republic of China. Recent books include: *Scientific Culture and Urbanisation in Industrialising Britain* (1997); *Technology and Industrialisation. Historical Case Studies and International Perspectives* (1998); (with Maureen Bryson) *Industrial Man: The Life and Works of Charles Sylvester* (1999); *Japanese Industrialisation 1603–2000. Historical and Cultural Perspectives*, (2000); and, with Fumihiko Satofuka, *Culture and Technology in Modern Japan* (2000).

Frank James is Reader in History of Science at the Royal Institution of Great Britain, London and was co-editor of *History of Technology*. Recent work includes *The Correspondence of Michael Faraday*, 3 vols and continuing (1991–), and (with G. Cantor and D. Gooding), *Faraday* (1991, 2nd edn published as *Michael Faraday*, 1996), and scholarly interests centre on the life and work of Michael Faraday and the wider scientific and technical milieu of the nineteenth century.

Gary Moses is Senior Lecturer in History at Nottingham Trent University. His research interests are in labour relations and popular culture in English rural society in the nineteenth century, on which he has written recent articles in *Rural History* and *Agricultural History Review*.

Catriona Parratt is Assistant Professor of Cultural Studies in the Department of Sport, Leisure and Health at the University of Iowa, where she teaches the history of sport and of gender. Her recent publications are on Victorian woman and sport, and include papers in *Journal of Sports History*, *Sport History Review* and *International Journal of the History of Sport*.

Harold Perkin, Founder and Life-President of the Social History Society, is Emeritus Professor of History and Higher Education, Northwestern University, and was Professor of Social History at the University of Lancaster for many years. He was editor of the well-known series of books *Studies in Social History*, and is one of Britain's best-known social historians and author of the now classic text, *The Origins of Modern English Society, 1780–1880* (1969).

Judith Rowbotham is Senior Lecturer in History at Nottingham Trent University; Secretary, Social History Society and Joint Director of the History and Law Interdisciplinary project based there (Behaving Badly, Socially Visible Crime and Legal Responses). Recent publications include '"Soldiers of Christ"? Images of Female Missionaries in the Late Nineteenth Century: Issues of Heroism and Martyrdom', *Gender and History*, 12:1 (2000); '"Hear an Indian Sister's Plea", Reporting the Work of British Women Missionaries c. 1870–c. 1914', *Women's Studies International Forum* (1998, summer); 'Only when Drunk': The Stereotyping of Violence in England, c. 1850–1900 in Shani D'Cruz (ed.) *Everyday Violence in Britain c. 1850–c. 1950: Gender and Class* (Longman's, 2000).

Kim Stevenson is Senior Lecturer in Law in the Department of Academic Legal Studies, Nottingham Trent University and Joint Director of the History and Law Interdisciplinary project based there. Recent publications include 'Ingenuities of the Female Mind: Legal and Public Perceptions of Sexual Violence in Victorian England 1850–1890' in Shani D'Cruz (ed.) *Everyday Violence in Britain c. 1850–c. 1950: Gender and Class* (Longmans, 2000); 'Observations on the Law Relating to Sexual Offences: the Historic Scandal of Women's Silence', *4Web JCLI* (1999).

Geoff Timmins is Principal Lecturer in History and History Subject Co-ordinator at the University of Central Lancashire. Recent books include *The Last Shift: the Decline of Handloom Weaving in Nineteenth Century Lancashire* (1993), *Four Centuries of Lancashire Cotton* (1996), and *Made in Lancashire: a History of Regional Industrialisation* (1998).

Nicola Verdon is Lecturer in History at Harlaxton College, British campus of the University of Evansville. Her 1999 PhD thesis was on 'Changing patterns of female employment in rural England, c. 1790–1890', and her conference papers include 'A Tale of Two Counties: Contrasting the fortunes of women employed in agriculture in Norfolk and East Yorkshire in the 19th century', Economic History Society Conference, April 2000; 'A Concealed Presence: New perspectives on women's employment in the 19th century countryside', Agricultural History Society Conference, April 2000.

Sarah Wilson is Lecturer in the Department of Law at the University of Leeds teaching Business Law, Business History and also Equity and Trusts. Her PhD thesis on the historical foundations of modern white-collar crime, focusing on the origins of financial crime in modern British society, is in its final stages. She is also co-author of Blackstone's *Textbook on Trusts*.

Events of the Golden Age

1847
July election: Liberals 338, Conservatives 227, Peelites 91. Factory Act, ten hour day maximum for women and thirteen- to eighteen-year-old children but 'relay system' survived. Poor Law Act made administration more responsible to parliamentary inspection.

1848
Public Health Act to establish local boards. Fall of the French monarchy, the 'geographical expression' Italy revolts against Austria.

1849
The opening of Liverpool's Albert Dock inspires the prince to promote a great exhibition of industry and maritime strength. Joseph Paxton builds flat-roofed Victoria Regia Lily House with glass curtain wall hung from girders, features to be incorporated in Crystal Palace in 1850.

1851
The Great Exhibition of the Works of All Nations in Hyde Park in the 19 acre Crystal Palace opened 1 May. During its six months of opening some 6.2 million people viewed over 14,000 exhibits of craft and industry. Foundation of Amalgamated Society of Engineers, first modern trade union to survive the period. Submarine telegraph links London and Paris and Baron von Reuter opens telegraphy agency in London. Paris *coup d'état* of December. Census shows that agricultural workers and domestic servants composed by far the largest group of workers, but with engine and machine makers at some 48,000 civil servants at 31,000. Total population England and Wales nearly 18 million, Scotland 2.8 million, Ireland 6.5 million (and falling with lagged emigration effects of the famine). Collapse of Whig government of Lord John Russell, leading to Aberdeen's coalition government and the increased influence of Palmerston. July election: Liberals 310, Conservatives 299, Peelites 45. Forty Dissenters returned to Parliament, when religious census showed that nearly half of active worshipers were Dissenters with particular strength in Wales, the West Riding, Staffordshire and Cornwall. Outdoor Relief Regulation Order: outdoor relief possible for able-bodied males in specific instances. Publication of Harriet Beecher Stowe's *Uncle Tom's Cabin.*

1853

Employment of Children in Factories Act caught up with younger eight to thirteen age group anomaly and established standard day. Vaccination compulsory for all infants. London exhibition of George Scheutz's working copy of Charles Babbage's difference (calculating) engine.

1854

Religious tests abolished at Oxford and Cambridge, extended in 1856, but only in 1871 that membership of university government allowed to Dissenters. George Boole publishes first form of symbolic logic in his *Laws of Thought*; perforated postage stamps introduced; the American Commodore Perry forces Japan to trade with the United States.

1854–56

Anglo-French Alliance March 1854 to aid Turkey, Triple Alliance December 1854, Austria joins; Crimean War, September 1855, fall of Sebastopol, March 1856 Paris Peace Treaty – Turkish independence, Black Sea neutrality, European Commission to control navigation of Danube.

1855

Collapse of coalition over mismanagement of Crimean War and Palmerston appointed Prime Minister heading a Conservative majority. Removal of penny stamp duty on newspapers, following removal of advertising taxes (1851) and presaging removal of excise duty on paper (1861), together meant speedy emergence of 1d newspapers, especially *The Daily Telegraph*, *The Morning Star*, *The Manchester Guardian*, and *The Daily News*. Paris International Exhibition, including among prizes one to Scheutz for above Babbage machine.

1856

County and Borough Police Act. By great majority of 376:48, MPs vote against the Sunday opening of Museums. Henry Bessemer introduces his process for cheap steel based on 1855 patent, blasting air into molten iron and coke to remove carbon, and incorporates Robert Mushet's 1856 patent for removing sulphur impurities, for which he pays the latter an annual pension for life of £300; Henry Perkin synthesizes first artificial (aniline) dye, mauve, heralding an alternative label for these years, 'the mauve age'; Friedrich Siemens developed the regenerative furnace as forerunner of open-hearth steel process.

1857

April election: Liberals 367, Conservatives 260, Peelites 27.

1856–60
The Arrow War (Second Opium War), the French and British defeating China in a series of chaotic skirmishes, and October 1860 Treaty of Peking allowed trade in Yangtze, eleven additional open ports, importation of opium legalized.

1857–58
Indian Mutiny principally against East India Company's Bengal army and against background of surge of railway, telegraph and school building. By India Act, East India Company lost control of India to the Crown.

1858
Local Government Act and Public Health Act whereby local authorities now empowered to establish their own sanitary authorities free from central control. The banker L.N. Rothschild finally allowed to sit in Parliament as MP for the City of London as a consequence of the Oath of Abjuration Bill. The Year of Communications, Foreland lighthouse first with electric-powered arc lights, laying of first (unsuccessful) transatlantic telegraph cable, Cromwell Varley improves pneumatic system for sending telegrams for Stock Exchange, balloon Nadir takes first aerial photograph, I.K. Brunel launches the *Great Eastern* driven by paddles and propeller. Tin-opener finally invented.

1859
May election: Liberals 325, Conservatives 306, Peelites 23. W.J. Rankine publishes *Manual of the Steam Engine*; Karl Marx publishes *The Critique of Political Economy*; George Eliot (Mary Anne Evans) publishes *Adam Bede*; Charles Darwin published the *Origin of Species* after he and A.R. Wallace had offered preliminary evolutionary theories in the previous year in the *Journal of the Linnaean Society*. The geologist Charles Lyell remained famously cautious in 1863, but by 1870 the new British journal *Nature* judged Darwinism the new paradigm.

1860
Winter and Christmas of the great frost, closure of London docks, massive rise of destitute subjected to wholly useless workhouse and labour tests. Charles Wheatstone invents printing telegraph. First patent for 'cracking' petroleum issued in Boston, United States.

1861
Census. Population of England and Wales 21 million, 3 million in Scotland, in Ireland 5.5 million, with famine migration continuing to the United States, migrants to gold of Australia and Canada. Foundation of Massachusetts Institute of Technology; the Belgian Ernest Solvay improves manufacture of

sodium bicarbonate, regular production by 1863, Brown and Sharpe develop universal milling machine. Italian unification with Victor Emmanuel II as king.

1861–63
American Civil War between Confederate and Union forces at a time when some 4 million blacks were held in slavedom. In May 1861 Britain declared official neutrality approximately – 80 per cent of Britain's raw cotton came from America, upon which was in some manner dependent the employment of some half a million workers. 1862 Cotton Famine hits industry hard. In America over 600,000 soldiers and 50,000 civilians were killed.

1861–70
Rise of Count Otto von Bismarck under William I of Prussia, the latter becoming German emperor in 1871. British incapacity and Palmerston's misreadings concerning Russian defeat of Polish rebellion and Danish defeat and occupation by Prussian-led German Confederation. *Realpolitik*.

1862
Revised Education Code in response to Newcastle Commission 1858c–61, grant sizes dependent on performance in the three Rs and instituted inspectorate. First practical application of electric generator in arc light of lighthouse in Straits of Dover, A.B. Nobel begins experiments with nitroglycerin, Alexander Parkes demonstrates uses of celluloid. Abraham Lincoln's Emancipation Proclamation of 23 July leads to freeing of remaining American slaves.

1863
The Frenchman Louis Pasteur introduces heating process to be known later as pasteurization; J.F.A. von Baeyer of Berlin develops the first barbiturate probably named after his girlfriend; first 'horseless carriage' using internal combustion engine by J.J.E. Lenoir; and coining of the word 'aviation'.

1865
Palmerston majority in general election (July: Liberals 370, Conservatives 288), his death in October, appointment of Lord Russell as Prime Minister and heightened expectations of a new Reform Bill to extend franchise and geography of representation. Defeat of this brings Derby as Prime Minister with Disraeli as his Chancellor and Gladstone in opposition. Disraeli leads the Conservatives in the Commons towards a Second Reform Bill, passed in 1867 (Scotland 1868). Women excluded, rural working class excluded, no secret ballot, forty-five new seats, electoral boundaries reformed. J.B. Lister introduces phenol as a disinfectant in surgery. Lewis Carroll wrote *Alice in*

Wonderland and Jules Verne of France published *From the Earth to the Moon*. First pay toilets introduced outside Royal Exchange in London on Joseph Bramah design. Gilbert Scott designed St Pancras station and Edward Whymper climbed the Matterhorn.

1866
Sanitary Act whereby local authorities compelled to remove nuisances to public health, these now embracing domestic property and authority to demolish slums. T.C. Allbutt develops clinical thermometer, Robert Whitehead the torpedo, William Budd demonstrates in Bristol that cholera epidemic might be halted by limiting water contamination. British Aeronautical Society founded. Kimberley diamonds uncovered.

1867
Factory Act Extension Act extended earlier legislation to non-textiles and stipulated minimum ten hours education weekly for child workers. Agricultural Gangs Act, prohibiting employment of women and children alongside men in field-gangs. Conference of London in May guarantees neutrality of Luxembourg. British North America Act establishes a federal Dominion of Canada. Karl Marx first volume of *Das Kapital*, William Rathbone publishes *Social Duties*. Remington Model 1 typewriter. The American C.S. Peirce recognizes that Boolean algebra (above) can describe switching circuits where ON/OFF represent TRUE/FALSE. United States barbed-wire patents.

1868
Disraeli succeeds to Conservative Prime Minister in February, followed in November by Gladstone as Prime Minister with a Liberal majority of 100, after a general election centred on the so-called Irish Question. John Bright as President of the Board of Trade and the beginning of an intense reform period. Composition of House of Commons: 280 county seats, 369 borough seats, 9 university seats. November election result: Liberals 387, Conservatives 271. Of 267 constituencies, 195 contested, the greatest contestation since 1832. Now in the Commons there were 63 Protestant Dissenters, 26 Catholics and 6 Jews, and for the first time a Nonconformist sat in Cabinet, John Bright at the Board of Trade. Abolition of flogging in the army, and of colonial transportation and of compulsory payment of church rates. Foundation of Trade Union Congress. James Clerk Maxwell specifies notion of 'feedback', the information sent back to a system to effect a correction or change; exhibition of the first blended margarine; Siemens brothers develop open-hearth process; Sir Francis Galton demonstrated bell curve distribution of mental abilities, and in 1869 published *Hereditary Genius*. Meiji Restoration in Japan initiated first case of non-Atlantic sustained industrialization.

1869
Successful struggle of Liberals to pass the Irish Church Bill, designed to disestablish the Church of Ireland. Cutty Sark built, Suez Canal opened, invention of chained back wheel bicycle.

1870
Land Act to protect tenants from unfair eviction, W.E. Forster's Elementary Education Act to begin a national system of primary schooling with remittance of fees in hands of school boards, latter elected by all ratepayers including women, and by Cowper-Temple Clause no specifically denominational religious education in public elementary schools. Edward Cardwell's overhaul of army abolishing commissions by purchase at a time when army and navy expenditures by government were over twice the expenditures on civil administration, and reforms of judiciary and universities. George Eliot's *Middlemarch* 1871–72. Paris declaration of the Third French Republic, Franco-Prussian war. Invention of chewing gum in US. Margaret Knight invents a machine which manufactures paper bags and is still in use today. Completion of Mont Cenis Railway Tunnel through the Alps.

1871
Repeal of 1851 Ecclesiastical Titles Act which had prevented Catholics from adopting titles already taken by Anglican clergy. Zenobe Gramme presents his dynamo before French Academy of Science, German chemist Hermann Sprengel shows explosive power of picric acid, American W. McGaffey constructs steam-powered vacuum cleaner. By this year, in Britain railway mileage reached over 13,000 miles, over twice that of 1852.

1872
(Secret) Ballot Act passed, which mobilized local Conservative organizations. Foundation of Agricultural Labourers' Union. England and Australia linked by submarine cable telegraphs. Siemens and Halske research laboratory founded in Germany. Jane Wells patents the Baby Jumper. First oil tanker built, and in France a steam-driven car reaches a speed of 25 miles per hour.

1873
Gladstone defeated in Commons over Dublin University Bill for sectarian reasons. Gladstone defeated by Conservatives in January 1874 election, main issue income tax. In France Hippolyte Fontaine demonstrates that generator can serve as a motor; first refrigeration plant for meat in Australia.

Preface and Acknowledgements

The present volume is the first result of a programme of historical research undertaken at Nottingham Trent University during the academic years 1997 to 1999. A series of the regular fortnightly seminars of the Nottingham Trent History Workshop were devoted to uncovering aspects of the Golden Age, and overall interpretative essays were requested from Harold Perkin and David Edgerton. These seminars, a programme of research on Golden Age patenting, and a one-day workshop of all contributors in June 1999 were financed by the internal research funds of NTU. We would therefore very much like to thank Richard Joyner as NTU's Dean of Research for supporting this program so readily and generously. Our editor John Smedley attended the workshop and offered helpful advice at all stages, and suggested the orienting value of a list of key events. Ian Inkster would particularly like to thank Judith Rowbotham for much early help and enthusiastic collaboration as well as for the continuing organizational skill which helped bring this stage of our project to a fruitful conclusion. The volume is dedicated to the work and spirit of Sidney Pollard, who was a brilliant contributor to the Workshop and an influence upon several of the perspectives offered, but who died quite suddenly on 22 November 1998. The editors and contributors would like to thank Janet Elkington for all the work in producing the scripts to the required standard.

Chapter 1

Introduction: A Lustrous Age?

Ian Inkster

In 1850 the Industrial Revolution came to an end. In 1851 the Great Exhibition illustrated to the whole world the supremacy of industrial England. For the next twenty years Britain reigned supreme. From around 1870 Britain began to decline. Britain is now a second-rate power with strong memories of its former glory.

Such first sentences may read like a passage from *Kings and Things* or *1066 And All That*, but they are not so very unrepresentative of either popular or academic judgement concerning England after its initial forging of the first industrial revolution. It is true that some writers, such as Wiener, seem to date the collapse of Britain as a drawn-out process beginning as early as 1850, whilst a smaller number of historians are sceptical about decline at any time much prior to 1914. Even more confusingly, academic economic historians tend to run the Industrial Revolution itself through to at least 1860.[1] Whichever of these somewhat contradictory and seemingly arbitrary positions is adopted, the years from around 1850 to around 1870 tend to be either denied the status of a coherent period, or are unproblematically assumed to be ones of success and dominance, the natural postscript to industrial revolution and technological superiority. By concentrating on central aspects of social and industrial change, the authors of the present volume of essays expose the underpinnings of supremacy, its unsung underside, its tarnished gold. Major themes cover industrial and technological change, social institutions and gender relations in a period during which industry and industrialism were equally celebrated and nurtured. Against this background it is difficult to argue for any sudden decline of energy, assets or institutions, nor for any significant move from an industrial society to one in which a hearty manufacturing was replaced by commerce and land, sensibility and artifice.

---

[1] Martin J. Weiner, *English Culture and the Decline of the Industrial Spirit 1850–1980* (Cambridge University Press, Cambridge, 1981); for strong counter-arguments to claims of meaningful decline or to the potency of faulty policy based on hostile culture, see the brilliant Sidney Pollard, *Britain's Prime and Britain's Decline. The British Economy 1870–1914* (Edward Arnold, London, 1989), especially ch. 5, where it is clearly concluded (p. 265) that there 'was no lack of commercial spirit in Britain; on the contrary, from the nobility downwards, all were keen to make money': R. Floud and D. McCloskey (eds) *The Economic History of Britain Since 1700*. vol. 1: 1700–1860 (2nd edn, 1994).

Of course, to label is to invite criticism. The 'Mid-Victorian Boom' or 'the Age of Equipoise' titles, do just that,[2] but for these particular twenty years we would judge that 'the Golden Age' is a term which might evoke new thoughts yet stall the damage done when, to steal the wonderful phrase coined recently by David Eastwood, 'titular hegemony forecloses interpretative debate'. Because these years have so often been located betwixt and between the supposedly epic conjunctions of industrial revolution on one hand and industrial decline on the other, or incorporated within such seemingly commanding long-terms as 'the age of improvement', they have not been foreclosed by labels so much as relegated to limbo by the successful applications of nomenclatures designed for adjacent conjunctures. We may generally affirm the notion that the years following the commercial depression of the early 1840s and the subsequent Peelite 'resolution of the crisis of early-Victorian Britain' laid something of the attitudinal and institutional foundations of a new beginning, a confirmation of 'Britain's future as an urban, manufacturing nation' based on hard-earned understandings of the frictions and the heat generated by both the machinery of industry and the partialities of good governance.[3] But if this was a Britain whose new confidence was based on a proven commercial supremacy, this was not because industrial culture now delivered its rewards more adequately, but because the social problems which suffused the era – as in Harold Perkin's due warnings – were now seen as residuals of the recent past rather than as harbingers of an imminent future. From Mill to Jevons,[4] these years witnessed the growth to certitude of modernist social, economic and political thinking, as well as innovations in literary form and the dissemination of information by urban association and the printed word. This was the time when British commercial and technological supremacy coalesced in the establishment of 'respectable society', a system of non-centralized adjustment to new industrial and urban processes and institutions. New concerns demanded new legal solutions to such problems as prostitution or juvenile crime or for reducing such inefficiencies and frictions as drunkenness at work, as part of a range of modern mechanisms devised to cope with a complexity of site and agency not yet reached by other industrializing systems.

---

[2] R. Church, *The Great Victorian Boom 1850–1873* (1975); W. Burns, *The Age of Equipoise. A Study of the Mid-Victorian Generation* (Allen and Unwin, London, 1964).

[3] The quotations of this paragraph are drawn from David Eastwood's witty and telling 'The Age of Uncertainty: Britain in the early Nineteenth Century', *Transactions of the Royal Historical Society*, 6th series, 8 (1998), pp. 91–115, quotations pp. 92, 114, 115.

[4] J.S. Mill (1806–73) published *Principles of Political Economy* in 1848, was employed by the East India Company until its dissolution in 1858 and was MP for Westminster 1865–68. W.S. Jevons (1835–82), Professor of Political Economy at University College London, published his *Theory of Political Economy* in 1871.

Nowhere was this any better seen than in London during the year of the Great Exhibition. Over six million visitors were recorded between May and October, a maximum of 28,000 entering the main building in one hour.[5] As Sue Barton shows below in her apposite account of the provincial organization of such metropolitan visits, much good work and preparation was involved. Many celebrations emphasized good order, civility and sobriety. Less attention was paid to the very substantial additional policing arrangements which had been put into place at an early stage.[6] In a year during which Benjamin Disraeli could preach of 'general prosperity and particular distress', in which riots might occur at the closure of a workhouse gate, in which the Unitarian MP for Oldham, William Johnson Fox was advocating education of the people as a superior instrument of 'police', so too Lord Brougham addressed the House of Lords to the effect that two million had attended the Exhibition without major mishap by early July, one quarter of whom were foreigners, 'the rest being, for the most part, the tradesmen and operatives of this metropolis, and those too, not of the lowest classes'.[7] Indeed, the Great Exhibition represented one outstanding venue wherein technicians mixed relatively freely with savants, and both moved easily from sites of elite culture to places of practice. 1851 saw a particular combination of technological display with social mixing and association in such large and public venues as Exeter Hall and the Whittinghton Club and Metropolitan Athenaeum (Arundel Street). Joseph Hume MP, the Scots radical politician and anti-corn law advocate, used both these metropolitan venues to address large audiences on franchise and tax reform, and as principal public platforms of the National Reform Association.[8] When, in 1852, the House of Commons debated the future utility of the Crystal Palace, ushering in 'one of the most dangerous and serious divisions that could agitate this country', Hume used meetings at Exeter Hall to press for a retention of the building and its conversion into a centre of technical education for the masses, a far cry from ancient universities or grand scientific associations, (although he was, it might be noted, an FRS).[9] Where conservative interests saw the Crystal Palace as an all too clear threat to a gentler, garden metropolis, an institution in which the people were 'trepanned, seduced, ensnared and humbugged', Hume and the London technicians argued

---

[5] J. Tallis, *History and Description of the Great Exhibition*, 3 vols (London, 1852), vol. 3, pp. 922–3.

[6] *Hansard's Parliamentary Debates*, 3 series, 185, 114 (February–March) House of Commons, 10 February, pp. 278–9.

[7] *Hansard*, 1851, 118 (July–August), House of Lords, 11 July, pp. 505–8. Fox supported Joseph Hume in Parliament and in 1850 had introduced a Compulsory Education Bill.

[8] *The Times*, 1 January, 1851, p. 1; 28 January, p. 8.; 4 February, p. 4.

[9] *Hansard*, 1852, 120 (March–April), 29 April, p. 1352.

with ebullience for the maintenance of the new site, symbol of both industrial progress and social change.[10]

For the great number of the skilled artizans and tradesmen whose journals, associations and civic activities were growing most healthily throughout these years, this was the period when their energies, knowledge, expertise and competitiveness had resulted in a successful challenge to big capital. Thus, the artizans took Professor Thorold Rogers of Oxford to task for his advocation of a removal of special protection for technical innovation, on the grounds that it was precisely the inventive technicians who effectively fought the great capitalist manufacturers, who utilized their patents to 'break down the monopoly of capital by a short-lived monopoly, which immediately stimulates others to invent other improvements'. If the preceding years had been ones of 'industrial capital', those following the 1851 exhibition were to be the years of 'artizan intellect'.[11] This was not a position so very different from that of the liberal intelligentsia, many of whom now, and perhaps for the first time in such numbers, believed that good governance had at its basis 'a consistent union in the nation'. As elaborated by Walter Bagehot, such a union was to be formed of a newly-increased intelligence and responsibility, which would yield a harmony of opinion in the midst of great variance of 'social circumstances and social habits'. The earlier 'democratic theory of government' was natural to a phase of great industrial and urban change, the years 1832 to 1860, but could only be superseded in an 'age of political intellect' by a wider franchise based on a nice combination of property and political intelligence.[12]

Perhaps, then, a novel attribute of the Golden Age was the impact of artizanal culture on middle-class liberalism through the spread of machinofacture and its desiderata of efficiency, good order, exact measurement and reproducibility. Manchester's civic centre was not merely formed of merchant capitalism, but of the artizanal business and comings and goings of St Ann's Square, the Victoria Buildings, Manchester Chambers and Deansgate. It was in such venues that Bagehot's 'civilities' were spread and verified. The conversion of the intellectual radicalism of the 1820s and 1830s into the municipal communalism of the Golden Age took place within just such specific sites, guided by common interests and expertise, fired by Nonconformism, embodied in both T.H. Huxley and, later, Joseph Chamberlain. In his famous 'Duties of the State' lecture before members of the Birmingham and Midland Institute in 1871, Huxley answered the *laissez-faire*

---

[10] Ibid., pp. 1354, 1336.
[11] *The Scientific Review and Journal of the Inventors' Society*, 1:8 (October 1865), p. 121.
[12] Walter Bagehot, *The History of the Unreformed Parliament and its Lessons. An Essay Reprinted from the National Review* (Chapman and Hall, London, 1860), p. 42 and *Parliamentary Reform, An Essay Reprinted with Considerable Additions from the National Review* (Chapman and Hall, London, 1859), p. 5.

of his friend Herbert Spencer through a focus on the civic role of technique, expertise and formal knowledge systems, and the need for expert interventions at the local level in such mundane areas as vaccination, roadmending and sanitation programmes, as opposed to the violent alternatives of European contemporaries. This was the stuff of Walter Bagehot's 'unconscious imitation', by far 'the main force which moulds and fashions men in society'. Thus, Coleridge's earlier fears that the extended franchise would open the political system to the sway of 'fools and knaves' seemed, amongst the many spokesmen of the public sphere, to belong firmly in the past.

In this volume of essays we treat these years as a coherent era in the history of industrial Britain, as a Golden Age in which free trade and the gold standard at most times spelt British success, and in which social and political advances were accepted as the rewards due to an early industrial modernity. The starting point is Britain's position as the foremost economy exploiting global expansion, new ideas and institutional changes. In these years exports to India increased five-fold, to Australia fifteen-fold. A rapid rise in capital exports towards the £1 billion mark carried with it an empire of railway building, and income receipts from such investments rose from an average of around £12 million at the outset of the Golden Age to some £50 million at its end. But advance was not measured only in commercial terms. Britain's economic position was linked to significant developments in the character of her social institutions and relations. The essays in this volume go beyond a celebration of success into the construction of an interpretation of industrial Britain which emphasizes social explanation as well as social effects and which unravels linkages between institutions, industries and technologies. The Golden Age project was interdisciplinary, involving historians and lawyers at Nottingham Trent University and historians of science, technology, gender relations and social change from a variety of academic institutions in a series of seminars during 1997–99. The five parts of the volume are devoted to an overall picture of the nation, a study of individual industries, examination of technological change, analysis of social institutions, and of gender relations.

From the outset, Harold Perkin is insistent concerning the social lacunae and contradictions of these years, and provides several of the arguments which have induced the question mark of our title. In terms of competitive commercial and technological performance, riding on the back of the speedily emerging international economy and gold standard, Britain experienced the greatest success. For the great numbers of the commercial and professional middle classes, for many landowners and farmers, for those tens of thousands of literate engineers, mechanics and clerks who gained the material rewards due to their talents and energies, these were golden years. Higher incomes could be spent on a far greater range of products than in the earlier years of the century, and for many the printed word, sources of information, new forms of association and greater opportunities for gainful instruction were all obtainable at falling rather than rising prices. For these same groups, local political power was beginning to move into their hands from 1835, when the Municipal Reform Act introduced into boroughs rate-payer control and permitted the emergence to civic authority of a range of business, professional and industrial interests and Dissenting groups.

The four editorial introductions to each part of this volume clearly show that the industrial and commercial gains were substantial, were increasingly shared by new groups and localities, but were always at once tenuous and exclusive – a raw material famine might bring a locale or an industry to quick ruination, systematic legislation might exclude an entire gender from entry onto the paths of individual achievement. Thus, there was much furore concerning Nonconformist entry to the ancient universities, but little was heard there of the matriculation or graduation of women, an omission hardly rectified by the reluctant allowances of the London colleges or of Durham, or of the otherwise liberal, civic traditions of Scottish higher education. There is much in this volume to suggest that in this period and in the following years, the major problem of British industrialism was not so much any overarching 'anti-industrial' culture, but rather a persistence and increased subtlety in the processes and mechanisms whereby vital energies, skills and ideas were systematically excluded from crucial areas of industry, commerce and knowledge production.

Much has been made of the supposed anti-industrialism of the years following the Great Exhibition, and something more has been made of Victorian non-industrialism. Neither label can stick, especially if we at once note the 'pre-industrialism' of German bourgeois culture or the 'aristocratic aspirations' of the American business magnates, or the very similar arguments given for the supposed retardation of French industry. Although in comparison to European later developers and to the United States, British governments failed to construe policies directly designed to increase technological progress or industrial training, there is little in the way of convincing evidence which proves that if government had indeed accomplished such, then the rate of industrial or economic growth would have been higher prior to, say, the Great War of 1914–18. Dr Lyon Playfair's famous letter from the Paris World Exhibition merely represented an existing tendency towards emphasizing the potency of formal industrial training on the Continent, whilst neglecting the importance of apprenticeship and informal training systems in the British industrial economy.[13] Such statist policies might only have shifted scarce resources and energies into areas of less immediate profitability. It may well be

---

[13] 'Letter from Dr Lyon Playfair to the Rt Hon. Lord Taunton', *Journal of the Society of Arts*, 15 (7 June 1867), pp. 447–51.

the case that a reason for the reluctance of British governments to more directly dabble in the industrial economy of the Golden Age, lay in the influence of militarism, colonialism and snobbish disdain over the policy-making environment. But it is highly unlikely that such attitudes left much of a mark on the underlying culture of industrialism, a complex of institutions and attitudes lying well beyond the reaches of any romantic intellectualism, as depicted in parts 2 and 3 of the present volume. Rather than Wiener's image of a high culture of anti-industrialism powering down the mechanisms of the workshop of the world, we might suggest Karl Mannheim's, one in which 'intellectuals are more or less regarded as a foreign body in the nation'. In his famous essay on 'Youth in Modern Society', Mannheim went on to note how in Britain more than elsewhere the intellectuals 'are either somehow looked down upon, spiritually isolated, or not really taken seriously.'[14] Indeed, for Mannheim, it was just such a disparagement of the culture of the intelligentsia which was removing 'the main source of fermentation and dynamic imagination' from early twentieth-century British society.

Assertions concerning the essentially non-industrial character of Britain are more peculiar and more easily surmounted. If in the 1870s, the benchmark decade for the end of our period, the proportion of the employed labour force engaged in industry was only 42 per cent, no other nation on earth could boast anything higher, and the two main commercial competitors, the USA and Germany, had reached levels of industrial employment of around 24 per cent and 29 per cent respectively. Even one hundred years later, the British figure of 38 per cent remained comparable with those of the USA (31 per cent) and Germany (44 per cent), and was measurably above those of such nations as the Netherlands (31 per cent), France (35 per cent) and Japan (35 per cent).[15] Such features of the British system as an overvalued currency, large net emigration or the export of capital overseas, are not so much measures of non-industrialism as reflections of the attractions of Empire and overseas trade within the multiplicity of choices offered by the most sophisticated economic system in the world. And whilst it is true that the British economy lost its lead in efficiency to the USA by the turn of the century, it yet retained its lead over the rest of Europe until the 1960s.

We might conclude that the aftermath of the Golden Age was not any secular decline, but rather a furtherance of capitalist complexity. Such a complexity certainly embraced a movement towards a far more sophisticated and broad-ranging goal structure, as evidenced in the public clamours of a huge range of interest groups as well as the formal political parties. At a time when

---

[14] Karl Mannheim, *Diagnosis of Our Time. Wartime Essays of a Sociologist*, (Routledge and Kegan Paul, London, 1943), quote p. 42.

[15] Angus Maddison, *Phases of Capitalist Development*, (Oxford University Press, Oxford, 1982), table C5, p. 205.

late industralizers such as Germany, Russia, Japan or Italy were aiming to devise workable schemes of statist invertentionism around a goal of industrial modernization, Britain was moving beyond a Golden Age towards one of enhanced securities and increased civic purpose set within a massive and complex social matrix of site and agency.

Chapter 2

'Nor all that Glisters ...': The Not So Golden Age

Harold Perkin

>Not all that tempts the wand'ring eyes
> And heedless hearts, is lawful prize;
>Nor all, that glisters, gold.
>   (Thomas Gray,
>   'Ode on the Death of a Favourite Cat')

>All that Glitters is not Gold, or, The Factory Girl.
>   (Play at the Little Theatre,
>   North Shields, 1854)

The mid-Victorian age, as this volume witnesses, was known to contemporaries and historians as a Golden Age. Sir George Young, that Victorian manqué, said that the 1850s was the best time for a young man to be alive – he meant of course a young man of his own educated upper middle class.[1] It was a period of social peace and class harmony, of unprecedented prosperity, an 'age of equipoise' between the two great secular economic troughs of the nineteenth century, the 'hungry forties' (and hungrier twenties and thirties) and the Great Depression of 1874–96. The fear of revolution which had clouded British skies from 1815 to 1848, the mass meetings and riots from Spa Fields and Peterloo (1816 and 1819) through the Parliamentary Reform riots of 1830–32 to the Chartist 'plug plot' riots of 1842 and the aborted march from Kennington Common to the Duke of Wellington's guns at the Thames bridges in 1848, had all melted away. The Great Exhibition of 1851 ushered in what I have called elsewhere the mature, viable, class society of mid-Victorian England, a society in which class conflict was transformed from open warfare into peaceful negotiation and mutual acceptance.

All this is true and incontrovertible, but it masks a great paradox. When historians look for the causes of this unexpected harmony, they discover a strange contradiction. The exceptional prosperity is hard to find, still more to explain. National income and average incomes – rents, profits and wages –

---

[1] G.M. Young (ed.), *Early Victorian England, 1837–65* (Oxford University Press, Oxford, 1934), Introduction.

were still rising in real terms, as indeed they had been since the Industrial Revolution began, since that is the nature of industrial revolutions. But they were not growing so much faster than before or, for that matter, after. Gross National Product accelerated slightly, from about 2 per cent per annum in the early Victorian age to 2.4 per cent in 1845–75, before returning to just under 2 per cent again in the Great Depression of 1874–96. Real GNP per capita, given the growth of population, increased even less fast: from 1.61 per cent per annum to 1.85 per cent, before falling back to 1.07 per cent. Money incomes were rising, undoubtedly, but mild inflation reduced their real benefit. Real wages – what average wages would buy – stagnated, rising by only 5 per cent, in the 1850s, slightly more, by one eighth (12.4 per cent), in the 1860s, and did not accelerate until the early 1870s, by about one fifth (21 per cent by 1874), too late to affect the 'feel good' factor during most of the period.[2]

The middle-class apologists Sir Robert Giffen and Professor Leone Levi argued in the 1880s that the working classes got most of the increase in national income over the last fifty years, doubling their average income while the middle-class average rose by less than a third. But they cooked the books by comparing all taxpayers – whose numbers increased to take in the lower middle class, thus depressing the average – with their own arbitrary estimates of average wages, based on fifty-two weeks' fully-paid work with no unemployment or sickness. The true facts were different: upper- and middle-class incomes increased by 66 and 60 per cent while manual wages, allowing for unemployment, rose by 24 per cent, or 39 per cent if we allow for the movement from unskilled to more skilled occupations. So the 'great mid-Victorian boom' was much less prosperous for the three-quarters of the population who worked for wages than for the complacent middle class.[3] All that glitters was not gold for the factory girl of the North Shields play, still less for those even poorer than herself.

I argued thirty years ago, not very convincingly I now think, that mild inflation was the answer: rising prices, as long as they are mild enough not to exacerbate conflict, lead to rising money incomes for all parties. Landlords find it easy to raise rents, employers to increase profits, so that both more easily concede wage increases, and everyone feels happier, even though the gains are not so great as they seem. It lowers the friction, like pouring oil in the cogs of the machine. Conversely, falling prices as in the earlier and later periods, while increasing real incomes and living standards, put grit in the machinery and increase the friction. But more recent research has shown that prices were not rising consistently throughout the period but enjoyed a couple of short bursts,

---

[2] R.A. Church, *The Great Victorian Boom, 1850–75* (Macmillan, London, 1975), ch. 2.

[3] Harold Perkin, *The Origins of Modern English Society* (Routledge, London, 1969), pp. 413–16 for sources and analysis.

in 1853–56 and 1869–73, piled up on a short shelf or plateau in the long nineteenth-century decline.[4]

So the inflationary 'feel-good' factor operated, if at all, only at the beginning and towards the end of the period. That still leaves us with the paradox. Why did most mid-Victorians think that the environment they lived in – social, economic and political – was less hostile than before and that things were going their way? That the free-market capitalism which had produced and resulted from the Industrial Revolution was working satisfactorily and giving them the benefits it promised?

When we look beneath the complacency of the – or should we say some? – mid-Victorians we find the question even more puzzling. The Golden Age seems to have been, if not 'fool's gold' (iron pyrites), at best gold painted over baser metal. Like Horace Walpole's over-ambitious cat at Strawberry Hill, drowned in a Chinese vase in pursuit of goldfish, they mistook the glitter for the gold. If things were getting better for some Victorians, they were still notoriously bad for others. Poverty and the harsh New Poor Law, low wages and unemployment, wretched slums, ill health and short life-expectancy, foul air, water and sewerage, long working hours and bad conditions especially for women and children, industrial accidents without compensation, and a paucity of public amenities other than pubs and gin shops were, if no worse than before, not all that far off what we would now call Third World standards.

The cities, notably London, were awash with filth and horse manure requiring the services of scavengers and crossing sweepers, teeming with beggars and ragged, shoeless street urchins, and grimy street-traders of everything from fruit and vegetables to song sheets and dog's dung (for leather tanning). In the Strand and the Haymarket and similar streets from Brighton to Glasgow, prostitutes stood shoulder to shoulder soliciting business, and pickpockets – Fagin and the Artful Dodger were not figments of Dickens's imagination – plied their trade, while 'mugging' was a term invented by contemporaries for a particularly painful kind of robbery with violence by garrotting the victim. In slum areas 'offcomers' were not welcome: there the new policemen went in threes where, as the Punch cartoon put it, the local hooligans cried 'Here's a stranger – heave half a brick'.

Poverty and the Poor Law

'The poor ye have always with you' (Matthew, 26:11), and in the mid-Victorian period it was as true as ever. The Poor Law statistics are misleading, counting only the mean number of paupers relieved on 1 January and 1 July,

---

[4] Ibid., pp. 342–5; Church, *Great Victorian Boom*, p. 16.

which fluctuated between 800,000 and 1 million, but even this was misleading. Contemporaries estimated that between 2.5 and 3 million were relieved during the year, between 13 per cent and 19 per cent of the population – an underestimate of those in poverty since every effort was made to discourage claimants.[5] In the 1860s craftsmen earned 25s. to 30s a week (when in full work), farm labourers 12s. to 18s. Even with wives and children working and earning perhaps half as much again, labourers' families, at a time when bread alone for a family of five might cost 7s or 8s, must have been constantly in poverty.[6] In the 1890s at least a third of the population lived in poverty by Charles Booth and Seebohm Rowntree's standards, who estimated that some 31 per cent of Londoners and 28 per cent of the inhabitants of York were in poverty (though by a somewhat impressionistic standard).[7] Given the fairly steep rise in real wages and living standards in the last quarter of the century, things were certainly worse in the mid-Victorian age.

Charles Dickens, Henry Mayhew, Douglas Jerrold and Gustave Doré, and contemporary working-class memoirs, put rather emaciated flesh on these poor dry bones. What did poverty mean in real terms? It meant small ragged children begging on the streets or sweeping a path through the mire for a halfpenny; a five-year-old boy trying to sell a bunch of wilting watercress begged at Covent Garden; John Ward, a weaver in the Cotton Famine earning 7s 3d a week on three looms, and only 3s unemployment relief in the winter; 2000 boys, Charles Kingsley's 'water babies', still climbing chimneys and risking scrotal cancer for a few pence in 1867; children of five to ten years of age working at gang-labour on farms several hours' walk from home; Lucy Luck, a little parish orphan girl turned out of doors at 3 a.m. by her foster parents to become a domestic skivvy on 1s 6d a week and then an apprentice straw-plaiter working for food only; small children sent by their mothers to cajole their drunken fathers to leave the pub before spending all their wages; the drunkard's pregnant daughter throwing herself off London Bridge into the Thames in Cruikshank's cartoon; a boy mould-runner carrying heavy pots in and out of the fiery pottery kilns for 1s a week; unemployed workmen breaking stones in the workhouse yard for 1s. a day; a printer, William Adams, working for Ruskin's *English Republic* at Coniston in the Lake District until it failed in 1855, walking 274 miles to London and finding 'not an odd job anywhere' on the

---

[5] Geoffrey Best, *Mid-Victorian Britain, 1851–75* (Panther, London, 1973), p. 167, citing Aschrott and Preston-Thomas, *English Poor Law System* (London, 1888), p. 282, n. 1, and S. and B. Webb, *English Poor Law History* (Longmans, London, 1929), pp. 1041–2. See also Michael E. Rose, *The Relief of Poverty, 1834–1914* (Macmillan, London, 1972), p. 535.

[6] E.H. Hunt, *Regional Wage Variations in Britain, 1850–1914* (Clarendon Press, Oxford, 1973), pp. 70 and 82.

[7] Charles Booth, *Life and Labour of the People of London*, vol. 1 (Macmillan, London, 1893), pp. 131–2, 165; B.S. Rowntree, *Poverty: A Study of Town Life* (Longmans, London, 1901), pp. 111, 117–18.

way; even a strong and well-paid stone mason like Henry Broadhurst, later Secretary of the Trades Union Congress, having to tramp 1,200 miles in an unavailing search for work in the slump of 1858–59. Small wonder that James Ireson, a stonemason born in 1856 in Whittlesea near Oundle on 5s a day, talked of the trials and difficulties of what he called the 'Hungry Sixties'.[8]

The official poor, one in five or six of the population, were worse off still. Those, about a quarter of the total, in the 'Bastilles' of the Poor Law Unions, which were intended to offer conditions 'less eligible' than those of the lowest-paid labourer outside, were incarcerated like convicts and allowed only occasional walks in the fresh air. Women, old and young, were separated from their menfolk – deliberately, to stop them breeding more paupers as Malthus feared – along with their children, and fed chiefly on bread, potatoes and gruel with a little fat mutton or bacon once or twice a week.

The irony was that, despite some well-meant improvements, their situation was actually worse than the New Poor Law of 1834 intended. The workhouse test and less eligibility were only meant to apply to the 'able-bodied poor', the unemployed, to test their willingness to work at free-market wages. The 'impotent poor', those too old to work, widows and orphans, the sick and the mentally and physically handicapped, were supposed to be treated in separate institutions – segregated from contamination with the wilfully idle – with better food and conditions. Although the sick and the orphans began, in the 1860s, to be housed in infirmaries and 'district schools' (orphanages), the rest were placed in the general mixed workhouse.

Charles Dickens described one such 'little world of poverty' in 1860:

> It was inhabited by a population of some fifteen hundred or two thousand paupers, ranging from the infant newly born or not yet come into the pauper world, to the old man dying on his bed ... Groves of babies in arms; groves of mothers and other sick women in bed; groves of lunatics; jungles of men in stone-ved downstairs day-rooms, waiting for their dinners; longer and longer groves of old people in up-stairs Infirmary wards, wearing out life, God knows how ... In all these Long Walks of aged and infirm, some old people were bedridden, and had been for a long time; some were sitting on their beds half-naked; some dying in their beds; some out of bed, and sitting at a table near the fire. A sullen or lethargic indifference to what was asked, a blunted sensibility to everything but warmth and food, a moody absence of complaint as being

---

[8] Examples taken from Henry Mayhew, *London Labour and the London Poor*, 4 vols (Griffen, Bohn and Co., London, 1861–62); George Cruikshank, *The Drunkard's Children* (1848); and John Burnett's Collections of working-class memoirs: *Useful Toil* (Allen Lane, 1974), pp. 68–74, 84–85, 298–9, 316–17; *Destiny Obscure* (Penguin, Hammondsworth, 1984), pp. 8–88, and *Idle Hands* (Routledge, London, 1994), p. 96. See also the many illustrations of slums and the poor in Douglas Jerrold and Gustave Doré, *London* (London, 1872).

> of no use, a dogged silence and resentful desire to be left alone again, I thought were generally apparent.

Though the pauper nurses were kind, they could not remember the names of the many old men 'who died in that bed'.[9]

These were the casualties of the new industrial system, the unskilful artisan, the sick labourer, the widows and orphans 'left to struggle for life and death', that Herbert Spencer, the great Victorian protagonist of social Darwinism and the free market, thought a price worth paying for 'the far-seeing benevolence' of the universal law of nature, backed up by Christianity, which sacrificed the individual for the benefit of society and the human race.[10]

The New Poor Law had in fact been enacted to deal with the wrong problem. It was intended to end the underemployment of farm workers of the South of England and their large, improvident families, living on starvation wages supplemented by out-relief. It was totally irrelevant to the problems of the growing industrial population, who suffered from the upswings and downswings of the trade cycle. The workhouse during booms was too large for them, during slumps far too small. It needed to be elastic-sided to meet the ever-changing need. Hence the reform forced on the system by the pressure of facts in the 1843 Act, to allow the unemployed during slumps to be offered the 'labour test', work in the workhouse yard in return for a pittance just sufficient to buy bread. The unemployed during the frequent depressions – apart from the Cotton Famine of 1861-64, there were deep slumps in 1858 and 1866, the first of which has been described as 'possibly the most profound depression of the century'[11] – were therefore rarely housed in the workhouse, which had in effect become a poorhouse, a dumping ground for the impotent poor.

Most able-bodied workers had a horror of the workhouse and either joined the many friendly societies and trade unions which offered sick and unemployment benefit, or used up their savings, pawned their possessions, or lived off their relatives until they found another job. Many, like Henry Broadhurst and William Adams, went on the tramp from town to town, looking for work, often with disastrous effects on their health, boots and clothes, and their precious savings. No doubt things were no worse than before, and perhaps a little better in good times when work and wages were available, but they were not golden for the poor, or even for most of the working class who met the contingencies of life with almost no recourse except the grim and grudging Poor Law or the strictly rationed handouts of the Charity Organization Society.

---

[9] Charles Dickens, 'A Walk in the Workhouse', *Household Words*, 1860; reprinted in *The Uncommercial Traveller* (Oxford University Press, Oxford, 1958).
[10] Herbert Spencer, *Social Statistics* (Williams and Norgate, London, 1851), pp. 322–3.
[11] Church, *Great Victorian Boom*, p. 76.

Slums, Disease, and Working Conditions

The Victorian period was, in the title of a contemporary book, an age of great cities. The author, the Congregational Reverend Robert Vaughan in 1843 saw them as the heralds of a glorious future, the source of culture and civilization, the home of middle-class moral leadership which would purge the corruption and selfishness of the feudal aristocracy and tame the fecklessness and violence of the working masses.[12] Two foreign visitors the very next year saw it differently. Henri Faucher saw both the centres of industrialism and the old commercial cities as seats of corruption.[13] Frederick Engels in 1844 looked out from Cathedral Bridge in the heart of Manchester over the River Irk and saw:

> piles of rubbish, the refuse, filth, and decaying matter of the courts on the steep left bank of the river. Here one house is packed very closely upon another ... All of them are blackened with smoke, crumbling, old, with bone window panes and window frames. On the right, low-lying bank stands a long row of houses and factories ... The background here is formed by the paupers' cemetery and the stations of the railways to Liverpool and Leeds. Behind there is the workhouse, Manchester's 'Poor Law Bastille'. It is built on a hill, like a citadel, and behind its high walls and battlements looks down threateningly upon the working-class quarter that lies below.[14]

Not all working-class housing in mid-Victorian towns and cities was old and crumbling. Much of it was brand-new, solid brick, and though monotonous a lot healthier than the damp country cottages and Irish turf huts from which many of Manchester's industrial workers came. But a great deal could only be described as slums, disease-ridden and verminous, and reinforcing the poverty and squalor in which many, perhaps most, of the unskilled working class lived.

Henry James first arrived in London in 1868 on 'a wet, black Sunday, about the first of March' and found it 'hideous, vicious, cruel, and above all overwhelming'. He drove in a greasy four-wheel cab from Euston, after dark, to Morley's Hotel in Trafalgar Square: 'The low black houses were as inanimate as so many rows of coal scuttles, save where at frequent corners, from a gin shop, there was a flare of light more brutal than the darkness'.[15] If this was the centre of London's 'charming immensity', imagine what the grubby and grimy outskirts, were like still more the squalid backstreets etched by Gustave Dore in 1872, with their crowds of wretched barefoot children, ragged women, shabby

---

[12] Robert Vaughan, *An Age of Great Cities* (Jackson and Walford, London, 1843).
[13] Leon Faucher, *Manchester in 1844* (Manchester: Heywood, Manchester, 1845), pp. 90–91.
[14] Frederick Engels, *The Condition of the Working Class in England in 1844* (Oxford University Press, Oxford, 1958)
[15] Henry James, *English Hours* (Heinemann, London, 1960), pp. 1–6.

street traders, and sleazy prostitutes; not to mention the slums of the old provincial cities and new industrial towns.

With squalor went disease, the ever-present typhus and typhoid, smallpox and tuberculosis, the 'white plague', which afflicted so many Victorian celebrities like the Brontes and John Stuart Mill, but was endemic in the slums, along with the other diseases of poverty. Most fearful was cholera, which repeatedly hit the cities from the 1830s to the 1870s. In 1866 Dr John Snow discovered the 'cholera morbus', the waterborne source of the disease, from the famous pump in Conduit Street in the West End fed from the Thames, but it would be years before the epidemic could be controlled. It could nearly wipe out whole families. In an earlier outbreak William Webb, son of the Parish Clerk of East Kennet, Wiltshire, was the only child out of five to survive, and only because a childless neighbour took him home so that 'Willie should not die'.[16] Child mortality was horrendous. A gravestone in a Yorkshire village shows eleven children in the same family who died, most under one but all under five years old, between 1838 and 1861.[17] The Victorian novelists' sentimental concern with children, like *Eric, or Little by Little*, 'making a good death' was firmly based in all-too-common reality.

Working conditions were also menacing to health, and even lethal at times. The Ten Hours' Act of 1847 applied only to textile factories, and was really a twelve hours' Act with one and a half hours for meals, for women and children under nine until 1874. Other factory industries came under regulation only in the later 1860s. The factory inspectors' reports record small children as well as adults being mangled by machinery and then turned away without any compensation. Walter Bagehot, editor of *The Economist* and guru of the free-marketeers, complained in 1858 of the 'sentimental radicalism' of Dickens and one of his contributors, Henry Morley. Their offence was to expose the fearful accidents still happening in mills despite the Factory Act of 1844 requiring the fencing of machinery. They had reported more than 10,000 victims, including 100 killed, in the three years 1851–54. A factory boy carelessly looking out of the window was caught by the arm and whirled to his death by the shafting. Another, sent up to dust the ceiling before whitewashing, was caught by his headcloth and hurled to his death. A man trying to fit the driving strap to the drum on the shaft was smashed to pieces.[18] Bagehot blamed the victims for their 'negligence' and excused the factory owners' evasion of the law. Mines and railways were even more dangerous: about 1,000 miners were killed each year, and in 1875, 767 railway workers were killed and 2,815 injured.[19]

---

[16] Burnett, *Destiny Obscure*, p. 80.
[17] Joan Perkin, *Victorian Women* (John Murray, London, 1993), illustration 1.
[18] Walter Bagehot, *The Economist*, 1858; Henry Morley, *Household Words*, 22 April 1854.
[19] Figures cited by Best, *Mid-Victorian Britain*, p. 139.

Outside the factories and mines working conditions were often no better. Landed aristocrats like Lord Ashley, the saint of factory reform, blamed the industrialists for their callousness, but his own father, the Earl of Shaftesbury, paid much lower wages and housed his workers in rural slums worse than in the towns. Gangs of tiny farm labourers led by brutal taskmasters trudged through the rain and cold to weed and pull turnips on faraway fields. Little domestic skivvies toiled twelve hours and more a day for tiny wages and their food, for lower middle-class families hardly more prosperous than themselves, often harassed by the master of the house or his sons into unmarried pregnancy and prostitution. The unregulated domestic workers, in weaving, straw-plaiting, dressmaking, box-making, nail-and-chain making, the later 'sweated industries', found themselves working twelve and more hours a day for starvation wages. The Victorian gospel of work as the salvation of society held few charms for most manual workers.

Drink, Crime and Prostitution

Drink, it was said at the time, was the quickest way out of Manchester. Gin and beer were the chief solace of the manual worker and often of his wife, at least when they could afford it. The mid-Victorian period holds the record for the consumption of alcohol. Average intake of spirits rose to 1.3 gallons and beer to 33.4 gallons in the mid-1870s. Drunkenness was seen as one of the main causes of poverty: the chief division in the working class was not between the skilled and the unskilled but between the respectables and the roughs. Drink could reduce a well-paid miner or iron-moulder's family to penury. One of the chief duties of a wife, and often of her children, in many industrial communities was to fetch her husband from the pub on pay-day before he drank all his wages. At the same time, this provided a wonderful alibi for the Pecksniffs of the sanctimonious middle class, who could blame the poor for their own poverty.[20]

Hence the power and influence of the temperance movement, which also reached its apogee in the period. George Cruikshank's cartoons and illustrations to Dickens's novels echo Hogarth's Gin Alley and Beer Street of 100 years earlier, with the difference now that beer, which anyone from 1832 could sell for a £2 licence, became as vitiating as gin. Temperance, which condemned spirits but allowed 'wholesome' beer, was trumped by teetotalism, which condemned both. The movement was, significantly, as much an attack

---

[20] Brian Harrison, *Drink and the Victorians, 1815–72* (Pittsburgh University Press, 1971), *passim*.

by the respectable working class on the feckless roughs as by the middle class on the inebriate lower orders.[21]

It was also, more covertly, a sideways blow at the aristocracy and gentry, who consumed far more than their share of the wine and spirits. The ladies left the gentlemen to their port and whisky after dinner as much for the relief afforded by the chamber pots hidden behind the window shutters as for the talk of sport and politics. 'Drunk as a lord' was a well-deserved Victorian epithet. But the rich could afford their vices, in this as in more reprehensible ways, while the working classes for the most part could not. Drink was responsible for many a fight at closing time, for many a marital rumpus and broken marriage, and for not a few families ending up in the workhouse. Pubs in Lancashire were categorized by the number of fights on Saturday nights reported in the local press.

The mid-Victorian period also saw the peak of the great crime wave that increased the number of indictable offences more than six-fold, from 4,605 in 1805 to 30,349 in 1848, and the number of convictions more than eight-fold, from 2,783 to 22,900.[22] The reduction in capital offences and the greater readiness of juries to convict led to more convictions, but the new police, extended from London to the towns in 1839 and to the counties in 1856, left prosecution to the victims, so the statistics are to that extent a minimum. *The Builder*, a trade journal, reported in 1867 that 100,000 Londoners lived by plunder, and the police had a hard time at every fire preventing looters breaking in.[23] Property crime began to decline in the 1850s, but assaults went on increasing, along with drunkenness, to reach their peak in 1876.[24] Much crime went undetected, and certainly unreported, since much of it was street crime, pickpocketing, mugging and the like, which was hard by its nature to prove. A minority, though only a fraction of property crime, was sheer mindless violence which rose and fell with the consumption of alcohol, and was only reported when it led to broken bones or manslaughter. The popular press bears witness to the dangerousness of urban streets at night; they were certainly unsafe for women alone, whose presence was taken as an invitation to sexual harassment: Victorian fathers' obsession with the protection of their daughters was thoroughly justified.

Yet crime was by no means a working-class, or even an underclass, monopoly. If street violence and public drunkenness were proletarian crimes,

---

[21] Anne L. Helmreich 'Reforming London: George Cruikshank and the Victorian Age', in Debra N. Mancoff and D.J. Trela (eds), *Victorian Urban Settings* (Garland, New York, 1996).

[22] Perkin, *Origins*, p. 167.

[23] *The Builder*, 3 August 1867.

[24] V.A.C. Gatrell and T.B. Hadden, 'Criminal Statistics and their Interpretation', in E.A. Wrigley (ed.), *Nineteenth-Century Society* (Cambridge University Press, Cambridge, 1972), pp. 369–75.

only the middle class could perpetrate white-collar crime, since it required access to money and positions of trust. Although it was always underexposed, because most banks and companies covered it up to preserve their good name, the cost of financial peculation far outweighed that of burglary and street robbery. In 1860 one offender alone, W.G. Pullinger, a cashier at the Union Bank of London (in which Lionel Redpath, the crooked Registrar of the Great Northern Railway, had already stashed his ill-gotten quarter of a million pounds) embezzled £250,000, over three times the £71,000 stolen by London's thieves that year. The few middle-class criminals, usually clerks and other small fry, actually prosecuted in the county of Surrey between 1855 and 1865 stole amounts nine times the average of lower-class thieves.[25] George Hudson the 'Railway King', who made a fortune by hyping up his companies and paying dividends out of the new capital thus raised, fled abroad to escape goal in 1854.[26] Readers of Dickens and Trollope well knew the ubiquity of white-collar crime from the depictions of Carker in *Dombey and Son*, Merdle in *Little Dorrit*, Uriah Heep in *David Copperfield*, or Alaric Taylor in *The Three Clerks* and Augustus Melmotte in *The Way We Live Now*.

Prostitution was the traditional female solution to lack of resources, and women's alternative to crime (not exclusively, since Victorian London had its 'Mollies', rent boys and obliging guardsmen). Some historians, like James Laver, believe that the 1850s and 1860s saw the high-water mark of prostitution, since the moral revolution and the cult of respectability deprived the unrespectable, or hypocritical, of the amateur supply and drove them to the professional market. The figures are necessarily esoteric, and Mayhew's 41,954 'disorderly prostitutes' arrested in London between 1850 and 1860 mark the barest minimum. Estimates varied from 7,261 (Sir Richard Mayne, Police Commissioner) to 80,000 (the Bishop of Exeter) – one in eight of the adult female population of London. The police in 1857 counted 2,825 brothels in London with 8,600 professional prostitutes.[27] These did not include the 'dollymops' or amateurs drawn from the domestics, shop assistants, milliners and dressmakers who were said to eke out their small wages with occasional sexual forays.

The profession ranged from the 'pretty horse-breakers', the high courtesans kept by peers and rich bourgeois, like 'Skittles', Catherine Walters, who seems to have been intimate with half the peerage, or Cora Pearl, 'the girl with the swansdown seat', who graced the *demi-monde* of London and Paris, down through the glamour girls of the 'best houses' like Kate Hamilton's in Mayfair frequented by the 'swells' of the London clubs, to the "gay" creatures (then the

---

[25] George Robb, *White-Collar Crime in Modern England* (Cambridge University Press, Cambridge, 1992), pp. 2, 7, 182.
[26] Ibid., ch. 2.
[27] Mayhew, *London Labour*, 4, pp. 215, 262–3.

euphemism for heterosexual prostitutes) lining the pavements of the Strand or Piccadilly, to the drabs who serviced the sailors and dockers of the East End. Some of them were child prostitutes of twelve or thirteen, much valued by lecherous old gentlemen for their supposed virginity and lack of disease. The Victorian myth, supported by Gladstone who rescued many from 'a fate worse than death' with his wife's approval, was that they were driven to the trade by poverty. Several told Mayhew, however, that they chose the life because it was far more rewarding than being a servant, a shop girl or a factory worker; they ate and dressed well, and they often married well too.[28]

Their customers were equally diverse, but they all had to have money, which gave the middle class the first preference. Dr Ryan thought that £8 million a year was spent on professional sex in London, some girls earning £20 or £30 a week, some as little as £1 or £5, but averaging overall £100 a year. They had considerable expenses, for dress, attending theatres and assembly rooms, paying rent and maid service, and some had 'bullies' (pimps) who took a large share of their earnings. A high-class courtesan like 'Skittles' might cost her aristocratic lover £2,000 a year plus a coach and pair of matched horses, but at the other end a low-class whore might charge 1s for a 'quickie' in an alley, and perhaps eke out her living by shaking down the customers with the help of her bully.

The Contagious Diseases Acts of 1863–71 were a product of the "social evil" and the fears of the government for the health of the armed forces, while Josephine Butler's campaign against them reflected the Victorian preference for sweeping it under the carpet. Both were evidence of the contemporary obsession with extramarital sex, which was reinforced by the verbatim reports of divorce cases after the 1857 Matrimonial Causes Act. The *Saturday Review* complained in 1864 at the time of the Public Health Commission:

> We want a Moral Sewers Commission. To purify the Thames is something, but to purify *The Times* would be a greater boon to society ... The unsavoury reports of the Divorce Courts, the disgusting details of harlotry and vice, the filthy and nauseous annals of the brothel, the prurient letters of adulterers and adulteresses. the modes in which intrigues may be carried out, the diaries and meditations of married sinners, these are now part of our domestic life.

The common belief that prostitution, promiscuity and adultery were at their height in the mid-Victorian age adds a spurious glitter, or tinsel, to its image.

---

[28] Ibid., section on Prostitution; Judith Walkowitz, *Prostitution and Victorian Society* (Cambridge University Press, Cambridge, 1980); Linda Mahood, *The Magdalenes* (Routledge, London, 1990).

Social Harmony and Class Conflict

If the scale of poverty, disease, slums, drunkenness, crime and prostitution all give the lie to the golden image of the mid-Victorian period, was the comparative social harmony also a myth? Certainly there was a decline from the cruder forms of social protest associated with the Chartist movement and the violent industrial disputes of the previous generation. After Feargus O'Connor doffed his cap to Police Commissioner Mayne at Kennington Common on 10 April 1848 and cancelled his massive march on Parliament with his million-signature petition, the demand for democracy and the fear of revolution seemed to disappear. The great mid-Victorian social peace was at hand. But once again, when we look closely at the disputes and disagreements of the day, there was far more conflict than the legend suggests.

The Chartist challenge may have dissipated, or have mutated into temperance Chartism, educational Chartism, and even Chartist churches, but the working-class Radicals had never given up their hopes of manhood suffrage, and they could mount impressive and effective mass demonstrations as they did at Hyde Park in 1866 over Parliamentary Reform. Political conflict was only muted because violence was no longer necessary, and both political parties came to accept the case for a wider franchise, only bidding with each other on how far down the social scale the vote should go. The compromise of 1867 admitted the majority of working-class householders, a leap in the dark which doubled the electorate without fear, it seems, of revolution or socialism.

Industrial conflict supposedly became muted by the 1850s with the coming of the new model unions, which officially believed in bargaining rather than strikes, but could not always dissuade the employers from provoking them. The most famous of the new national skilled unions, the Amalgamated Society of Engineers, began with a famous dispute, the engineering lock-out of 1852 by which the employers tried to end its existence. The builders' dispute of 1859 and the iron workers' clash of 1866 were similar lock-outs, with the employers trying to force the workers to sign the "document" renouncing membership on penalty of dismissal. The new agricultural worker's union of 1872 provoked a lock-out by the farmers of East Anglia which lasted from the spring of 1873 to August 1874. Strikes, despite the repeal of the Combination Laws in 1824–25, could still be prosecuted as illegal conspiracies under common law until 1871, and even then picketing was illegal until 1875.[29]

The employers had an even more powerful weapon in the Master and Servant Acts under which any worker leaving work unfinished was guilty of breach of contract and could be summarily convicted by a magistrate and

---

[29] Cf. Perkin, *Origins*, pp. 393–407; Trygve Tholfsen, *Working Class Radicalism in Mid-Victorian England* (Croom Helm, London, 1976), esp. ch. 8.

sentenced to three months in the house of correction. Until their amendment in 1867 there were 30,000 convictions a year under these Acts, which did not apply to breach of contract by masters, and 10,000 a year until their final repeal in 1875.[30] As James Hole, an unreconciled Owenite cooperator, put it in 1851, 'The relation between master and servant approaches slavery in the degree in which the servant is deficient in counteractive force'.[31] Or, as another malcontent put it in 1861, 'You make the laws in your own favour; you lay burdens on our shoulders that you will not touch with your own fingers; you overtask us; you underpay us; and when we receive our miserable pittance of wages, you would have us make our obeisance and say, 'Thank you, sir'.[32] These unequal laws challenge the notion that class and class legislation were no longer important in the mid-Victorian period.

Class versus Hierarchy

David Cannadine has recently argued that class has three different meanings in English which have been in competition throughout history: class as a seamless hierarchy of individuals knowing their place and not resenting it; class as a three-story pyramid of upper, middle and lower; and class as conflict, 'us' versus 'them'. Despite the tendency of historians and sociologists, especially Marxists, to emphasize the last two (which may in fact be the same, depending on whether you think the 'haves' are divided between landlords and capitalists or united against the 'have-nots'), he thinks that hierarchy, the acceptance of 'degree above degree' in a 'live and let live' society, has been the dominant experience of British history. The Victorian age in particular, he thinks, saw a 'viable hierarchical society' in which conflict and even negotiated accommodation between collective identities were overlain by a patriotic sense of community extending outwards from Britain to the Empire.[33]

This formulation offers what might be called a critical echo of my 'viable class society' in which the classes had accepted each other's right to exist and to share in the responsibilities of government and economic relations. My analysis never implied that the classes were isolated groupings of self-consciously allied people, armies of antagonists aligned against each other in mutual warfare. On the contrary, the class ideals were aspirational versions of

---

[30] Daphne Simon, 'Master and Servant' in John Saville (ed.), *Democracy and the Labour Movement* (Lawrence and Wishart, London, 1954).
[31] James Hole, *Lectures on Social Science and the Organisation of Labour* (Chapman, London, 1851), quoted in Tholfsen, *Working Class Radicalism*, p. 255.
[32] *Social Science: Being Selections from John Cassell's Prize Essays by Working Men and Women* (Cassell, Petter and Galpin, 1861), quoted in ibid., p. 249.
[33] David Cannadine, *Class in Britain* (Yale University Press, London, 1998), chs 1 and 3.

themselves as ideal citizens making the most important contribution to society. These aspirations were held, not necessarily by the whole class, but by activists who vigorously propagated the ideal and tried to get others to join them in pressing for recognition. Only insofar as they could persuade their fellows on their own level and, more importantly, the rest of society, to accept their view of the identity and role of the class could they hope to impose their values upon society and so dominate its politics and culture. The mid-Victorian age, I argued, saw the triumph of the entrepreneurial ideal, the domination of public policy and the economy by the values and culture of the business middle class.[34] Enough contemporaries accepted their leadership to ensure their domination of the culture and values of the age, though not, as we shall see, to the entire exclusion of competing ideals.

In practice, both class and hierarchy always exist in every society. As I have argued elsewhere, they are like the warp and weft in the social fabric, the vertical and horizontal threads without which the whole cloth could not hold together.[35] At different times warp or weft comes to the surface and dominates the face of the cloth. In eighteenth-century British hierarchy predominated in the shape of patronage and the vertical landed and mercantile interests which competed for government favour. In the early nineteenth century, class came to the fore in the Radical movements and the industrial conflicts of the Regency and Chartist periods. In between, transitional movements occurred which reflected both: machine-breaking, for example, as a means of pressurising the agricultural and textile industries, and the Anti-Corn Law League as a conflict between the manufacturing interest and the agricultural, and as an attack by the industrial middle class on the landed aristocracy. At such times the social fabric was like shot silk, which could be seen as hierarchical or class-based according to the point of view of the observer.

This way of looking at the period does not preclude any of the three definitions of Cannadine's tripartite formula. Whether or not class conflict, between two or three classes, still survived from the preceding turbulent half century, there can be no doubt that the mid-Victorians were obsessed with class in one or more of the three senses. It could be seen on the streets of every town: in the dress of the ladies and gentlemen, so different from the manual workers even on Sundays; in the horses and carriages of the gentry and the horse-buses of the middle class; in the church and chapel pews with their differential rents, the rich at the front and the poor if there at all at the back; in the inns and taverns of the gentry and middling people and the pubs and gin shops of the workers and the poor; in the West End gentlemen's clubs and the branches and lodges of the trade unions and friendly societies; in the children's schools,

---

[34] Perkin, *Origins*, ch. 8.
[35] Perkin, *The Rise of Professional Society* (Routledge, London, 1990), pp. 2–3.

carefully matched to each class by the Clarendon, Taunton, and Newcastle Commissions; in the law courts where white-collar criminals got off with a fine or a nominal sentence while petty thieves were sent to hard labour and the treadmill; in the first-, second- and third-class carriages of the new railways; in the size and appearance of the houses, from terraced brick boxes to the semis and villas of the better-off; in the very topography of the cities, the increasing segregation between the classes who inhabited the inner-city slums, Georgian squares, lower middle-class inner suburbs, and upper middle-class outer ones; even in death, in the serried slate slabs, angelic statues and imposing mausoleums of the cemeteries. Whether such differences were resented to the point of open conflict or rebellion depended on how successful the triumphant entrepreneurial ideal was in selling its belief in a fair, prosperous and upwardly-mobile society to the working masses.

An Answer to the Paradox

This view of class, whether as viable hierarchy or viable class society, suggests an answer to the paradox with which we began, social harmony in the midst of a not so Golden Age. After the turbulence of the previous generation the business middle class was able, with the help of a reformed landed elite and a reforming professional class, to convince the working classes that their ideal society offered something for everybody. The heart of that ideal was the belief in free trade as the means to economic growth and prosperity and in self-help and upward mobility as the means to a fairer and more equal society.

It could not be sold, however, without some concrete evidence of practical reform. One great achievement was the ending of Old Corruption and the unwarranted privileges and monopolies, sinecures and pensions of the old elite. *The Black Book; or Corruption Unmasked*, the bible of Parliamentary Reform, went through successive editions in 1820, 1831 and 1847.[36] By the third edition, it could claim that most of the unearned government pensions and sinecures of Old Corruption had come to an end. In the same way, the old municipal corporations had been cleaned up and new local authorities appointed to deal with police and public order, paving and lighting, public health and sanitation, water and sewerage, educational and library, and other urban services. The Poor Law, though still draconian, was somewhat ameliorated, with workhouse diet and clothing often better than the poorest labourer's family outside. Factory legislation, food adulteration, air and water pollution, and acid-rain laws were promulgated, especially in the 1860s.

---

[36] [John Ward], *The Black Book, or, Corruption Unmasked* (John Fairburn, London, 1820, 1831, 1847), 1847 edn, p. vii.

Education for the working classes was expanding with government subsidies for three decades before the Elementary Education Act of 1870. Free trade in corn had stabilized the price of bread and by 1860 most of the petty import and excise duties had been abolished. Even the great hope of the Radicals and Chartists, male household suffrage, was achieved in the towns in 1867. There was a sense that things were moving in the right direction.

In this atmosphere, the ideal of the business class, the belief in self-help and progress, could make headway. But it did so in a way that did not compromise working-class self-respect. The working class, or at least the more skilled, respectable part of it, transformed it into a version of their own ideal. Self-help to them was not individual but collective, it meant mutual aid, in the form of their own institutions of defence and security: the trade unions, friendly societies, and cooperative movement. All three burgeoned and thrived in the mid-Victorian period as never before. The new model unions grew to a quarter of a million members, persuaded the Royal Commission on Trade Unions in 1867 that they were a respectable force for good industrial relations, and created their own national body, the Trades Union Congress in 1868, whose Parliamentary Secretary, George Howell, became the Liberal Party's organizer of the new working-class vote. The friendly societies, whose function was to provide the workers and their families with social security in sickness and old age, or at the very least a decent burial, came to number forty-one million members by 1874. And in the biggest challenge to the competitive system, the cooperative movement by 1872 created 927 local societies with over 300,000 members, their own English and Scottish wholesale societies, and in due course their own factories, overseas tea and coffee plantations, and shipping to import their own supplies. We should also add the mechanics' institutes, with 1,200 branches and 200,000 members in 1860, which certainly offered technical education for upwardly-mobile workers, though many of them were under middle-class leadership and used by aspiring clerks and business managers. These mutual aid organizations served only the skilled and more respectable sections of the working-class, it is true, but they were the most intelligent and active part, the natural leaders, and without them the class could not achieve anything. Their acceptance of the middle-class ideal in a form suited to their own needs ensured that social peace and harmony would prevail despite the provocations of some of the more hostile and short-sighted employers.[37]

---

[37] See, *inter alia*, E.J. Hunt, *British Labour History, 1815–1914* (Weidenfeld and Nicolson, London, 1981), ch. 8; A.E. Musson, *British Trade Unions, 1800–75* (Longmans, London, 1972); P.H.J.H. Gosden, *The Friendly Societies in England, 1815–75* (Manchester University Press, Manchester, 1961); Beatrice Potter, *The Cooperative Movement in Britain* (Longmans, Longmans, 1891); Sidney Pollard, 'Nineteenth-Century Cooperation', in Asa Briggs and John Saville (eds), *Essays in Labour History*, vol. 1 (Lawrence and Wishart, 1967); J.F.C. Harrison, *Learning and Living, 1790–1960* (Routledge, London, 1961), ch. 5; and for other, especially contemporary sources, Perkin, *Origins*, pp. 380–407.

If Britain had, for better or worse, a more peaceful transition to democracy than most other industrializing countries, it was due to the accommodation between a hegemonic middle class and an independent and creative skilled working class which turned revolution into evolution. That happened, crucially, in the mid-Victorian period. To that extent it was after all, if not a Golden Age, an 18-caret gold-plated one.

Part I

Industry

Introduction to Part I: Industry

Colin Griffin

A quarter of a century ago, Roy Church posed the question: 'Are historians justified in referring to the period between 1850 and 1873 or thereabout, as the Great Victorian Boom?' or mid-Victorian Golden Age in the parlance of this volume. His answer was 'a severely qualified affirmative', in part because 'the labels ... contain sufficient truth to conceal their several defects',[1] and he added 'that an understanding of these years would be enhanced if historians were to concentrate less on national macroeconomic aggregates and more on microeconomic, regional developments' and other more specific dimensions of the economy.[2]

These four brief microeconomic studies of British industries confirm, rather than contradict, Church's warning that labels such as 'Golden Age' hide as much as they reveal about the character of the third quarter of the nineteenth century. In agriculture, for instance, a transient combination of favourable market circumstances produced a Golden Age of prosperity for the capitalist food producers which did not filter down to the farm workers who 'did not see even an illusion of gold'.[3] Similarly, in the coalmining industry a period of unprecedented growth, which culminated in the El Dorado of the so-called 'coal famine', stretched the availability of human capital of both management and worker to the limit and beyond, with terrible consequences for the health and safety of the largest subterranean workforce on the planet.[4] The cotton industry, too, made greater progress during the 1850s and 1860s as a whole than in the previous quarter century but was, nonetheless, dogged by overinvestment in new technology and overproduction which constantly threatened the profitability of factory masters and the livelihood of the cotton worker, and not only during the notorious years of the 'cotton famine', 1862–65, which foreshadowed the lean years of the so-called post-1875 'Great Depression'.[5] Unlike the 'staple' industries of agriculture, coal and cotton, the electrical engineering industry was created in the form of the submarine telegraph cable and telegraph communication industry in the two decades after

---

[1] R.A. Church, *The Great Victorian Boom 1850–73* (Macmillan, London, 1975–76).
[2] Ibid., p. 78.
[3] See chapter 4 in this volume.
[4] See chapter 3 in this volume.
[5] See chapter 5 in this volume.

1850. During these years 'undersea telegraphs [developed] from an experimental technology into a significant science-based industry', though as Gillian Cookson demonstrates this was a Golden Age shot through with paradox and inconsistency. In the 1850s, for instance, high public expectations, fuelled by the fulsome company prospectus and the story-grabbing press, were dashed and fortunes lost as inadequate technology resulted in cable failure. Thereafter, general investors, not anxious to get their fingers burnt twice, could not be tempted into investing in second-generation technology that worked, and missed out on what soon proved a low risk, high profitability industry. Moreover, public cynicism resulted in the industry being dominated by a few large limited companies which, contrary to the spirit of the so-called age of laissez-faire, indulged in oligopolic practices that maximized their profits at the expense of society at large.[6] George (later Sir) Elliot MP, a prominent figure in both the coalmining and submarine cable industries was, indeed, the embodiment of mid-Victorian inconsistency.[7] In his role of mining and mechanical engineer and colliery owner he told the Institute of Mechanical Engineers in a lecture on the future training and education of mining engineers 'It will be useful to remember that it is an Englishman's pride to do for himself that which the citizens of many countries have provided for them by their Governments'.[8] Yet in the 1860s and early 1870s he 'repeatedly pressed the Government to offer a premium to stimulate investors to apply their minds' to the development of coal-cutting machinery.[9] Nor was he unhappy to see the government underwrite the shares of private enterprise in the cable-laying industry in which his company was a major player. Nor was this rags-to-riches self-made man averse to being left alone by government to pursue monopolistic business practices.[10] Similarly, coalmining was an industry in which capitalists could profess to be 'deeply interested in any apparatus, or system, which would tend to prevent the present sacrifice of human life, and at the same time effect economy in the working of coal' and yet employ management that endangered both of these objectives.[11]

---

[6] See Chapter 6 in this volume.

[7] Elliot appears in both Griffin and Cookson's chapters in this volume and his career is summarized in Colin Griffin, 'Sir George Elliot' for the new *Dictionary of National Biography* (Oxford University Press, Oxford, forthcoming).

[8] George Elliot, 'President's Inaugural Address', *Proceedings of the North England Institute of Mining Engineers* (1868), p. 32.

[9] Samuel Parker Bidder, 'On Machines Employed in Working and Breaking Down Coal as to Avoid the Use of Gunpowder', *Proceedings in the Institute of Mechanical Engineers*, 28 (1868–69), p. 43.

[10] Cookson, chapter 5.

[11] Bidder, 'On Machines Employed', pp. 143–5 where he also notes the support for coalmining machines by the South Staffordshire and East Worcestershire Association of Mine Agents criticized by the HM Inspector of Mines for the area for the excessive accident rate in mines which he managed. Griffin, coalmining chapter

Gladstone proclaimed in the late 1860s that the prosperity of the country 'went forward, not by steps, but in "leaps and bounds"'.[12] But as this limited study of a range of vital industries suggests, his peroration was based on a partial reading of the national economy and society. In what sense mid-Victorian British industry might be said to have experienced a 'Golden Age' depends on whether the cursor searching the desktop alights on the entry for tenant farmer or farm worker, or the statistics for coal production or accident rate. Since the label 'Golden Age' or 'Golden Years', like that of 'Industrial Revolution', has sufficient basis in a selective reading of the past and the evidence on which it is constructed, it is likely to remain a candidate for reappraisal and reinterpretation well into the new millennium.

---

[12] Walter Bagehot, '*The Economist*, 4 January 1873' in W.H.B. Court, *British Economic History 1820–1914. Commentary and Documents* (Cambridge University Press, Cambridge, 1965), p. 12.

Chapter 3

Coalmining in Mid-Victorian Britain: A Golden Age Revisited?

Colin Griffin

A newspaper editor declared from the heart of the Black Country in January 1867 that the

> coal trade is so intimately bound up with the greatness of this empire, affording, as it does, the very primary cause for our motive and mechanical power ... The question of the probable exhaustion of the coalfields of this country is one which we may leave to be solved by coming generations. No matter what the speculative theories of merely idealists, or perhaps the deliberate opinion of scholars, based, however, upon an unnecessarily imperfect aggregate of facts, may advance in support of the doctrine that our coalfields are well-nigh exhausted, it is, at least, a well ascertained fact that for many years to come the coal mines of Great Britain will continue to yield their rich treasures to uphold the supremacy and augment the resources of this vast empire.[1]

This oration was both an expression of the belief, in the greatness of the mid-Victorian coal industry and a critique of Stanley Jevons recently published *The Coal Question*, with its ominous sub-title 'an enquiry concerning the progress of the nation and the probable exhaustion of our coal mines'.[2] Only killjoys and cranks questioned the coal industry's ability to underpin the workshop of the world for the foreseeable future since Britain's pre-eminence among the coal- producing nations of the globe in the new age of statistics was beyond dispute and a source of commendable self-congratulation. This chapter will analyse the salient features of this mid-Victorian dominance and consider whether the epithet 'Golden Age' is a fitting testament to it.

---

[1] *Dudley Guardian*, 26 January 1867.
[2] For some perceptive observations on Jevons and the place of the coal industry in Victorian society see Asa Briggs, *Victorian Things* (Penguin, London, edn, 1990), ch. 8.

Economic Progress

Table 3.1 Output of the Principal Coal-Producing Countries, 1850–70
(million tons)

Year	Belgium	UK	USA	France	Germany
1850	62.5	7.5	4.4	5.0	5.7
1860	87.9	17.9	8.2	12.2	9.5
1870	115.5	36.1	13.1	26.0	13.5

Source: Roy Church, et al., *The History of the British Coal Industry*, vol. 3 Victorian Pre-eminence (Oxford University Press, 1986) p. 773.

Britain's coal output doubled between 1850 and 1873[3] and the output of the world's second-largest producer had in 1870 only just passed half that of Britain's some twenty years earlier. It was in the two decades after 1850 that Britain also came to dominate the world's coal trade. Exports increased from 3.2 to 11.2 million tons and at 28.7 million tons in 1890, comprised 80 per cent of the world's coal trade.[4]

The British also had grounds for self-congratulation on another score: the efficiency of their collieries. If this is measured in terms of labour productivity (output per man year) Britain's superiority is striking since on average in the mid-1870s it was 270 tons per man employed compared to 209, 154 and 135 respectively in Germany, France and Belgium. Only that of the USA exceeded Britain's at 341 tons per man; the gap reflecting the superior geological conditions prevailing in the New World compared to those of the old countries of Europe rather than any failure on the part of colliery management.[5] The basis of this striking record of labour productivity was the colliery owners' commitment to technological innovation which has earned the epithet 'coalmining's Industrial Revolution'.[6] The quarter century after 1850 was the

---

[3] UK coal output was 128 million tons in 1873. Roy Church, et al., *The History of the British Coal Industry, vol. 3. Victorian Pre-eminence* (Oxford University Press, Oxford, 1986), p. 86. According to Crouzet 'It is apparent that the most rapid advance was achieved during the mid-victorian period, the 1850s and 1860s and the boom at the beginning of the 1870s, when output doubled in twenty years', Francois Crouzet, *The Victorian Economy* (Methuen, London, 1982), p. 264.

[4] Church, *Coal Industry*, p. 36.

[5] Neil Buxton, *The Economic Development of the British Coal Industry* (London, Batsford, 1978), pp. 95–97; Donald N. McCloskey 'International Differences in Productivity? Steel and Coal in America and Britain Before world War 1' in Donald McCloskey (ed.), *Essays on a Mature Economy: Britain after 1840* (Methuen, London, 1971), pp. 289–95.

[6] A.R. Griffin, *The British Coalmining Industry. Retrospect and Prospect* (Moorland, Buxton, 1977) p. 106; Colin Griffin, 'An Industrial Revolution in the East Midland Coalfields

period when the most advanced forms of mining technology pioneered in particular collieries or the more advanced coalfields became fairly standard practice throughout the industry, and when new technologies were developed and diffused with unprecedented rapidity in the more advanced enterprises. There was a general acceptance of the superiority of the long-wall system of mining over alternative methods and this increased the extraction rate by as much as a third and also produced a lower percentage of small coal which was difficult or impossible to sell at a profit. There was also a more general introduction of best practice ventilation systems. In the 1850s this was furnace ventilation at the bottom of upcast shafts though by the 1860s it was mechanical ventilation using centrifugal force, such as the Waddle Fan, which required less maintenance and was generally safer than furnace ventilation. Shaft work and underground haulage were transformed which enabled the advantages of long-wall extraction to be fully realized. More powerful steam winders raised cages held rigid in shafts by conductors. Haulage along the main roadways to the shaft bottom was increasingly by steam or compressed air powered tub rope haulage, through conveyance on side roads from the coalface, or on difficult gradients on main roads, was still the province of ponies and horses. The increasing use of safety lamps instead of naked lights permitted the extension of mining into gaseous seams and areas of mines, though this practice continued to be fraught with danger as the continued loss of life in explosions testified.[7]

Technological innovation on this scale produced an increase in labour productivity between c. 1850 and 1870[8] which was all the more remarkable given that coalmining, like other extractive industries, was subject to diminishing returns with collieries having to be sunk to ever greater depths with more extensive underground working. These characteristics progressively increased the amount of capital required to be invested in them before coal could be efficiently mined. In 1843 the cost of sinking and opening out underground of a typical south Staffordshire colliery was £3,000–£4,000, whilst an investment of £50,000 was not uncommon in the North East. Thirty years later these outlays had probably about doubled while the most advanced mines in the East Midlands represented an initial investment of about £50,000

Between c. 1850 and c. 1880? The Case of the High Park "Superpit", Nottinghamshire', *Transactions of the Thoroton Society* (1990), pp. 75–82.

[7] Buxton, *Economic Development*, pp. 100–112; A.R. Griffin, *Coalmining*, pp. 106–11 and A.J. Taylor, 'The Coal Industry' in R.A. Church (ed.) *The Dynamics of Victorian Business. Problems and Perspectives to the 1870s* (Allen and Unwin, London, 1980), pp. 53–5 provide excellent summaries of the process of technological innovation.

[8] Labour productivity measured in terms of output (tons) per man year and output (tons) per man shift increased according to the most authoritative account from 280.6 and 1.36 in 1842–49 to 310.8 and 1.40 in 1860–69 respectively. Church, *Coal Industry*, pp. 474.

in the 1850s and £70,000 by the later 1870s.[9] Moreover, the most recent estimates suggest that the 'historic capital' of the industry grew from £17.64 million in 1854 to £54.14 in 1873,[10] a three-fold increase. Returns on invested capital were heavily dependent upon the level of coal prices in the short term and the outcome of the constant struggle against diminishing returns in the long term. Coal prices followed a broad upward trend in the 1850s and 1860 and exploded between 1871–73.[11] Mean gross profit per ton averaged about 1s in the 1850s, 1s 6d during the 1860s and rose spectacularly from 1870 to reach 6s in 1873 when the most efficient companies were earning 25 per cent or more per annum on their assets.[12]

Given this scale of investment in response to the massive increase in the demand for coal it is hardly surprising that the demand for labour was intense for most of the period and reached desperate proportions during the coal famine of the early 1870s as management scrambled after any labour they could recruit or poach. The trend of wage rates was relentlessly upward and average weekly earnings for hewers increased from 19s 2d in 1854 to 25s 10d in 1866, and 32s 9d in 1873.[13] Earnings peaked in 1873–74 by which time it was being asked of trade union leaders: 'Is there much champagne drunk in your district?' to which they were inclined to reply, tongue-in-cheek, 'I do not think that the colliers knew about champagne until you gentlemen began to talk about it and since it has been talked about so much, I have known them meet together and subscribe for the purpose of having a bottle to see what it was like'.[14]

Total coalmining employment increased from 218,230 in 1851 to 295,810 in 1861 and 386,560 in 1871.[15] The pressure on management to recruit was intense given the industry's deserved reputation for offering dangerous, dirty and arduous work which produced a high wastage rate through incapacitating injuries, early retirement and labour turnover. It could take up to eighteen months or more to convert a 'green hand' into an efficient, safety-wise, coal-getter equivalent to an AB in the navy, though only a matter of weeks for him

---

[9] Taylor, *Coal Industry*, pp. 55–7; C.P. Griffin, 'Some Comments on Capital Formation in the British Coalmining Industry during the Industrial Revolution', *Industrial Archaeology Review*, 1:1 (1976), pp. 81–3.
[10] Church, *Coal Industry*, p. 103.
[11] The price of coal in the London market was 17s 6d in 1850, 19s in 1860, 18s 6d in 1870, rising to 33s in 1873. The pithead price of coal in South Yorkshire increased from 6s to 19s between 1871 and 1873. *Royal Commission on the Depression of Trade and Industry 2nd Report*, Cd4715, 1886, Q3009. Evidence of John Ellis, colliery owner.
[12] Church, *Coal Industry*, pp. 521–2.
[13] Church, *Coal Industry*, pp. 561; 575.
[14] *Select Committee on the Present Dearness and Scarcity of Coal*, C313, July 1873, Q7386. Evidence of John Normansell, Secretary of the Miners' National Association.
[15] Church, *Coal Industry*, p. 189. It had increased to 514,100 by 1873 and peaked at 538,800 in 1874. B. Mitchell and P. Deane, *Abstract of British Historical Statistics* (Cambridge, Cambridge University Press, 1962), p. 118.

to become competent at other operations, such as loading coal.[16] Both sides of industry preferred to recruit and promote labour through an informal apprenticeship system and juvenile recruitment comprised 77 per cent of the total in the 1850s and 81 per cent a decade later.[17] During periods of the greatest demand for labour, most notably 1870–73, agricultural labourers, framework knitters, railway servants, tailors, shoemakers, sailors, carpenters and policemen to name but a few sources were attracted into the industry, though these recruits subsequently left in large numbers during the so-called Great Depression that followed.[18]

Rapid expansion not only required an ever larger labour force but a growing body of managerial and supervisory personnel to run the industry. The character of mine management varied, like the scale of mines and quantity of investment, from coalfield to coalfield and even from colliery to colliery within a single coalfield. It was equally uneven in its quality and competence.[19] Mine safety legislation from 1850, and the appointment of a government inspectorate to enforce it, exerted pressure on colliery owners to reform management practice and raise its quality to at least a minimum standard and, at best, to that employed in the most progressive mines. In the 1850s and 1860s the mines of the North East and the most heavily capitalized mines elsewhere in the North were managed by practical men with scientific and technical knowledge often following the advice of a consultant mining engineer, or viewer as they were commonly known. Some mines in many coalfields, and most mines in others, such as the West Midlands, were under the management of contractors or 'butties', working miners who had acquired a little capital and considerable practical knowledge and who provided their own working capital and labour and produced coal on a piece rate for the colliery owner who had invested in the fixed capital of the mine. The contractors might be loosely supervised by the owner's agent, who would be responsible for many or several mines, or

---

[16] S.C, *Coal* 1873, Q904 Evidence of Thomas Evans, HM Inspector of Mines and Q3792–3. Evidence of J.T.Woodhouse, colliery viewer.

[17] Church, *Coal Industry*, p. 227.

[18] S.C, *Coal* 1873, Q905, Thomas Evans, Q3792, J.T. Woodhouse and Q1451, J. Willis, HM Inspector of Mines.

[19] S.C. on Accidents in Coal Mines, 2nd and 3rd Reports, C258, 1854, Q1024, evidence of T.J. Taylor, colliery viewer; Q1892, J.T.Woodhouse, colliery viewer; Q2951, S. Dobson, General Manager, Duffryn Collieries, South Wales who argued that colliery management 'from head viewer downward' was generally 'far superior' in the North-Eastern coalfields than Lancashire, South Wales, Yorkshire, East Midlands and South and West Midlands. Thomas Wynne, HM Inspector of Mines for Staffordshire, opined, Q2987–90, that 'the coal district of Durham and Northumberland is managed in a much more scientific way than any other part of England' and that 'The South Staffordshire district is as bad as it can be'.

entirely, or largely, left to his own devices.[20] The expertise of management at mines with a resident manager could also be more apparent than real since many of them were knowledgeable in local mining practice but lacking in scientific and technical education, unlike their counterparts on the Continent.[21] Legislation from 1855 required all collieries to observe a code of increasingly extensive rules governing planning, working practices and mine closure, and to nominate a person responsible for enforcing them. It was not until 1872, however, that these responsible persons (who titled themselves variously, and confusingly, managers, under-managers, underground bailiffs and agents) were required to have a statutory certificate of competence acquired through public examination in order to manage a mine, though even then this only applied to large mines (employing thirty or more underground) and exemptions were possible on the basis of long service and accident-free record.[22] Colliery owners also ideally required their collieries to be operated with increasingly complex and expert management systems (including consultant mining engineers acting as advisers and disseminators of best practice) to protect their investment and remain competitive in an industry in which business casualty rates were high.[23] The finest general managers were multi-talented, having both entrepreneurial drive and flair combined with expertise in all branches of mine management.[24]

The existence of institutions devoted to providing this expertise through training in scientific, technical and commercial education was limited before 1870. Schools of 'viewers' and other forms of managerial apprenticeship, including dynasties of mining engineers, provided a steady stream of talent, and knowledge was disseminated and obtained informally from the trade press, the annual reportage of the Mines Inspectorate, government inquiries into mining accidents and coal supplies and above all, perhaps, through the activities of professional societies such as the 'North of England Institute of Mining

---

[20] The most expert discussion of the 'Butty System' remains A.J. Taylor 'The Sub-contract System in the British Coal Industry' in L.S. Presswell (ed.), *Studies in the Industrial Revolution* (Athlone, London, 1960).

[21] S.C. of 1854, Q1023. Evidence of T.J. Taylor. Joseph Dickinson, HM Mines Inspector, Lancashire, claimed in 1860 that 'a great number' of managers 'could never pass the examination which has to be passed in France, even before the candidates are admitted to the Polytechnic School'. Thomas Evans, HM Mines Inspector, argued that 'the difficulty is to get both qualities, a theoretical, scientific man and a man of practical experience'. Few were to be found in the British coal industry before 1872. S.C. *Appointed to Enquire Into The Operation of the Mines Acts*, Report 1866, C398, Q606 (Dickinson), Q945 (Evans).

[22] Sir Andrew Bryan, *The Evolution of Health and Safety in Mines* (Ashure Publishing, Letchworth, 1975).

[23] Taylor, *Coal Industry*, p. 59; H.A. Shannon, 'The Limited Companies of 1886–1883' reprinted in E.M. Carus-Wilson (ed.), *Essays in Economic History* (Arnold, London, 1966), pp. 396–401.

[24] Church, *Coal Industry*, pp. 409–22. For a typical example see C.P. Griffin, 'Robert Harrison and the Barber, Walker Co: A Study in Colliery Management 1850–90', *Transactions of Thoroton Society*, 82 (1979), pp. 51–62.

Engineers, and others interested in the prevention of Accidents in Mines, and in the advancement of Mining Science generally', established in 1852.[25] The inspectorate and many distinguished members of the mining engineering profession complained that the supply of management expertise and personnel was inadequate during these years of hectic expansion in output and labour force and so 'in many instances, through a mistaken notion of economy by proprietors, men are employed as responsible managers who are totally ignorant of the most elementary principles of mechanics, the laws of gas, the principle of the safety lamp, the geometry of figures formed by faults and heaves and the theory and accurate practice of surveying and levelling [which produced] excessive commercial failure and loss of life'.[26] The following exchange in 1866 said it all: 'Is not the most foolish economy that can be practised by the owners of a colliery to employ an inferior manager? It is, but the owners say, on the other hand "we get the best men we can employ, under the circumstances"'.[27] Mine management felt the twin pressures of the need to improve its technical experience and implement mine legislation most acutely after c. 1850 and reacted by forming voluntary associations for mutual support in responding to them. Unlike miners' trade unions these managerial associations have remained largely unstudied. The purposes and activities of the Incorporated Association of Mine Agents of South Staffordshire and East Worcester may be typical. Formed in October 1866, its objectives were 'the improvement of the practice of mining ... and the protection and aid of the Members in any matters connected with their practice'.[28] The association provided legal aid for members charged with an infringement of the Miners Inspection Acts and this service was soon required to support, for instance, James Cope, prosecuted for failing to prevent three miners entering an area of fatal choke damp. Cope argued that it was the underground bailiff (a butty) who was responsible for day-to-day management at the mine, which he visited once a fortnight. His defence was rejected since he was the only official with both the technical expertise and practical knowledge to ensure the safety of the mine, and he was given the maximum fine of £20 plus costs, less than £10 per

---

[25] Church, *Coal Industry*, pp. 422–34. Other district societies soon followed.
[26] *Dudley Guardian*, 25 January 1868.
[27] S.C. of 1866, Q14,495. Evidence of J.P. Baker, HM Mines Inspector. Managers dismissed following fatal accidents and prosecution by the inspectorate had little difficulty finding alternative employment in which to reoffend. *HM Inspector of Miners Annual Report 1876*. C.P. Griffin, *The Economic and Social Development of the Leicestershire and South Derbyshire Coalfield 1550–1914*, unpublished PhD thesis, University of Nottingham, 1969, pp. 425–7.
[28] *Memorandum and Articles of Association of the Incorporated Association of Mine Agents of South Staffordshire and East Worcestershire October 1866 Minute Book* (copy in the author's possession).

head for the victims who had walked unwittingly to their deaths.[29] The association's reaction was to challenge the regulations which made their members responsible for safety rather than the chartermaster (or butty) who employed the men, despite the latter's reputation for maximizing their short-term gains at the expense of mine safety and the long- term economic interests of the mine. The association claimed that the new 1865 regulations 'would encourage an amount of carelessness and recklessness among all the subordinate officers of the mine which would result in an increase in mine casualties and destruction of property'.[30] Moreover, making mine managers responsible for safety simply would not work since 'the amount of carelessness and recklessness on the part of the miners involved in two recent explosions at Talk of the Hill and Oakes ... defied the most skilful and attentive supervision. Science in mining was completely frustrated by such acts of the workmen'.[31] The Association conducted a protracted battle with the district Inspector, J.P.Baker, accusing him, for instance, of using 'Mediaeval practices ... of fines, pains and penalties' following his report that in 1868 '111 lives had been lost, one third preventable ... if ordinary care, ability and supervision had been exercised by colliery managers'.[32]

Social Cost

The social cost in terms of mortality, occupational disease and human suffering of the rapid expansion of the industry and its economic benefits must do much to mitigate the latter. Coalmining safety legislation produced the first authoritative statistics on mine accidents and confirmed what informed contemporaries had long suspected; that coalmining was an exceptionally dangerous occupation (only possibly exceeded by that of deep-sea fishing). Given the large number of miners of working age who died in accidents at work, and that there were 100 serious non-fatal accidents to every fatality, the chance of a career miner working underground avoiding a severe accident were slim indeed and many miners must have experienced several serious injuries during their working lives.[33] Although the long-term trend in the incidence of fatalities was downward, numbers remained stubbornly high and far from flattering compared to Continental experience before the Mines Regulation Act of 1872, as the following statistics indicate:

---

[29] *Dudley Guardian*, 24 April 1867.
[30] *Mine Agents of South Staffs*, M.B., 23 December 1866.
[31] *Mine Agents*, 23 January 1867. p. 582
[32] *Mine Agents*, 5 October and 3 November 1868.
[33] Church, *Coal Industry*, pp. 582–87; John Benson, *British Coalminers in the Nineteenth Century: A Social History* (Gill and Macmillan, Dublin, 1980), pp. 37–43.

Table 3.2 Death Rates in Mines from Different Causes, 1851–71
(per 1,000 employed)

Year	Underground workers					On the Surface
	Explosions	Falls of Roof	Shaft Accidents	Misc	All Causes	
1851–55	1.28	2.02	1.30	0.56	5.15	1.01
1856–60	1.23	1.85	0.90	0.66	4.63	0.99
1862–65	0.62	1.71	0.67	0.79	3.79	1.11
1866–70	1.16	1.58	0.53	0.73	4.00	1.26
1871–75	0.52	1.21	0.44	0.57	2.74	0.90

Source: H. Stanley Jevons, *The British Coal Trade* (1915, reprinted 1969) p. 371.[34]

Miners must share the responsibility with management for the carnage, and it was the pressure to produce that was so often the undoing of both. The most common form of payment underground was the piece rate, and miners and hauliers were tempted to cut corners to increase output and earnings as in Leicestershire in the 1850s and 1860s, when the inspector placed great emphasis on 'the men's own recklessness [he had] many times in the last year seen places unsafe; the men, however, take no heed of warning but, to save themselves a little trouble in propping, will run any amount of risk ... sometimes these accidents on roadways arise from people being deaf; sometimes from being too venturesome, and trying to get as far as possible before the tram overtakes them; instead of going into a refuge'.[35] The excessively rapid expansion of the labour force exacerbated the situation since 'the rapid extension of mining operations over the last four years [1859–63] has caused a demand for miners far in excess of the growth of the mining population and therefore other labourers have been taken into the mines. Some of these make excellent miners, but they are less numerous than I would like to see ... and especially those who have gone into the pits from surface labour at middle age, are very indifferent miners and careless as regards their own

---

[34] Death Rates in Mines From Different Causes in Selected Countries (per 10,000 employed):

Country	Explosions	Falls of Roofs	Other Accidents	All Causes
Belguim (1851–2)	11.4	10.4	14.2	36.0
Westphalia (1841–52)	1.0	10.0	5.0	16.0
G.B. (1851–52)	12.4	14.4	5.1	41.9

Source: T.J. Taylor, S.C. of 1854.

[35] S.C. of 1866, QQ8816–24, Evidence of J.T. Atkinson, HM Inspector of Mines. *Annual Report, 1865* (hereafter *Ann. Report*). In 1881 it was still being emphasized that 'The system in vogue in the English, Welsh and Scottish coalfields generally, is that the propping at the face of work is made over to the hewers, pikemen, or butty colliers ... In cases in which they are not specifically paid for setting props it would appear that they are apt imprudently to delay it too long'. *Preliminary Report of HM Commission Appointed Enquire into Accidents in Mines*, c.3036, 1881, IX.

safety'.³⁶ The massive demand for labour also encouraged miners to migrate between different coalfields in response to higher earnings and better opportunities, and according to the inspectorate this increased 'the difficulty of enforcing mine discipline and adversely influences the great loss of life annually experienced'.³⁷ These tendencies came to a head during the coal and labour famine of the early 1870s it at least in areas of phenomenal growth like the Midlands, where roof-fall mortality increased from twenty-two in 1870 to thirty-six in 1872 whilst the labour force was mushrooming from 28,810 to 39,265.³⁸ The lack of mine discipline exhibited by former framework knitters, who displayed 'excessive independence' produced by their previous domestic employment situation, and Irishmen, not noted for being 'regular in their habits', made them particularly accident-prone,³⁹ particularly given the 'difficulty in meeting with a sufficient number of good officers [supervisors] whose precept in the mine will, in the course of time, have good effect'.⁴⁰

Coalmining was a particularly dangerous occupation not only because of its horrendously high accident rate but also because of the range and level of occupational diseases from which its workforce suffered. Two of the most significant were a variety of lung diseases and nystagmus and there is considerable evidence that both rapid expansion and technological innovation impacted on both. Coal-dust and stone-dust were created in massive quantities in coal-getting and extending underground workings and roadways, especially since explosives were being used in greater quantities. The quantity of coal-dust present in working-places could be substantially reduced by efficient ventilation systems. Before the introduction of the more stringent regulations of the 1872 Mine Act this situation prevailed in only the best-managed mines, since the average mine would have a perfectly adequate ventilation capacity but an inadequate system of maximizing air flows to the more distant workings away from main roadways, or as H.F. Mackworth put it 'there is hardly a mine in the country where that [the volume of air] is carried throughout the working faces, as far as my own observation has gone'.⁴¹ In the poorest mines in coalfields like north and south Staffordshire, totally inadequate natural forms of ventilation persisted and miners continued to tell the nation that 'want of ventilation ... brings on asthma; that is the greatest complaint amongst

---

³⁶ *Ann. Report*, 1864.
³⁷ *Ann. Report*, 1859.
³⁸ *S.C. on the Present Dearness*, 1873, QQ1381–7 and Appendix 4. Evidence of Thomas Evans.
³⁹ *S.C. of 1873*, QQ920, 1451, and 3813
⁴⁰ *Ann. Report*, 1856.
⁴¹ S.C. of 1852–53, Q7135. Evidence of H.F. Mackworth, HM Inspector of Mines, South Wales.

colliers'.[42] Working in inadequately ventilated places was, ironically, made possible and indeed encouraged by the introduction of the safety lamp for illumination, since this enabled work to continue when fire-damp accumulated due to poor ventilation.[43] Mackworth declared that 'carbonic acid and other gases "silently numbering their victims", killed and incapacitated many more men than explosions', since both fire-damp and choke-damp thrived in poorly-ventilated places.[44] More generally, safety lamps were increasingly considered a necessary insurance against the risk of explosions and were relentlessly replacing naked lights despite opposition from miners who complained that the resultant poorer illumination had adverse effects on their productivity, increased the danger from accidents and damaged their eyesight, particularly increasing the incidence of nystagmus 'a distressing disease of the eye arising from strain due to poor standards of illumination'.[45] Since the mines inspectorate gave a high priority to the prevention of explosions in their early decades of activity, it was inclined to refute the miners' claims and the disease remained misunderstood, misrepresented and unrecorded.[46] Both lung disease and nystagmus resulted in the premature retirement of miners at a time when recruitment was at a premium and helped to guarantee the comparative youth of the workforce for which the industry was noted.[47]

Industrial Relations

The colliery owners of mid-Victorian Britain were left in no doubt as to the importance of their industry to the achievement of 'the workshop of the world':

---

[42] S.C. of 1866, Q5494, Evidence of B. Owen, working miner, Bilston, Staffordshire. See also Q2641, Evidence of J. Ackersley, working miner, Kearsley Hall, Lancashire who also answered in the affirmative to the question: 'Do you think it affects them [miners] in after life, and shortens their lives'. The oxygen deficiency in poorly -ventilated working places 'produces a pressure on the chest, a strange fatigue and feebleness, the breathe becomes quick and heavy; the labour is perfomed with great effort and thirst and drowsiness' (Mackworth, S.C. of 1852–53, Q1553).

[43] The practice of substituting safety-lamps in lieu of ventilation was still common practice, for instance, in the East Midlands in the 1860s. *Ann. Report* 1864.

[44] G.M. Macdonagh 'Coal Regulation: The First Decade 1842–1852' in Robert Robson (ed.), *Ideas and Institutions of Victorian Britain* (Bell, London, 1967). p. 83.

[45] Bryan, *Health and Safety*, p. 108. As late as 1890 the HM Inspectors of Mines, Midland District, attributed nystagmus to 'the sudden change from dark to light when the miners reached the surface, exacerbated by a tendency for them to rub their eyes and scratch their eyeballs with dust'.

[46] Poor ventilation also led to adverse effects on labour productivity and horse power since 'where the stalls are well ventilated [miners] can do one fourth more work in the course of a day'. S.C. of 1852–53 Mackworth Q638.

[47] Church *Coal Industry*, pp. 198–200.

> To a nation like England coal is only another name for gold; and we
> might even say that the presence of gold in Kent would be of far less
> importance to London than the existence of coal.
> *Standard*, 1878.[48]

They also certainly believed that their achievements were dependent upon managerial prerogative in the workplace. They felt themselves free, as we have seen, to appoint good, bad or indifferent mine management and they equally insisted on their right to employ labour of their choosing at wage rates and conditions of labour determined by their management. The formation of miners' trade unions, with their demands for the right to work for 'a fair day's work for a fair day's pay' and State intervention to protect their members from the excesses of capitalist exploitation, was a direct assault on managerial prerogative which the owners determined to resist through concerted action if all else failed since 'the system of organisation which the men have got amongst them, unless it be met by some counter-combination, must practically be eating up piecemeal the colliery owners and capitalists'.[49] They adopted a policy of non-recognition of unions and a stubborn refusal to compromise in negotiations with groups of men or their representatives.[50] The owners justified their position by insisting that they were obeying the laws of classical political economy, a not so dismal science which maximixed economic gain for both sides of industry and which trade unions destroyed as in the following case:

> At the time the delegates entered into South Derbyshire ... the whole district was in a state of prosperity. The men were regularly at work. The competition was such as to keep up the wages to the top price ... The continued strikes caused by the delegates makes them counterbalance any advantage they may get ... the effect of the policy of trade unions upon the workmen has been to produce mistrust and irregular habits, and upon the employers to cause a loss of capital and a check to enterprise.[51]

Trade unions were not only unnecessary to maximize the interests of the workforce but were also destructive of the paternalism that existed in many enterprises such as the Moira collieries where:

---

[48] Quoted in Briggs, *Victorian Things*, p. 288. A major coalfield was discovered under Kent in 1890 after many years in which 'coal in Kent' had been a standard joke on the Stock Exchange. It was producing 110,000 tons in 1914. H Stanley Jevons, *The British Coal Trade*, (1915 edn reprinted, David and Charles, Newton Abbot, 1969), pp. 155–75.

[49] S.C. of 1873, Q7549. Evidence of George Elliot MP, colliery owner and manager.

[50] Church, *Coal Industry*, p. 664; C.P. Griffin, 'Colliery Owners and Trade Unionism: The Case of South Derbyshire in the Mid-nineteenth Century', *Midland History*, 6 (1981), pp. 109–23.

[51] R.C. on Trade Unions and Employers' Associations, Final Report, C4123, 1868–69 Appendix 18. Answers to questionnaire by Arthur Higginson, manager Stanton Colliery, South Derbyshire.

> We had a very good trade and we made from time to time alterations, and advanced wages as we found wages advancing in the country. There was no necessity for a union to have wages advanced ... the union, and the doctrines they have taught ... teach them independence of action from their employers. It breaks entirely the old lines of good faith as between master and men ... and from that moment all good feeling is lost. The master is soured, loses his temper, and will do nothing for his men.[52]

Typically, the owners in the inland coalfields (and isolated ones like South Wale) used an extensive armoury of weapons in their offensive against nascent coalfield trade unionism in the 1850s and 1860s. To join a union was to invite dismissal as Rule 13 of Snibston colliery, Leicestershire stated: 'That if any man or boy engaged at these collieries shall be a member of the union or association called "a Miners' Union or Association", or contribute thereto, not being a member, he or they should be subject and liable to instant dismissal from his or their employment at these collieries'.[53] The owners compiled blacklists of union members which was highly effective as Alfred Hall, for instance, found to his cost when he moved to Cinderhill colliery, Nottinghamshire following his 'victimisation' at Snibston.[54] Union members were locked out by employers in one coalfield after another and replaced wherever possible by local 'free labour' augmented by imports from other poorly union-organized coalfields, such as the West Midlands in the case of East Midlands.[55]

Employer domination was, however, increasingly challenged by trade unionism in the larger coalfields from the mid-1860s[56] and to a limited extent proved its worth to both sides of industry in settling disputes before they became serious during the coal famine of the early 1870s. In the North East, for instance, it was claimed 'that the action of the union is very beneficial in avoiding strikes ... They send for the delegate now almost in the way that we would send for the doctor, to get the thing [threatened strikes] put right'.[57] As the great boom, ended several coalfields were on the brink of adopting the institutionalized conciliation and arbitration procedures tied to a sliding-scale of prices and wage rates that, on balance, served the owners and miners well

---

[52] Quoted in Griffin, 'Colliery Owners' p. 110.
[53] S.C. of 1866, Q5050. Evidence of A. Hall, Leicestershire coal-miner.
[54] S.C. of 1986, Q4979.
[55] J.E. Williams, *The Derbyshire Miners* (Allen and Unwin, London, 1962), pp. 105–7, 111–17.
[56] For a short, valuable summary of these developments see H.A. Clegg et al., *A History of British Trade Unions Since 1889 Vol. 1. 1889–1910* (Clarendon Press, Oxford 1964), pp. 15–20.
[57] S.C. of 1873, Q7560, G. Elliot, colliery owner and mining engineer.

once they became nationally firmly organized following the establishment of the Miners' Federation of Great Britain in 1889.[58]

Conclusion

The colliery owners' association representing the East Midlands advised the Royal Commission on the Depression of Trade and Industry in 1886 that 'The downward tendency commenced towards the close of 1874, and became a "Profitless depression" in 1877, and has remained so, with one or two brief intervals ever since'.[59] Colliery owners looked back with amazement, tinged with nostalgia, at the preceding quarter century that had culminated in the 'very great excitement in the coal trade' of 1871–73 which generated the mistaken belief that there was going to be a 'famine of coal forever'.[60]

Britain's coal output underpinned the country's temporary mid-Victorian economic ascendancy and was able to do so through instigating significant technological and organizational changes. These changes were not unproblematic and were only achieved at considerable social cost as human resources and knowledge struggled to meet the manifold challenges posed by the insatiable demand for 'black diamonds' in a society, which the industry epitomized, whose business philosophy was grounded on the pursuit of self-interest tempered only marginally by more altruistic motives. If massive investment, increasing profitability, rising real wages and technological innovation are indicative of a 'Golden Age' of coalmining, then the third quarter of the nineteenth century merits that image. It was, nonetheless, an image severely tarnished by the bankruptcy of less skilful or foolhardy investors, excessive mortality and ill health among the workforce, and class warfare in the coalfields. Yet another example of paradox for the history books.

---

[58] A.R. Griffin, *Mining in the East Midlands 1550–1947* (Cass, London, 1971), pp. 131–144, 152–7. Though whether the procedures prevented the miners from maximizing their earnings remains a controversial issue. See, for instance, J.H. Porter, 'Wage Bargaining under Conciliation Agreements, 1860–1914', *Economic History Review*, 23:3 (1970), pp. 460–75.

Chapter 4

A Golden Age of Agriculture?

Stephen Caunce

I

It may seem strange to seek a Golden Age of agriculture after 1850, when industrialization was taking a firm hold on England. However, more people then worked on farms than a century before, more land was cultivated, and far more English farm produce was consumed. Previously, when the population had moved above 5 million the food supply became unreliable, but in 1801 it reached 9 million, by 1851 18 million, and in 1871 23 million. The two decades after 1850 were therefore good for farmers, leading to incomes in 1873 being at their highest point in the nineteenth century.[1] They could afford high rents, and so landlords had little to complain about. The countryside generally seemed to be at ease with itself after a prolonged period of turmoil, an almost inchoate amalgam of resistance to enclosure, Captain Swing, food riots, and resistance to reform of the Poor Law.[2] Little sign remained of an old-established peasantry, but historians now generally do not see this as a recent demise. On the surface, the two central decades were a Golden Age indeed.

Technically as well, everything seemed excellent. Landlords provided specialized buildings and began to invest in field drainage as well as enclosure. They promoted 'high farming', a high-cost, high-return system dominated by substantial farmers, who increasingly formed a hereditary caste.[3] Productivity and the reliability of supply both stood higher than ever before, though their farming was inevitably organic and still conducive to many kinds of wildlife. It had evolved primarily through discovering and combining the best available

---

[1] F.M.L. Thompson, 'An Anatomy of English Agriculture 1870–1914', in B.A. Holderness and M. Turner (eds), *Land, Labour and Agriculture 1800–1929, Essays for Gordon Mingay* (Hambledon, London, 1991), p. 213. The tables of county statistics in this excellent survey illuminate much of the argument below.

[2] G.E. Mingay (ed.), *The Unquiet Countryside*, 1989; A. Charlesworth (ed.), *An Atlas of Rural Protest in Britain 1548–1900* (Croom Helm, London, 1983) are general surveys. See also E.P. Thompson, *The Making of the English Working Class* (Gollancz, London, 1963), ch. 7.

[3] B.A. Holderness, 'The Origins of High Farming' in B.A. Holderness and M. Turner (eds), *Land, Labour and Agriculture 1800–1929, Essays for Gordon Mingay* (Hambledon, London, 1991), 149–64; P.J. Perry, 'High Farming in Victorian Britain: Prospect and Retrospect', *Agricultural History Review*, 55 (1981), 156–66.

practices contained within the age-old pattern of European farming, though new crops from America, and especially the potato, had contributed.[4] The progressive integration of marketing systems had also gradually encouraged farmers to adapt cropping to terrain and environment instead of aiming for self-sufficiency. Science and the government had made little contribution, and the urge to tabulate the life of the nation only reached agriculture in 1866, when official statistics were first collected.[5]

Numerous agricultural machines appeared in the 1851 Great Exhibition, but they presaged rather than represented mechanization, for hand tools and muscle-power remained fundamental.[6] In 1851 the 1.8 million farmworkers slightly exceeded the entire mining and mechanized manufacturing population, and equalled that of domestic servants.[7] The health and size of most rural communities, including most market towns, therefore depended on agriculture and its support activities. It also remained a vital source of income for the Church, via tithes, and for the gentry and aristocracy, via rents. While some of this money returned to the farmers through landlords' investment, most did not. Given the extensive voluntary element in English government at all levels, this funding of Church and State was of great significance, but it constituted a heavy burden with only a distant historical justification now that the economy was no longer based almost exclusively around the land.

Between 1850 and 1870 hindsight, gives us a sense of still water at the turn of the tide. The 1870s saw the advent of mass trade unionism, renewing social conflict.[8] The rural population ceased to grow, and between 1861 and 1901 the agricultural labour force contracted by a quarter, shrinking from 19 per cent of the labour force to only 9 per cent.[9] The repeal of the Corn Laws in 1846

---

[4] The various volumes of the *Agrarian History of England and Wales* Cambridge University Press, Cambridge (hereafter *AHEW*) now provide a comprehensive work of reference. Unfortunately volume 7, *1850–1914* is still in preparation, so G.E. Mingay (ed.), *The Victorian Countryside* (Routledge, London, 1981), 2 vols, remains essential for our period. Its thematic chapters provide a context for most of what follows. *AHEW*, vol. 6, *1750–1850*, (ed.) G.E. Mingay, also contains much of value. See also M. Overton, 'Re-establishing the English Agricultural Revolution' *Agricultural History Review*, 44 (1996), pp. 1–20; J.A. Chartres, 'Market Integration and Agricultural Output in Seventeenth-, Eighteenth-, and Early Nineteenth-Century England', *Agricultural History Review*, 43 (1995), pp. 118–37.

[5] The Board of Agriculture, *Agricultural Returns for Great Britain*, published annually as Parliamentary Papers from 1866.

[6] E.J.T. Collins, 'Agricultural Hand Tools and the Industrial Revolution', in N. Harte and R. Quinault (eds), *Land and Society in Britain, 1700–1914: Essays in Honour of F.M.L. Thompson* (Manchester University Press, Manchester, 1999), pp. 57–77.

[7] J.D. Chambers, *The Workshop of the World: British Economic History 1820–1880* (Oxford University Press, Oxford, 1968), p. 15.

[8] A. Howkins, *Reshaping Rural England, A Social History 1850–72* (HarperCollins, London, 1991), J.P. Dunbabin, *Rural Discontent in Nineteenth-Century England* (Holmes and Meier, New York, 1974).

[9] P. Deane and W. Cole, *British Economic Growth 1688–1959* (Cambridge University Press, Cambridge, 2nd edn, 1967), pp. 142–4.

symbolically confirmed the end of agriculture's near-sacred status, but the feared flood of foreign wheat did not materialize immediately. By 1871 imports met about 40 per cent of British wheat consumption, but this almost exactly equalled the population rise and they topped up home production rather than competed with it.[10] This stabilized prices of the basic bread grain below crisis levels even in difficult years, ensuring that economic growth was not threatened, something previously unknown in an advanced economy largely reliant on its own farms. Ireland's awful experience of the 1840s would not be repeated on the mainland. Imports also supplied raw materials for industry on a scale beyond home production, but other cereals, meat and perishables were not affected. By the 1880s, however, farmers and landlords were complaining loudly but ineffectually about seemingly limitless and very cheap imports, first from the USA, but then from Russia and elsewhere. High farming was destabilized, and though historians have mostly abandoned the concept of a generalized economic depression at this time, agriculture forms an exception. Thus it is not surprising that F.M.L. Thompson has said that farmers came to look back on the central decades as 'a lost paradise before the fall'.[11]

Yet farmworkers' experience was very different. They are a neglected group, too often treated by historians as a largely inert factor of production whose only power of independent action was to obstruct and delay the inevitable through a fear of the consequences.[12] Without a peasantry, they formed the vast majority of the rural population and now depended mostly or entirely on selling their labour. They got very little for it, and if protest was muted, both the start of mass migration away from the countryside in the 1860s and the ferocious strikes of the 1870s indicate that peace had not stemmed from satisfaction. If rent survived without question, and if tithes survived in a modified form, the battles had largely been lost that had been fought to preserve the corresponding social entitlements of the poor, embodied in the customary concept often now called the 'moral economy'.[13] Even the franchise was denied them when it was extended to other working-class groups in 1867. The punitive diet of the workhouse was better than many families in work enjoyed, and the mood of the workers has been characterized as 'bleak fatalism', hardly the stuff of which golden ages are made.[14]

---

[10] Ag. Returns, *1871,* Parl. Papers, 1871 LXIX, p. 26; *Ag. Returns 1881*, Parl. Papers, 1882 CIX, pp. 72–3.

[11] Thompson, 'An Anatomy', p. 218.

[12] A. Armstrong, *Farmworkers in England and Wales: A Social and Economic History 1770–1980* (Batsford, London, 1988) is the only general study.

[13] See E.P. Thompson, 'The Moral Economy of the English Crowd in the Eighteenth Century', 185–257; 'The Moral Economy Reviewed', pp. 258–351, in *Customs in Common* (Merlin Press, London, 1991).

[14] D. Jones, 'Rural Crime and Protest', in Mingay, *Victorian Countryside*, p. 575; J. Burnett, 'Country Diet', in ibid., p. 561. See also B.S. Rowntree and M. Kendall, *How the Labourer Lives* (Nelson, London, 1913).

The lack of factories in the South, the heavy preponderance of arable in the South East, and the continuing reliance on bread as the staff of life has encouraged a belief that economic specialization had led an industrializing North into symbiosis with a South ever more focused on farming.[15] Farmworkers may therefore seem to have suffered for the sake of affordable food for the cities, with the suffering spreading to their employers in the 1880s. Rural separation is thus often seen as central to rural prosperity and well-being. However, recent research has undermined this vision, not by saying it is entirely untrue, but by showing it to have been part of something much more complex. It was based overwhelmingly on the experience of the South, and especially the South East, with other regions reduced to a secondary status, or ignored. Too often, the technical has been sundered from the social, or social change has been agonized over without sufficient consideration of economic context. The intention here is to seek a fresh perspective by concentrating on the rural North of England, where farming remained both more vital and more stable, where farmworkers' experience was in direct contrast with accepted stereotypes, and where links with industry seem to have been beneficial rather than a threat.[16]

II

To take specialization first, rural manufacturing was certainly in a poor state. Originally overwhelmingly based south and east of a line from the Exe to the Humber, where three-quarters of the English population and most of its wealth were also found, it had been one of the most striking features of early-modern England. Little new industry grew up in the rural South to compensate as it decayed, but farming did not fade away in the north as a simplistic specialization models suggests that it should have. Factories took up little land, and much of that was of poor quality. The good land, and there is a great deal of it, was developed rather than given up. It was a lack of demand, isolation, thin populations, and poor drainage that led to agrarian underachievement in the early-modern North, and the nineteenth century ended all that. Tables 4.1 and 4.2 suggest that the result was diversity and dynamism, and the north had more types of farming and of employment structure than any comparable

---

[15] See, for instance, M. Berg, *The Age of Manufactures 1700–1820: Industry, Innovation and Work in Britain* (Routledge, London, 1994), pp. 106–7.
[16] P.J. Perry (ed.), *British Farming in the Great Depression, 1870–1914: An Historical Geography* (David and Charles, Newton Abbott, 1974), was seminal in establishing the variety of regional experiences. See also A. Howkins, 'The Marginal Workforce in British Agriculture', *Agricultural History Review*, 42 (1994), pp. 49–62.

English area.[17] By 1871 the combined corn acreage of Yorkshire and Lancashire exceeded that of Norfolk and Suffolk. Most of the land, as elsewhere, was controlled by great estates, and large farms of 1,000 acres or more were not unusual in places. However, very small farms of 50 acres or less were also common in several areas. This included Lancashire, whose agriculture arguably became the most successful in England after 1850 as mossland was reclaimed to create an isolated but highly effective western zone of high-yield, high-wage, arable farming.[18]

Table 4.1 Comparative Corn and Wheat Acreage, 1871–1901, Selected English Counties

County	% Mountain and heath, 1901	Acreage of corn, 1871	Acreage of wheat, 1871	Av. wheat yield, 1891–1900, bushels per acre
Northumberland	36.6	152,108	41,716	32.42
Yorks., N. Riding	23.3	229,762	74,374	29.65
Yorks., W. Riding	12.7	262,372	103,910	28.43
Lancashire	7.8	107,057	36,765	31.71
National average	7.1			29.93
Suffolk	3.4	390,441	157,915	29.24
Norfolk	3.4	458,527	206192	32.05
Yorks., E. Riding	0.3	281,548	116,139	30.93
Lincolnshire	0.1	624,185	306,238	33.65

Source: *Ag. Returns*, Parl. Papers. Wheat and corn acreage, 1871 LXIX, pp. 40–45; remainder, 1902 CXVI, Pt. 1, pp. xiv, 34, 52; these figures not available for previous years.

Armstrong has pointed out that in 1841, 'the West Riding employed more farmworkers than Wiltshire, while those of Lancashire approached twice the numbers in, say, Berkshire, Dorset or Oxfordshire'.[19] Moreover, whereas a prime cause of poverty in the south was the effective casualization of almost all labourers by 1850, casual labour in the north remained a matter of choice for men, and a mutually satisfactory way of employing women, tramps and Irish

---

[17] A. Edwards and A. Rogers (eds), *Agricultural Resources: An Introduction to the Farming Industry of the United Kingdom*, (Faber, London, 1974), fig. 9.12, p. 207. Contrast the map of earlier regional farming in *AHEW*, 6:2, p. xx.
[18] T.W. Fletcher, 'Lancashire Livestock Farming During the Great Depression', *Agricultural History Review*, 9 (1961), pp. 17–42; J. Walton, 'The Agricultural Revolution and the Industrial Revolution: the Case of North-West England, 1780–1850', in C. Bjorn (ed.), *The Agricultural Revolution Reconsidered* (Landbohistorisk Selskab, Denmark, 1998).
[19] Armstrong, *Farmworkers*, p. 85.

harvesters at peak times.[20] Before 1850, threshing machines had mostly been welcomed, and after 1870 Joseph Arch testified that 'about Newcastle and those northern districts the men were much better paid and they said "the union is a good thing, but we are well off and can get along without it"'.[21] All wage data suggests that the northern high-wage zone identified by Caird in 1851 persisted throughout the nineteenth century.[22] This helps explain a striking feature of northern life everywhere except in south Lancashire, the hiring of farm servants who were paid yearly or half-yearly and who lived in the farm houses, boarded and lodged as part of their wages.[23]

Table 4.2 Paid Male Farmworkers, Selected Counties

County	Wages, 1850	Agric. acreage, 1866	Workers, 1851	Workers, ratio per hundred acres	Workers, 1901	Change, % 1851–1901
Lancs.	13s 6d	708,827	29,388	4.2	21,057	-28.4
Lincs.	10s 0d	1,387,826	38,713	2.8	38,332	- 1.0
Norfolk	8s 0d	1,009,087	38,644	3.8	36,418	- 5.8
N'land	11s 0d	656,989	11,939	1.8	7,781	- 34.8
Suffolk	7s 11d	775,404	35,305	4.6	29,810	- 15.6
E. Riding	12s 0d	612,084	14,162	2.3	18,365	+29.7
N. Riding	11s 0d	760,778	14,062	1.9	20,944	+48.9
W. Riding	14s 0d	1,094,152	29,360	2.7	25,319	- 13.8

Sources: Wages, Lord Ernle, *English Farming Past and Present*, 6th edn. 1961, app. IX; acreage, *Ag Returns*, Parl. Papers, 1866, LX, 1; remainder, *Census of England and Wales*: 1851, *Occupations of the People*, summary table XXVIII, pp. ccxlv–ccxlvii; 1901, *Occupations of the People*, table 32.

Usually seen as a relic of early-modern times, service in the North (and Scotland and Wales) flourished and diversified in the mid-nineteenth century. It secured the willing labour of the young and unmarried, a large group in any rapidly growing population, and used their time to the full by keeping them on the farm. A sixteen-hour day was not uncommon, in return for good food and very high pay by southern standards. In Northumberland, service even drew in

---

[20] S. Caunce, *Amongst Farm Horses: The Horselads of East Yorkshire* (Alan Sutton, Stroud, 1991), pp. 25–9, 194–6, 223–4; A. Howkins, *Poor Labouring Men, Rural Radicalism in Norfolk 1870–1923* (Routledge, London, 1985), pp. 8–10.
[21] J. Arch, *The Story of his Life, Told by Himself* (Hutchinson, London, 1898), p. 221.
[22] J. Caird, *English Agriculture in 1850–51* (Longman, London, 1852), frontispiece map. E.H. Hunt, *Regional Wage Variations in England* (Clarendon, Oxford, 1974).
[23] Caunce, *Horses*; S. Caunce, 'Farm Servants and the Development of English Capitalism', *Agricultural History Review*, 45 (1997), pp. 49–60. See also A. Kussmaul, *Servants in Husbandry in Early Modern England* (Cambridge University Press, Cambridge, 1981).

married men, with cottages provided around the farmstead, and single men were unsuccessfully pressured to hire 'bondagers', that is female servants who would work in the fields at peak times. Hereabouts farming had expanded dramatically even by northern standards, labour and cottages were both in very short supply.[24] It is the most extreme case of the generally tight northern rural labour markets, which everywhere gave workers substantial bargaining power. Table 4.2 shows that this translated into relatively high wages, though they were still below those paid in industry.[25] Industrial employment grew less in the Midlands and was more concentrated, and neither here nor in the South could farming expand as it did in the North. Moreover, while we undoubtedly see a less fractured rural community than in the South, this did not detract from profits. East Yorkshire farm servants were officially pronounced the best-fed English working-class group in the 1860s, just after their employers were described as 'the richest men of their class in the country'.[26]

In the North migration was common, especially among those leaving service. Few northern farming districts were without an obvious industrial counterpart even where they seemed utterly rural, like north Northumberland or the East Riding of Yorkshire. However, Redford long ago confirmed a contemporary view that 'no appreciable movement had taken place from the southern counties into the manufacturing districts'.[27] The Poor Law settlement system did not stop intra-regional migration, but it did cripple long-distance mobility. In so far as southerners moved, it was along the tried and tested road to London, not the novel and unknown one to the North. Despite enclosure, the southern farm labour force therefore continued to increase up to the 1850s even though the extra work was not there to employ them usefully. Wages were fixed by dividing up what money remained after those with powerful entitlements had taken the share they felt they needed. By 1850, jobs were found for nearly everyone who needed one, partly to avoid any return to the disorder many farmers remembered very clearly, and partly to maintain enough workers for the increasing needs of harvest-time.[28] Mechanization made little sense in this setting until after 1870, when migration finally began to make

---

[24] Dunbabin, *Rural Discontent*, ch. 7.

[25] Extensive but differing patterns of payments in kind make comparing real income and welfare levels hard. J.P. Dunbabin, 'The Revolt of the Field: The Agricultural Labourers' Movement', *Past and Present*, 26 (1963), p. 73, noted that 'the greater welfare of the north must have been ascribable as much to a different pattern of life ... as simply to higher wages'. See also Armstrong, *Farmworkers*, p. 138.

[26] *Sixth Report of the Medical Officer of the Privy Council*, Parl. Papers 1864, XXVIII, p. 238; Caird, *English Agriculture*, p. 310.

[27] A. Redford, *Labour Migration in England 1800–1850* (Manchester University Press, Manchester, 1968), pp. 62–7, See also tables 4 and 5, p. 127, and maps D and E.

[28] E.J.T. Collins, 'Harvest Technology and Labour Supply in Britain, 1790–1870', *Economic History Review*, 22, (1969), pp. 453–73. When wages later rose, this was more than counterbalanced by the loss of jobs, Thompson, 'An Anatomy', pp. 215–16.

inroads, whereas in the North its advantages had always been worth considering. There, towns and machines did not go hand in hand with destitution, but with prosperity.

III

Similarly, while it may seem paradoxical to associate peasant farming with urbanization, small family farms multiplied and flourished in the north. This is readily comprehensible on poor hill-land far from a potential market, where cash for wages was difficult to find. Families too small to manage hired a servant or two, but that was all. However, the same attitudes were evident in south Lancashire on prime arable land, as well as along the Pennine edge of that county and of West Yorkshire. In fact, in 1871 Lancashire and the West Riding of Yorkshire had 14.4 per cent of all holdings under 20 acres and 17.2 per cent of land in such holdings, compared to 9 per cent of national acreage.[29] Pennine pasture farms were located on rough terrain, but they mostly lay very close to at least one town, and often to several due to the complex and unconventional pattern of northern urbanization.[30]

Thus, whereas London was always a single, highly centripetal conurbation, northern towns spread out as separate and independent entities. Industrial development built towns into cities, but it also turned villages and even hamlets into towns as well, partly due to intense local specialization of activity. This is true even around Newcastle, the most concentrated cluster, but it reached the ultimate in the textile district of modern West Yorkshire. To this day this conurbation has no centre, and many constituent towns remain physically separate. Mining in this period was rarely associated with towns except by geographical accident, and many miners lived a thoroughly rural life, often growing a substantial part of their own food. Intensive inter-urban canal and railway systems brought steadily increasing levels of dispersed employment to the countryside between. Travellers may have been appalled by the pollution and environmental degradation associated with areas like south Lancashire and the Black Country, but even here the first Ordnance Survey maps show how limited urbanization was. Textiles provided a lot of organic waste for manure, and pollution did not impede farming except in the immediate vicinity of really

---

[29] *Ag. Returns, 1871*, p. 21. Note that while Suffolk had 1.78 per cent and 1.58 per cent respectively, Norfolk was well above the national average with 4.84 per cent and 3.64 per cent. See also D. Grigg, 'Farm Size in England and Wales, from Early Victorian Times to the Present', *Agricultural History Review*, 35 (1987), pp. 179–89.

[30] S. Caunce, 'Urban Systems, Identity and Development In Lancashire and Yorkshire: A Complex Question', in N. Kirk (ed.), *Northern Identities* (Ashgate Publishing, Aldershot, forthcoming).

obnoxious processes. Occasionally it even helped, inhibiting pests and diseases among the market gardens of south-west Lancashire, for instance.

These small farms on the urban fringes usually depended absolutely on their proximity to a big and reliable market.[31] Halifax's eighteenth-century local historian observed that there were no real farming families in this Pennine parish, an area nearly as big as Rutland.[32] Much land was then waste, and the rest was mostly occupied by clothiers who managed holdings primarily as an adjunct to business, though they grew some crops for their own use. By 1850, factories were rendering the rural clothier's lifestyle untenable, but the descendants of clothiers were turning into highly-commercialized microfarmers and land was being brought in from the waste on a large scale. Our period was one of constant change as the railways began to open up town markets to distant suppliers, but strong wholesale systems took some years to develop. In the interim even cereal production became commercially possible on freshly-enclosed land in places like Erringden, a Pennine township west of Halifax with no village centre and an exposed situation over 820 feet high, especially as many local townsfolk still willingly ate oatbread.[33] As wheat became the rule and distant supplies became competitive, production shifted towards speciality crops like rhubarb, and especially to liquid milk for direct sale to consumers.[34] Landlords encouraged all this because they got more rent from a mass of small farms, typically about 15 to 20 acres, and they actively developed moorland by enclosure and by building large numbers of laithe houses (literally, barn houses), the specialist farmstead of the mid-Pennines. They housed the farmer, the crops and the animals in one range under one roof, though with separate quarters, unlike the traditional longhouse of further north.[35]

In south Lancashire a similar process could be observed on newly-drained mossland. In 1879 William Hornsby, an ex-miner, brought his family to 'a treacherous bog with a few ridges of firm clay for pathways. They had come to

---

[31] M. Winstanley, 'Industrialization and the Pastoral Economy: Family and Household Production in 19th Century Lancashire', *Past and Present*, 160 (1996), pp. 157–95.

[32] J. Watson, *The History and Antiquities of the Parish of Halifax* (T. Lowndes, London, 1775), pp. 8–9. See also R. Brown, *A General View of the West Riding of Yorkshire, Surveyed... 1793* (Edinburgh, 1799), pp. 223–9.

[33] W.B. Crump, *The Little Hill Farm* (Halifax Antiquarian Society, Halifax, 1949). See R. Scola, *Feeding the Victorian City: The Food Supply of Manchester, 1770–1870* (Manchester University Press, Manchester, 1992), chs 2–5 for the development of market systems in the North West.

[34] W.H. Long, *A Survey of the Agriculture of Yorkshire* (HMSO, London, 1969), pp. 60–71. This was quite different from urban cow-keeping.

[35] Royal Commission on Historical Monuments of England, *Rural Houses of West Yorkshire* (1986), pp. 178–83.

live in a bare wooden shack in a misty trackless wilderness'.[36] This land at Rixton had never been cultivated before, but it was only 8 miles from the very centre of Manchester. The industrial areas were thus surrounded and interpenetrated by farmers who responded eagerly to incentives to produce all sorts of crops, and who were free of the constraints imposed by well-established systems and agricultural communities. Wage labour was scarce because industry was everywhere, and paid better than any farmer.[37] Indeed, it was widely accepted that small farmers probably lived worse than a labourer and worked harder. Even the laithe house, impressive as it is as a whole, usually included only a one-up, one-down cottage as the farmer's share. It was the willingness of families to exploit themselves in search of independence and a rural life which sustained them, and like the clothiers of old they combined social conservatism with innovation and enterprise. On arable farms they specialized in meeting town demand and sold almost as soon as crops were picked, or even grew on contract. The dairy farms pioneered an industrialized approach, where the farmstead was simply a production base, with the milking herd and much of the feed bought in, and imports proved invaluable in cheapening production costs.[38] This was quite different from the patterns that developed around London, with their heavier emphasis on hay production and market gardening, for instance.

Agriculture was thus not withering in the North, and neither was it the only big market for farm products. London stood squarely at the functional centre of the South, with a superb system of wholesale markets and well-established communications with all the rural areas, often water-borne. It was the world's largest and wealthiest city, and its population was roughly equivalent to that of all the manufacturing areas added together.[39] Moreover, convenient assumptions that almost all arable production was concentrated in the South East, and that the link with low wages was one that reflected the nature of arable farming, are clearly not true. The high-wage zone was far from being entirely pastoral, indeed it was less so than the South West, which was at least as badly-paid as East Anglia. The Midlands saw most conceivable combinations of farming systems, employment systems, and wage levels.

---

[36] K. Fryer, *The Farmers and the Rest: Rixton in the Nineteenth Century* (Hollinfare Publications, Rixton, 1995), pp. 9–10.

[37] Margaret Penn's father was a farmworker in Glazebrook early in this century, but despite enjoying the work he (and his son) were drawn out by the much higher wages available, M. Penn, *Manchester Fourteen Miles* (Cambridge University Press, Cambridge, 1981).

[38] The oil-seed milling industry of the Humber provided plenty of cattle cake from an early date, see Holderness, 'Origins', p. 154.

[39] B.R. Mitchell and P. Deane, *Abstract of British Historical Statistics* (Cambridge University Press, Cambridge, 1962), p. 19; E.A. Wrigley, 'A Simple Model of London's Importance in Changing English Society and Economy 1650–1750', *Past and Present*, 37 (1967), pp. 55–8

Lincolnshire straddled the North–South divide in all these respects, and despite its enormous agricultural acreage it receives far less attention than it should.

IV

It thus seems that far from agriculture, and those who made their living by it, thriving when separated from industry, taken as a whole it suffered and ossified even though the farmers themselves might prosper. East Anglian agriculture was most dynamic before 1800 when the region was a hive of manufacturing activity, and a local Golden Age defined for the community as a whole probably lay well in the past in 1850. Third World experience today shows that specialization is rarely beneficial when it concentrates resources into only a few productive sectors, still less into one primary sector. That usually reflects a core-periphery relationship with a highly unequal power balance rather than rational economic specialization.[40] The concentration of so much southern economic activity in London and the lack of enterprise shown by southern manufacturing both suggest that this was the fate of the rural South, and especially East Anglia. The North as a whole diversified rather than narrowing its economic range, and there was little sign of real agricultural depression after 1870 in places like East Yorkshire.[41] The Royal Commission on Labour's report in 1893 on a visit to Driffield, one of its leading market towns, shows that despite a clear expectation that structural problems would be found, the rural economy there was still working well.[42]

The English countryside could not be isolated from change. The rail network spread out and the best-situated market towns prospered, but often at the expense of neighbours less well placed. Villages lost non-agricultural functions because people could get around more, but village shops thrived as previously unknown goods came in. Craftsmen seemed certain to lose out to factory competition, but they mostly adapted and developed links with the new producers. Local customers valued local styles of tools and vehicles, and true mass production was a long way off for most things. Transport systems remained too expensive for national markets to develop in items like farm carts and wagons, by far the most common items found on farms, so wheelwrights prospered as farming intensified and more and more goods were shuttled back and forth around farms and between farms and the outside world. Cheap iron

---

[40] For the strength of ties between East Anglia and London, see *AHEW*, 6, pp. 215–22. On core–periphery relationships, see, for instance, C. Stoneman and J. Suckling, 'From Apartheid to Neocolonialism?', *Third World Quarterly*, 9 (1987), pp. 515–44.

[41] S. Caunce, 'Complexity, Community Structure, and Competitive Advantage Within the West Yorkshire Woollen Industry', *Business History*, 39 (1997), 26–43.

[42] *The Royal Commission on Labour,* Parl. Papers 1893–94, XXXV, B-II, Driffield, pp. 53–4. See also Thompson, 'An Anatomy', pp. 230–37.

benefited the blacksmith, and there was no reduction in the numbers at work in any of the three ridings of Yorkshire, for instance.[43] Rather than making everything from scratch, saddlers could use the products of Walsall factories to assemble harness according to local taste.[44] Other products did threaten traditional local manufacturing, but only where people saw a superior, cheaper product that often made life easier and more convenient for ordinary people.

This all contradicts many well-established views, notably Lee's regional specialization model which compared the share of employment in agriculture, mining, textiles and services for English counties at this time.[45] Agriculture then appears to be concentrated in the South, but only because relative figures were used. An absolute measure of labour produces quite a different result, as table 4.3 shows.

Table 4.3 Employment in Agriculture, 1871 Census Regions, per 1,000 acres

Region	'000 acres	Numbers	Ratio
London	78	27,918	35.8
South East	4,066	236,283	5.8
South Midlands	3,201	224,257	7.0
Eastern	3,214	211,578	6.6
South West	4,994	277,162	5.6
West Midlands	3,865	228,628	5.9
North Midlands	3,541	187,557	5.3
North West	2,000	147,217	7.4
Yorkshire	3,655	171,290	4.9
Northern	3,492	98,828	2.8***
England	**32,106**	**2,050,401**	**6.4**
Wales*	5,219	190,806	3.7***
Scotland	N/a**	239,683	

* includes Monmouth
** too much poor land for meaningful result
*** high proportion of poor land
Source: *Census, 1871*, tables 8 and 10.

There were relatively small differences in the numbers per hundred acres, and if any areas stand out as unusually committed to agriculture, it is the urbanized regions and especially London. In Lee's calculations they were dwarfed by the industrial and service sectors in these areas, whereas in the rural South there

---

[43] A. Raithby, 'Yorkshire Blacksmiths and Industrialisation', unpublished research report, Kirklees Museums Restoration Unit, 1983, in author's possession.
[44] S. Caunce, *Oral History and the Local Historian* (Longman, London, 1994), pp. 34–9 and 161.
[45] C.H. Lee, 'Regional Growth and Structural Change in Victorian Britain', *Economic History Review*, 33 (1981), pp. 438–52.

was little else. All Lancashire's agricultural population crowded together equalled the inhabitants of Salford, Oldham and Preston, but even such numbers did not constitute an imposing presence compared to the number of towns and cities in the North West. Similarly, reclaiming the mosses, the Pennine moorlands and the Northumbrian plain involved investment, sometimes heavy investment, but compared to that in industry, it does not impress as it would in a rural county.

V

If we are simply judging in terms of production and human survival, mid-nineteenth-century rural England generally compared well to the Middle Ages, or even to contemporary Ireland. There was no mass starvation despite an unprecedented rural population and an even bigger urban one. The manufacturing and service economy of London and the manufacturing towns could just absorb those who were determined to leave the land. This unprecedented wealth and the most benign commercial environment agriculture had ever seen, or was ever likely to see, could constitute a case for declaring a Golden Age. Yet endurance was the dominant theme for most of those living in the countryside, so any overall assessment has to be complex and conditional.

The aristocracy and gentry got high rents and maintained their lifestyles, but the end of protection also ended their guaranteed claim, as a group, to a sufficient income to maintain themselves at a level commensurate with their historic role. They had to invest in their estates as never before and they still suspected, rightly, that the days of the great estate as a basis for political dominance were drawing to a close. Certainly, their authority was already challenged in ways not seen before. For them this was a last fling as they started reworking their place within a nation that was changing fundamentally. Landowners' incomes declined by a quarter or more by 1914, and influence drained away, but it was a slow process, not a cataclysm that would make them see the twenty years before 1870 as their apogee.[46]

Everywhere, substantial farmers established themselves as a genuine part of the middle class. A yeomanry gave way to a group of employers and capitalists who took unprecedented risks, with heavy investment and unprecedented dependence on markets. The central decades could be seen as a time of fool's gold built on illusions, for the markets that had brought prosperity were to let many down badly after 1870, especially where social expectations had developed that could not be maintained in a more normal business

---

[46] J.V. Beckett, *The Aristocracy in England 1660–1914* (Blackwell, London, 1986), pp. 456–67.

environment.[47] Few modern farmers can afford the detachment from their enterprises that characterized these decades, and the later American term 'gilded age' seems most appropriate.

Farmworkers did not see even an illusion of gold. They were usually deferential and many fully accepted a hierarchical society, but they did not accept the arbitrary control that was all too often their lot in the south, living in tied cottages and instead of sufficient wages, receiving essential goods and services as charity. They were often threatened with dismissal if they combined, or even attended chapel instead of church. Everywhere, what they wanted was the right to live their own lives with the security that came from plenty of work; at least a living wage; and perhaps the chance of access to farms of their own. In the North, they got these things in moderation, though no one could pretend that they lived well in any absolute sense and the farming ladder was weak in arable areas dominated by large farms.[48] Southern labourers could hardly keep body and soul together, and their resistance before and after the central decades was a logical and sometimes successful response to their situation, which was primarily aimed at righting wrongs rather than tearing down for a fresh start. Howkins and Wells have shown that the countryside could identify firmly and fiercely with Chartism and other aspects of the urban working-class struggle, but the majority never showed real political radicalism.[49] What they did with most effect was simply to walk away. People had always left the countryside for the towns, but the rising flood that was developing in the south by 1870 signifies a real change in attitudes.[50] Dire poverty was no longer accepted as inevitable and they had no regrets about the new environment of the 1870s. If employment fell thereafter, at least wages stayed higher for those in work and the rest could leave more easily. In the light of northern experience, their plight was contingent rather than a stage in a determinist drama.

Urban–rural interactions were complex and agriculture was already an integral part of an industrializing and urbanizing socio-economic system in 1850, not a self-contained unit threatened and corroded by an aggressive industrial system. Taking England as a whole, what really emerges is neither a sunny lost paradise, nor a simple living hell. There had never been more people

---

[47] In the depressed years, northern and Scottish small farmers made bankrupt southern farms viable simply by changing cropping patterns away from approved local norms and by working themselves, Perry, *Farming*, pp. 98–101.

[48] A. Mutch 'The "Farming Ladder" in North Lancaster, 1840–1914: Myth or Reality?', *Northern History*, 27 (1991), pp. 162–83.

[49] Howkins, *Reshaping*, p. 180. R. Wells, 'Rural Rebels in Southern England in the 1830s', in C. Emily and J. Alvin, *Artisans, Peasants and Proletarians, 1760–1860* (Croom Helm, London, 1985).

[50] Farmworkers' unions always supported migration and emigration, see for instance Armstrong, *Farmworkers*, pp. 116–17.

in the countryside but economic activity could not rise in line with population because of the near-universal poverty. Farmers and landlords took what they regarded as their due regardless of the consequences for the rest, but the real cause of general misery was the structural failure of the southern rural economy. Where previously there had been manufacturing jobs in large numbers, now the overwhelming majority had to depend on farming, or go to London. This was not development, but regression.

To declare it a Golden Age is to allow the experience of the farmers to overrule that of other groups, and to celebrate production as an end in itself. The 'Golden Age' concept actually tells us more about the structure of rural society than anything else, with perception at least as important as fact. This period stood out in much higher relief in the South than in the North, and what was really at issue was a sense of lost power that had never really developed there. The Golden Age for southern farmers existed only as a short-lived and inevitably transient combination of circumstances in a country where long-term stability was simply not a possible outcome. It was unsustainable because the levels of coercion needed to maintain so exploitative a system no longer existed, and could not exist in the England of the time.

Chapter 5

The Cotton Industry in the 1850s and 1860s: Decades of Contrast

Geoff Timmins

That the history of the cotton textile industry during the mid-Victorian years can be accommodated comfortably within the notion of a Golden Age might at first sight seem highly problematical. After all, this period can be viewed as one dominated by the much-rehearsed events of the Cotton Famine (1862–65), with an over-supply of cotton goods causing output and employment in the industry to plummet and a massive and sustained relief effort being required to minimize distress and ease the threat of social unrest. The Cotton Famine era also saw sharp increases in the price of raw cotton, adding substantially to production costs and hence to the price of cotton goods, and heightened anxiety about future supplies, the more so because of the high dependency that had emerged on American sources.[1] Additionally, trade cycle downturns brought periods of recession in 1858 and 1868–69, causing pronounced falls in the profitability of cotton firms.[2] And there are indications that the long-term growth rate of the industry, which had generally shown an increase each decade since the early eighteenth century, was slowing down, a reflection of the intensity with which overseas competition was being generated.[3]

Yet an alternative and altogether less gloomy interpretation of the cotton industry's history during the mid-Victorian years can be offered. It emphasizes both the remarkable progress achieved during the 1850s and early 1860s when consumption of raw cotton reached unprecedented levels and the sustained, albeit protracted, recovery the industry made from the Cotton Famine, culminating in the high prosperity of the early 1870s. This progress arose from

---

[1] On economic change during the Cotton Famine years, see especially W.O. Henderson, *The Lancashire Cotton Famine, 1861–65* (Manchester University Press, Manchester, 1969, reprint of 1934 edition), chs 1–3 and D.A. Farnie, *The Cotton Industry and the World Market, 1815–1896* (Clarendon Press, Oxford, 1979), ch. 4. The bibliographies given in these works can be updated from T. Wyke and N. Rudyard, *Cotton: A Select Bibliography on Cotton in North West England* (Bibliography of North West England, Manchester, 1997), pp. 50–54 and G.R. Boyer, 'Poor relief, informal assisance, and short time during the Lancashire cotton famine', *Explorations in Economic History*, 34 (1997), pp. 56–76.

[2] R. Boyson, *The Ashworth Cotton Enterprise* (Clarendon Press, Oxford, 1970), pp. 67–9.

[3] Farnie, *Cotton*, pp. 89–96.

long-term developments taking place within the industry, including technological innovation, as well as from expansion of sales both at home and abroad, despite foreboding about the crucial Indian market. The industry continued to benefit from the enhancement of external economies with which both its long-term growth and its increasing localization in Lancashire were strongly associated. Moreover, as has often received comment, the mid-Victorian years saw a general transformation of labour relations in British industry, cotton production included, as the intense disruptions of the Chartist era were left behind.[4]

This chapter argues that the 1850s and 1860s comprise more a period of progress than of setback for the cotton industry, despite the profound difficulties encountered during the Cotton Famine years. A start is made by assessing the extent to which the industry grew during these decades and the structural change which accompanied this growth. A second section discusses supply-side influences on the industry's expansion, not only those arising within the industry, but also those from which industrialists more generally were able to profit. The final section considers how the growth of cotton manufacturing was influenced by changes in market conditions.

The Extent and Nature of Growth

Though far from ideal for the purpose, figures of retained raw cotton imports indicate the rate at which the mid-Victorian cotton industry expanded.[5] In table 5.1 the first column for each year gives the total volume, the second column the volume obtained from the United States and the third column the volume of re-exports. A key feature of the series is the strong upward trend during the 1850s and early 1860s, with supplies from the United States predominating. It was a trend that culminated in a period of remarkably high activity for the cotton industry, the *Manchester Guardian* reporting that 1859 had been the most profitable year for spinners and manufacturers 'of any known for a very considerable period' and that, at the end of 1860, two years 'of almost unexampled prosperity' had been achieved.[6] Not only were the cotton-makers able to benefit from record supplies of cheap raw cotton from the United States, but also from a heavy, though temporary, demand for cotton goods in the Far East.[7]

---

[4] See, for example, J. Belchem, *Industrialisation and the Working Class* (Scolar Press, Aldershot, 1990), pt 2.
[5] B.R. Mitchell, *British Historical Statistics* (Cambridge University Press, Cambridge, 1988), p. 334.
[6] *Manchester Guardian*, 2 January 1860, 1 January 1861.
[7] Farnie, *Cotton*, p. 138.

Table 5.1 Volume Retained Raw Cotton Imports, 1850–59

	Total vol.	Vol. from USA	Vol. re-exports
1850	664	493	102
1851	757	597	112
1852	930	766	112
1853	895	658	149
1854	887	722	123
1855	892	682	124
1856	1024	780	147
1857	969	655	132
1858	1034	833	150
1859	1226	962	175
1860	1391	1116	250
1861	1257	820	298
1862	524	14	215
1863	670	6	241
1864	893	14	215
1865	978	136	303
1866	1377	520	389
1867	1263	528	351
1868	1328	574	323
1869	1221	457	273

Yet the raw cotton import figures indicate that British cotton manufacturers were already beginning to encounter difficulties in 1860. In that year, re-exports of raw cotton rose to 17.9 per cent of total raw cotton imports, having not exceeded 14.5 per cent during the latter half of the 1850s, an indication that sales were becoming increasingly difficult to make to British cotton spinners. During 1861, the re-export figure rose to 23.7 per cent and at the start of 1862, the *Manchester Guardian* drew attention to the cotton industry's deteriorating situation:

> In the early part of the past year, the much talked of prosperity of the Lancashire mill owners was found to be fast declining and before the end of four months to be giving place to an adverse state of trade, in some degree, actually experienced by many, and, at least, felt to be rapidly approaching by the great body.[8]

Part of the problem was the loss of sales to the southern states of America with the outbreak of the Civil War, which also threatened to reduce raw cotton supplies and to increase their price. Indeed, as is evident from Figure 1, raw cotton imports from America all but ceased between 1862 and 1864, whilst the

---

[8] *Manchester Guardian*, 1 January 1862.

average price of the middling Orleans grade that America supplied rose from an average of 9d per pound in 1860 to 27.75d per pound in 1864.[9] But Eastern markets were also reported as being glutted and most others 'at least fully supplied', so that, not surprisingly, Britain's exports of cotton yarns and cloths fell sharply in 1862.[10]

Whether glutted markets (bringing a fall in demand for cotton goods) rather than inadequate supplies of cotton (resulting from the civil hostilities in America) were responsible for the cotton industry's crisis between 1862 and 1865 remains a matter of dispute. Current orthodoxy stresses the former, though David Surdam has recently questioned its validity, arguing that the world demand for cotton goods did not weaken significantly until the price of raw cotton rose sharply late in 1861 as the Union blockade on southern American ports began to bite. During 1862 increased prices of cotton textiles reduced demand for them, and this in turn reduced demand for raw cotton.[11]

The high price of raw cotton encouraged non-American producers to raise their supply to the British market and from 1863 raw cotton imports into Britain gradually recovered. Additionally, the cessation of the southern ports blockade helped to revive American supplies during the mid-1860s, though they remained relatively scarce throughout the decade. As demand for cotton goods also revived, production began to recover, though it did so in an uncertain business climate when prices of both raw cotton and of cotton manufactures, whilst gradually falling, remained far higher than in the pre-Famine era. Furthermore, falling prices created insecurity for cotton merchants and manufacturers, since they risked having to sell at low prices in relation to the amount they paid for their stocks of raw materials or finished goods, hence squeezing profit margins and increasing the risk of bankruptcy. By 1866, however, the volume of raw cotton imports had been restored to pre-war levels, as had raw cotton prices by the closing months of the following year.[12] Even so, 1866–68 proved to be years of uncertainty for the cotton industry, bringing 'great trial and heavy loss for the spinners and manufacturers of cotton' and it was not until the second half of 1869 that they moved into a profit earning position, with modest price rises in cotton manufactures exceeding those in raw cotton.[13]

Despite its fluctuating fortunes the mid-Victorian cotton industry expanded appreciably, as the raw cotton import figures indicate. Taking two fairly similar

---

[9] Henderson, *Famine*, p. 147.
[10] *Manchester Guardian*, 1 January 1862.
[11] D.G. Surdam, 'King cotton: monarch or pretender? The state of the market for raw cotton on the eve of the American civil war' *Economic History Review*, 51(1998), pp. 113–32.
[12] J. Kelly, 'The End of the Famine: The Manchester Cotton Trade, 1864–67 – a Merchant's Eye View' in N.B. Harte and K.G. Ponting (eds), *Textile History and Economic History* (Manchester University Press, Manchester, 1973), pp. 358–80.
[13] *Manchester Guardian*, 1 January 1870; Farnie, *Cotton*, pp. 164–5.

points on the trade cycle, retained cotton imports rose from an annual average of 621 million lbs between 1849 and 1851 to one of 1,155 million pounds between 1869 and 1871, an increase of 86 per cent. Moreover, using the raw cotton import figures to calculate the value of output suggests an even more impressive growth record. According to Deane and Cole's estimates, the value of the cotton industry's gross output at current prices rose from £45.7 million between 1849 and 1851 to £104.9 million between 1869 and 1871, a rise of 130 per cent. Whilst such figures are no more than approximations, they do suggest a much higher growth rate for the industry compared with previous decades. Thus, between 1829 and 1831, the annual value of gross output averaged £32.1 million, rising by only 30 per cent during the 1830s and 1840s. Furthermore, the same series of figures shows that the gross output value attained in 1869–71 was not exceeded throughout the remainder of the century, a testimony to the strength of the trade cycle upturn in the early 1870s.[14]

Cotton import figures apart, the growth of the mid-Victorian cotton industry is evident from employment data obtainable in the 1851 and 1871 census returns. However, because the data relate to a highly mechanized industry, they are likely to understate considerably the extent of growth that occurred. The figures for Lancashire, by far the most important area of production, are given in table 5.2.[15] Those employed in the spinning and weaving branches are grouped together as cotton manufacturers.

The table reveals a reduction of 12 per cent in printing and dyeing jobs. This trend may be more a reflection of growing mechanization than of declining output, though exports of unfinished (grey) cloths were high as in the Indian market where consumers preferred the designs executed by local printers.[16] More generally, however, employment in the Lancashire cotton industry rose to a marked degree, a net gain of no fewer than 80,974 jobs being achieved, a rise of 26 per cent. This figure is all the more impressive since it occurred despite the loss of at least 13,000 jobs in cotton handloom weaving.[17]

That the great majority of the new jobs absorbed female labour suggests that they mostly arose in weaving rather than in spinning. Part of the reason for this view is that the spinning branch, in which the mule predominated, largely confined women to piecing and preparatory work (carding and roving). Spinning itself remained almost exclusively a male occupation. Weaving, by

---

[14] P. Deane and W.A. Cole, *British Economic Growth, 1688–1959* (Cambridge University Press, Cambridge, 1962), pp. 186–7.
[15] 1851 Census, *Occupations of the People*, pp. 632, 635; 1871 Census, *Occupations of the People*, pp. 423, 426.
[16] Farnie, *Cotton*, p. 101.
[17] G. Timmins, *The Last Shift* (Manchester University Press, Manchester, 1993), p. 136.

Table 5.2 Numbers Employed in the Lancashire Cotton Industry, 1851 and 1871

Year	Employment Category	Males	Females	Total
1851	Cotton Manufacturing	136,313	150,763	287,076
	Cotton packer & Presser	1,570	-	1,570
	Calico Printing	8,634	866	9,500
	Calico dyeing	2,802	-	2,802
	Fustian Manufacturing	2,954	1,988	4,942
	Other cotton and flax workers	3,732	551	4,283
Totals		156,005	154,168	310,173

Year	Employment Category	Males	Females	Total
1871	Cotton Manufacturing	148,813	227,234	376,047
	Calico Printing	6,451	676	2,951
	Calico Dyeing	1,781	-	6,451
	Fustian Manufacturing	2,275	3,390	5,171
	Others	225	302	527
Totals		159,545	231,602	391,147

contrast, made use of female labour at every stage of the process.[18] Also at issue is the varying rate at which mechanization was taking place in the two branches. The growing use of the self-actor mule virtually mechanized spinning, though piecers – often females – were required in some numbers to mend threads that broke as spinning took place. Improvements to the powerloom also continued but not to the extent that one weaver could always manage four looms, the number that was to become the industry norm for non-automatic looms by the late Victorian era. Accordingly, labour requirements in weaving remained high, especially in the manufacture of fine and fancy cloths.

---

[18] M. Winstanley, 'The factory workforce' in M. Rose (ed.), *The Lancashire Cotton Industry* (Lancashire County Books, Preston, 1993), pp. 123–34.

Indeed, hand-weaving persisted in this section of the industry, albeit to a declining extent.[19]

Table 5.3 Spinning and Weaving Capacity in the UK Cotton Industry, 1850–70

Capacity	1850	1856	1861	1868	1870
Number of Spindles (millions)	21.0	28.0	32.0	32.0	38.2
Numbers of Powerlooms (thousands)	250	299	400	379	441

Data on spinning and weaving capacity provides a final measure of the cotton industry's growth in the 1850s and 1860s[20] (table 5.3). Again, appreciable expansion is evident, with the number of spindles rising by 17.2 million (more than four-fifths) and the number of powerlooms by 191,000 (more than three-quarters). Since the additional capacity was likely to have been more efficient than that installed before 1850, its significance is all the greater. Moreover, the number of new spindles comfortably exceeded the eleven million spindles that the industry installed during the previous twenty years. Not unexpectedly, the figures show that much of the additional capacity was created in the prosperous years before the Cotton Famine, though they also suggest a sharp upturn in investment during the late 1860s.

Discussion so far has taken place at industry level, but comment may also be made on the experience of individual cotton firms. Inevitably, many of them were forced out of business during the difficult years of the 1860s, with perhaps as many as 200 to 300 failures occurring between September, 1864 and January 1865. Weaving concerns were affected more than spinning concerns, being caught out by temporary price rises which were higher for raw cotton and yarn than for finished cloth.[21] Indeed, spinners generally fared better than manufacturers, perhaps being sheltered by the high profits they had earned in the pre-Famine years. For instance, J. and P. Coats of Paisley realized an impressive 21 per cent rate of return on average during the 1850s compared with a 16 per cent return in the 1860s. (These figures compare with a rate of 18 per cent in the 1840s and 7.5 per cent in the 1830s.)[22]

---

[19] Timmins, *Last Shift*, pp. 141–2, 147–50.
[20] Deane and Cole, *Economic Growth*, p. 191; E. Helm, *A Review of the Cotton Trade of the United Kingdom, during the Seven Years, 1862–1868* (Manchester Statistical Society, Manchester, 1869), p. 94.
[21] Kelly, 'Famine', pp. 365, 367.
[22] A.K. Cairncross and J.B.K. Hunter, 'The Early Growth of Messrs J. and P. Coats, 1830–83', *Business History*, 29 (1987), p. 169.

Other developments associated with the cotton industry's mid-Victorian growth were the growing concentration of production in Lancashire and the move from integrated to specialist production, the integrated firms probably reaching their zenith during the second quarter of the nineteenth century. The former development is associated with the continued exploitation of the county's natural advantages for cotton manufacturing, especially its humid climate; the enhanced external economies becoming available to Lancashire firms, which are noted in the following section; and the stagnation of cotton manufacturing in the Clyde valley as the local engineering industry, especially shipbuilding, grew to prominence.[23] The latter has been explained in terms of changing market conditions. Prior to the mid-nineteenth century, spinners gained by moving into weaving because they secured a guaranteed market for at least part of their output, whilst they also opened up additional markets through cloth sales and spread their fixed costs over a greater volume of output. Thereafter, the marked rise in demand for cotton goods greatly stimulated each major branch of the industry, so that the need to integrate was reduced.[24] It has also been suggested that firms had less need to diversify because of the ready availability of the industrial and commercial facilities they required, which also reduced their ability to do so.[25]

Developments in Production

During the mid-Victorian period, technological change in the cotton industry arose largely through the wider adoption of existing machinery, such as self-actor mules, to which improvements were continually made. Yet notable technical innovation in the industry was by no means absent. In the preparatory processes the opening of tightly-packed bales of raw cotton and the subsequent process of scutching (the loosening and cleansing of raw cotton) began to be made automatic during the early 1860s. At the same time, steam jets were adopted to increase humidity in spinning rooms, thereby allowing short-stapled Indian cottons to be spun with fewer breakages and to produce finer threads than had been anticipated.[26] In cotton printing, the ageing process was improved by the invention of the steaming machine in 1849. This used rollers to pass printed cloth through a chamber containing water vapour, so that the colours became faster and more intense. The rollers made the process

---

[23] G. Timmins, *Made in Lancashire* (Manchester University Press, Manchester, 1998), pp. 181–2.
[24] C.H. Lee, 'The Cotton Textile Industry' in R. Church (ed.), *The Dynamics of Victorian Business* (George Allen and Unwin, London, 1980), pp. 161–80.
[25] A. Marrison, 'Indian Summer' in Rose, *Cotton Industry*, p. 242.
[26] Farnie, *Cotton*, p. 153.

continuous and hence far quicker than could be achieved with traditional 'cottage steamers'. And in dyeing, the discovery of synthetic dyes, which date from 1856, was especially important.[27]

The impact of these advances should not be exaggerated, however. In some instances, a good deal of adaptation might be required before improved technology could be successfully applied. The use of the powerloom and the self-actor mule in manufacturing finer cotton goods provide examples, the harshness with which they operated causing frequent thread breakages. Even when the advantages to be derived from improved machines had been demonstrated, not all producers would be impressed enough to invest, even if they could meet the costs arising. Meanwhile, hand-weaving continued to attract labour, especially in country areas, partly because alternative work could be difficult to find, but also because domestic manufacture was still preferred to other types of work.[28]

Douglas Farnie's figures on labour productivity in spinning and weaving give some idea of the impact that technological advance made in the cotton industry during the 1850s and 1860s. They reveal continued improvement, though at a slower rate compared with the second quarter of the century. In spinning, the increases are estimated to have averaged 4 per cent per annum from 1820–50, but no more than 2.1 per cent from 1850–73. In weaving, the corresponding figures were 7 per cent and 2.7 per cent. Even so, the amount of yarn spun per hand rose from an annual average of 3,079 lbs in 1849–51 to 5,008 lbs in 1872–74, a rise of 63 per cent. During the same period, the annual average amount of cloth woven per hand increased from 2,438 yards to 4,489 yards, an increase of 84 per cent.[29]

Aside from technical progress within the various cotton manufacturing processes, the cotton industry was able to benefit from more general technological advance. For instance, from 1845, the beam engines traditionally used in industry could be converted into more powerful compound engines by the addition of a high-pressure cylinder. Appreciable fuel economies and smoother running resulted. Further advantages of this type were obtained from the adoption of the American Corliss valve, patented in 1849 and first used in England during the early 1860s. To cope with the improved steam engines, stronger and more efficient boilers were developed, a major advance being the introduction of the Lancashire boiler from 1844. It was equipped with two flues rather than one, which not only added to its strength, but also encouraged

---

[27] G. Turnbull, *A History of Calico Printing in Great Britain* (Sherratt, Altrincham, 1951), pp. 44–9, 64–5.
[28] Timmins, *Last Shift*, pp. 161–70.
[29] Farnie, *Cotton*, p. 199.

water circulation. Further improvement was achieved with the invention in 1849 of Galloway tubes, which enhanced steam output.[30]

Both the extent to which the cotton industry grew and the not-inconsiderable (albeit diminishing) dependence it continued to make on hand technology or hand-assisted technology, meant that it required large amounts of additional labour. As already noted, this was principally met by females, despite the constraints of motherhood. The numbers coming forward were influenced by rising wages and the more stringent controls on the hours females could legally work. But labour supply generally was enhanced as population in the cotton districts continued to rise, especially in the Lancashire towns. This rise was mainly the result of natural increase – Lancashire's urban populations had high proportions of young adults – but non-Lancastrians taking up residence in the county also made a contribution, especially the Irish in the wake of the potato famine disaster of the mid-1840s.[31]

In addition to attracting sufficient labour, the expansion of cotton manufacturing depended on securing adequate amounts of capital and raw materials. With regard to the former, the early application of joint-stock financing assumed importance during the investment boom of 1859 and 1860, when 92 cotton companies were formed in Lancashire.[32] As to the latter, anxiety intensified during the 1850s when crop failures and the actions of cotton speculators brought a sharp rise in raw cotton prices, reversing a long-term downward trend. This price rise limited the benefit that spinners and manufacturers were able to obtain from the higher prices they charged as demand for their products grew. In an attempt to resolve the problem, the Manchester Chamber of Commerce established the Cotton Supply Association in 1857. This was a pressure group which aimed to encourage cotton cultivation in various parts of the world, but especially in India. The Association received comparatively little support from cotton spinners, more than adequate supplies of raw cotton being available during the boom of 1858–61. Moreover, despite shortfalls in American cottons during the Cotton Famine and later in the 1860s, the business community expected that supplies would eventually be restored.[33] There was concern, too, that Indian cotton was much inferior to that of America and even though its use would be confined to producing coarser yarns and cloths, it would greatly reduce quality and create an enormous amount of waste fibre.[34]

---

[30] Timmins, *Lancashire*, pp. 216–17.
[31] Timmins, *Lancashire*, pp. 225–7.
[32] Farnie, *Cotton*, pp. 215–16, 226.
[33] Farnie, *Cotton*, p. 137.
[34] A. Howe, *The Cotton Masters* (Clarendon, Oxford, 1984), pp. 199–202; A. Silver, *Manchester Men and Indian Cotton, 1847–72* (Manchester University Press, Manchester, 1966); Farnie, *Cotton*, p. 137.

As already noted, the cotton industry was becoming increasingly concentrated in Lancashire during the mid-Victorian years, influenced by enhanced external economies. They included the continued growth of subsidiary industries, especially textile engineering and coal mining, and by improvements to the transport system, the completion of the railway network being particularly significant. Not only did it reach into parts of the textile districts that canals had by-passed, but it also added to the lines of communication between towns, a consideration of great importance given the extent to which the volume of trade between them was expanding. Moreover, the faster travel to which railways gave rise was highly significant, not least for the thousands of local businessmen who had traditionally relied on stage coaches for their regular journeys to Manchester and for travelling further afield. Major extensions to warehousing and dock facilities should also be noted, including the opening of Liverpool's Wapping Dock in 1858 and Herculaneum Dock in 1864.[35]

Developments in Markets

A key feature of the sales of British cotton goods during the Victorian period was the growing proportion of exports. In the earlier part of the nineteenth century, a fairly even balance was attained between the value of sales in home and overseas markets. But by the end of the century, it is estimated that more than three-quarters of sales value was generated overseas. The mid-Victorian period saw a continuation of this trend, with exports as a percentage of final product value being thought to have risen from an annual average of 60.8 per cent between 1849 and 1851 to one of 67.1 per cent between 1869 to 1871.[36]

The extent to which exports of cotton goods rose in the mid-Victorian years is shown in table 5.4.[37] As might be anticipated, faster expansion occurred in the 1850s than in the 1860s, the Cotton Famine production falls taking their toll. The 1859–61 figure is taken at a higher point on the trade cycle than the other two, favouring to some extent the figures for the earlier period. Even so, the differences between the two decades are striking. Cloth sales grew particularly strongly in the 1850s, being 85 per cent higher at the end of the decade than at the beginning and even during the 1860s a rise of 21 per cent was returned, a figure half as big in absolute terms as the 1850s rise.

---

[35] For further details and references, see Timmins, *Lancashire*, pp. 232–3.
[36] Deane and Cole, *Economic Growth*, p. 187.
[37] Mitchell, *Statistics*, pp. 356–7.

Table 5.4 Export of Cotton Goods from Britain, c. 1850–70

Years	Yard (million lb)	Piece Goods (millions linear yds)
1849–51	142	1,413
1859–61	189	2,634
1869–71	183	3,183

Yarn exports were less impressive, however, growing by a third in the 1850s but falling marginally in the 1860s. Plainly, cloth exports were of crucial importance in fostering the growth of the mid-Victorian cotton industry.

As to export destination, the industry's strong dependence on European markets waned during the second quarter of the nineteenth century as competition from domestic producers intensified, though the absolute volume of sales there expanded appreciably. At the same time, both the volume and value of sales to the American market (the United States and Latin America) grew strongly, leading to a slightly higher market share. But it was the Asian markets which grew most strongly, especially those in India. Sales to this area were fostered by the termination of the East India Company's trading monopoly in 1813; by the preference for cheap cotton clothing in a sub-tropical country where population and probably per capita income were both rising; and by the extension of British power in India which encouraged the purchase of British goods. By 1850, America took 34 per cent of Britain's exports of piece goods by volume, compared with Asia's 31 per cent and Europe's 20 per cent. In value terms, and taking cotton manufactures as a whole, the European share was higher (34 per cent) whilst the American and Asian shares were lower (29 per cent and 24 per cent respectively).[38]

During the mid-Victorian years, the proportion of cotton piece goods exported to India showed only a small rise, however, from an annual average of 19 per cent (by volume) between 1845 and 1849 to one of 24 per cent between 1865 and 1869.[39] Such limited progress did not escape comment by contemporaries, with J.C. Ollerenshaw linking annual fluctuations in the sales of cotton goods to India with variations in the price of grain that Indians had to pay. He pointed out that many districts in India were in continual danger of famine and demonstrated how exports of cotton goods to India fell away as crop failure occurred. Accordingly, whilst praising the efforts of the Cotton Supply Association in helping to increase cheap raw cotton supplies from India, he remarked that such a policy alone would 'not create a good and remunerative trade for the Lancashire manufacturer.' Also required, he insisted was a means by which the growing of grain crops in India could be 'so

---

[38] Farnie, *Cotton*, pp. 91, 96–104.
[39] Farnie, *Cotton*, p. 117.

encouraged and developed as to find a surplus of food, and ready means of distribution, for the districts more peculiarly fitted for the growth of cotton'. He fully recognized that income levels of consumers were by no means the sole determinants of demand for Britain's cotton exports, but he showed convincingly that they could have a marked influence.[40]

Sales of British cottons in India were further threatened when Indian import duties were increased in 1859 and 1860. The measure, much criticized by the Manchester Chamber of Commerce, was seen as a means of coping with India's deteriorating budget deficit in the wake of the 1857 Rebellion. The rate of duty on both piece goods and yarn was raised to 10 per cent, from 5 per cent in the case of the former and from 3.5 per cent in the case of the latter. In line with free trade principles, the rate was not seen as high enough to offer commercial advantage to Indian cotton manufacturers and, following an improvement in the budget position, the previous levels of duty were restored in 1862. The higher duties may have had some impact on Britain's ability to sell cotton goods in India, with quite sizeable falls occurring during the period the duty was imposed.[41]

As well as benefiting from rising overseas sales, Britain's mid-Victorian cotton producers were also able to take advantage of expanding domestic demand. Indeed, boom conditions developed in the 1850s, brought about by a series of hot summers and a rise in effective demand for dress goods, the result of rising real wages and the establishment of a statutory half-holiday on Saturdays. During the crisis years of the early 1860s, however, home demand for cotton goods fell away sharply, with consumers turning more to cheaper linens and woollens. And despite the growing availability of cheap food imports, high wheat prices could still exert an adverse influence on the domestic demand for cotton goods, as was evident in 1867 and 1868.[42]

Conclusion

Taking the 1850s and 1860s as a whole, the cotton industry made considerable progress. Output may virtually have doubled in terms of volume and may have more than doubled in terms of value. Probably, too, a much bigger increase in output occurred than in the previous twenty-year period. The impact of this growth was strikingly evident in landscape terms, both the manufacturing and

---

[40] J.C. Ollerenshaw, *Our Export Trade in Cotton Goods to India* (Manchester Statistical Society, Manchester, 1870). See also Farnie, *Cotton Industry*, pp. 104–8; P. Harnetty, *Imperialism and Free Trade* (Manchester University Press, Manchester, 1972), ch. 3; and J.A. Mann, *The Cotton Trade of Great Britain* (Frank Cass, London, 1968 reprint of 1860 edition), pp. 69–79.
[41] Harnetty, *Imperialism*, ch. 2.
[42] Farnie, *Cotton*, pp. 131–2: Helm, *Review*, p. 76.

commercial branches of the industry erecting stylish and, in design terms, distinctive buildings to accommodate their needs.[43] The industry continued to gain stimulus from a range of supply-side influences, by no means all of which were generated within the industry itself, as well as from rising demand for its products, with exports exerting a growing influence.

However, the progress made by the industry was far greater in the 1850s than in the 1860s, the boom that occurred between 1858 and 1861 giving way to the depressed years of the Cotton Famine and several unsettled years thereafter. How far overproduction was responsible for these difficulties has provoked disagreement amongst historians, at least with regard to the Cotton Famine years. Yet the idea of periodic overproduction was strongly voiced in contemporary explanations of the cotton industry's difficulties during the 1860s and was applied even to the years when cotton supplies had recovered. As the *Manchester Guardian* remarked at the start of 1870:

> The excess of machinery, however, is undoubtedly the present evil of the trade. During several years preceding the American war, and even, indeed, in subsequent years, a strong impulse was imparted year after year in the erection and increase of new mills and machinery, which, requiring the absorption of a vast amount of new capital, has since acted not only to its own detriment, but also antagonistically to the welfare of the trade at large, and which, enhancing the value of the raw product by an excessive demand, and, upon the other hand, depreciating the value of yarn and cloth by an equally excessive supply, has thus been the means of forcing the trade into its present inert and unsatisfactory position.[44]

---

[43] See, for example, E. Jones, *Industrial Architecture in Britain, 1750–1939* (Batsford, London, 1985), chs 4 and 5.

[44] *Manchester Guardian*, 1 January 1870.

Chapter 6

The Golden Age of Electricity

Gillian Cookson[*]

The development of undersea telegraphs from an experimental technology into a significant science-based industry falls exactly into the period of the 'Golden Age'. In 1850 there were no international submarine cables, the science of electricity was little understood, and each projected submarine line represented a gamble with high stakes and unfavourable odds. By 1870 submarine telegraphy had become almost a routine procedure. Besides being a major industry in its own right, it had a wider significance for mid-Victorian society. The most obvious sign of this was the transformation in global communication which undersea cables brought. Lines spanned the world, crossing the Atlantic, connecting North and South America with Europe and North Africa, linking with cables to India and across the empires of Russia and China, with an Australian connection almost complete. In accomplishing this communications revolution, British engineers had achieved an early advantage in electrical technology, which was to be central to a new wave of industrialization during the closing years of the nineteenth century.

While the industry experienced dramatic expansion over the two decades, its development was uneven. Financial obstacles, as much as technological problems, dogged its growth. By the early 1860s there was a consensus among electrical engineers and company projectors – following an unexpected but highly constructive scientific intercession by the British government – that the technology was viable and potentially very profitable, a confidence which did not transmit itself to smaller investors. The self-assurance of electrical engineers was gleaned from knowledge which had emerged from a succession of expensive failures; these same failures undermined the faith of a general investing public which had lost large sums while the engineers experimented. With the public reluctant to risk more on submarine telegraph shares, new means of money-raising had to be devised. The result was that cable manufacture and laying, and the operation of international telegraphs, became concentrated upon a few large limited companies, themselves controlled by a small group of entrepreneurs with interests spanning the whole industry.

---

[*] Acknowledgement: The research upon which this chapter is based was generously supported by the Leverhulme Trust. Thanks to Colin Hempstead and Colin Griffin for comments on a draft version.

Through this struggle to organize and finance new schemes, the submarine telegraph industry played a key role in the evolution of modern commercial practices. Its development over the two decades also reflects a changing business disposition, from the idealism and inspired amateurism surrounding an immature technology, to an uncompromising commercialism once technological questions were settled and large profits beckoned. The vision of the original projectors – some of them principled internationalists, others foreseeing a means of ruling a scattered global Empire or of using the telegraph for business purposes – had been an essential element in launching exceptionally ambitious and expensive schemes when large questions still hung over their technical viability. By 1870, the industry had been forced to abandon any vestiges of innocence in order to achieve its Golden Age.

After Wheatstone's first experimental telegraph, opened in 1837 between Euston and Camden, an overland telegraph system was quickly established in Britain. Within a few years, a network extended over much of Europe and the telegraph was spreading west across the United States. Undersea telegraphs, technically much more difficult than landlines, were especially important to nations separated from the continental mainland. With a commanding position in world trade and international politics, Britain had a particular need to communicate with her continental neighbours in the way that they could communicate with each other. Some far-sighted Americans also saw that their nation could be increasingly sidelined in politics and commerce without a transatlantic link.

The earliest successful submarine cable, laid in 1851 between Dover and Calais after a failed effort the previous year, established a design which formed the basis for future cables: a copper conductor, the cable's core, was insulated with gutta-percha, a kind of latex from Malaya preferred to indiarubber for underwater use. The cable was armoured with iron wire, thicker at the shore ends which needed extra protection from anchors and tides. The 1851 line continued to work well, was followed by others around Europe, and within five years serious plans were under way for a transatlantic telegraph. But the differences between short cables laid in shallow seas, and ones which could span the Atlantic, weighing thousands of tons and with a shortest route of 1,660 nautical miles in depths of up to two miles, went beyond problems of scale and handling. A range of issues relating to cable construction and the best methods of managing long submarine cables remained matters of fierce contention between electrical engineers. It was suspected that poor management had been responsible for the failure of some early deep-sea cables, but as the science of electricity was imperfectly understood and there were no agreed electrical standards and measurements, this remained largely a matter of unsubstantiated opinion. Furthermore, instruments for testing lines and for sending and receiving messages were not sensitive enough for commercial use on long cables, which were plagued by feeble signals – there was no way of amplifying or relaying in mid-ocean – and the phenomenon of 'smearing', where the sharpness of transmitted signals was lost.

It became clear during the 1850s that long deep-sea cables carried disproportionately large problems and heavy risk, and the Atlantic project brought the technical shortcomings into sharp focus. The Atlantic Telegraph Company's attempts to lay cables in 1857 and 1858 cost more than £500,000, which was ultimately a complete loss.[1] Between 1851 and 1866, when the Atlantic crossing was finally achieved, about 120 submarine cables had been laid, meeting with mixed fortunes: some of these were long, some short, some highly successful, others disastrous and expensive failures.[2] An undersea line could not partly succeed; any minor problem in construction or small accident during laying could prevent it from working at all, and because of the huge cost of cables this was likely to result in a shattering loss for the companies and individuals involved. But where a line worked, there was immense potential for profit-making, and this remained the case after the risk factor had fallen dramatically in the mid-1860s.[3]

Developing Electrical Technology

In confronting the electrical problems of submarine telegraphy – designing new instruments and techniques, coming to terms with the underlying science, measuring and standardizing units of electricity – the profession of electrical engineering came of age. Electrical science and the technology of undersea cables developed in tandem, each informing the other, through the late 1850s and early 1860s, until vastly superior instruments had been developed and a consensus was established on the best means of testing and running long cables.[4]

---

[1] See Gillian Cookson, 'Atlantic Crossing: the first transatlantic telegraph', *History Today* (forthcoming).

[2] The National Maritime Museum, TCM/7/2, gives a list of submarine cables which is almost complete although not entirely accurate. For general histories of submarine telegraphy, see Charles Bright, *Submarine Cables: Their History, Construction and Working* (Arno Press, New York, 1974 [1898]); B.S. Finn, *Submarine Telegraphy: the Grand Victorian Technology* (Science Museum, London, 1973); Hugh Barty-King, *Girdle round the Earth: The Story of Cable and Wireless* (Heinemann, London, 1979).

[3] This was widely known by 1870, when guides for investors strongly recommended submarine telegraph shares. See for instance J. Wagstaff Blundell, *Telegraph Companies Considered as Investments* (Effingham Wilson, London, 1869); W.L. Webb, *The Public Telegraph Companies* (Webb, London, 1869).

[4] See Gillian Cookson and Colin A. Hempstead, *A Victorian Scientist and Engineer: Fleeming Jenkin and the Birth of Electrical Engineering* (Ashgate Publishing, Aldershot, 2000), especially chs 3 and 5.

Central to this process of consolidating electrical science and technology was an investigation sponsored jointly by the Board of Trade and the Atlantic Telegraph Company following the breakdown of the 1858 Atlantic cable. The joint committee's deliberations marked a turning point for the industry. Chaired by Captain Douglas Galton of the Board of Trade, the committee listed all cables laid before 1859, and from this survey identified the most common mechanical causes of failure, usually breakages. The inquiry went on to interview all the leading electricians involved in the development of submarine telegraphy, a small and close-knit group yet one which held conflicting opinions about the working of long cables. It became apparent that early successes with cross-Channel and North Sea cables had concealed serious shortcomings in technical and scientific understanding. Galton's report was a legitimating tool for already known 'best practices'. The dominant ideas which Galton accepted came from a small coterie of engineers, but because his enquiry was exhaustive and wide ranging, he was able to settle unequivocally some of the main issues in contention. Galton's report proved to be decisive in shaping the future progress of deep-sea cables, and in consolidating the reputations and position of such leading electrical engineers as William Thomson (later Lord Kelvin), Sir Charles Bright, Latimer Clark and Fleeming Jenkin.

Galton conclusively decided issues of electrical management and cable construction. His committee established disciplined and methodical procedures, and brought a recognition, in the words of Thomson, that there could be no 'double set of natural laws'.[5] Not only was theory accepted as an integral part of the development of telegraphy, but telegraphy's role in informing scientific theory became clear. The success of deep-sea cables, as a technology and as a business venture, was achieved through a symbiosis of scientific understanding with best practice. Immediately following Galton, the British Association for the Advancement of Science took up the challenge of setting electrical standards, in a move instigated by Thomson, the experimental work carried on largely by Jenkin and James Clerk Maxwell through the 1860s. By 1865 an official resistance standard was offered for sale to telegraph companies and other interested parties around the world, a solid manifestation of the way in which electrical practice had been reconciled with electrical theory.[6]

---

[5] C. Smith and M.N. Wise, *Energy and Empire: a biographical study of Lord Kelvin* (Cambridge University Press, Cambridge, 1989), p. 666.

[6] Cookson and Hempstead, *Victorian Scientist*, ch. 5.

The State and International Cables

The Galton inquiry was arguably the most important contribution made by the State to the fledgling British electrical industry, although there are other instances of government assistance to early telegraph projects.[7] Overland telegraphs, used to deploy troops during Chartist disturbances, had already demonstrated their value to the state. During the Crimean War, a temporary unarmoured cable was laid between Bulgaria and Balaklava by the British government, expressly for military purposes and to good effect. Successive political administrations were reluctant to commit public money directly to peacetime telegraph projects, but were alive to the strategic potential of such schemes and maintained a consistent although slightly detached interest in developments. There was a particular focus upon Mediterranean telegraphs, partly for reasons of defence and trade, but above all because it was a route to India.[8] When, during the Indian Mutiny of 1857, an emergency request for troops had taken forty days to reach London, ideas of a submarine cable – speedier and more private and secure than any overland line through foreign territory – were rekindled. There was a further demonstration of the use of international telegraphs to government during the single month that the 1858 Atlantic cable worked, when an unnecessary mobilization of troops in Canada was averted, saving over £50,000.[9]

During the 1850s government support of undersea cables progressed from issuing landing concessions – guarantees of monopolies on certain routes which quickly acquired a value of their own – to practical assistance with ocean surveys and loans of naval vessels. By the end of the decade there had been several instances where the State underwrote schemes, assuring a level of return on investment which was conditional upon a line working. Such guarantees were of limited use in attracting investors, as they did not insure against the prospect of a total loss, and experience was beginning to show that once a cable was successfully installed, high dividends were to be expected and subsidies were thus irrelevant.

In one case a British administration made a massive misjudgement in guaranteeing a telegraph scheme. Such was the urgency of establishing a secure line to India under British control that, despite the recent failure of the 1858 transatlantic line, the government agreed to underwrite a privately promoted Red Sea cable, which led in six sections from Egypt to the west coast of India.

---

[7] Gillian Cookson, 'Government and Telegraphs: the role of the British government in promoting submarine telegraphy', paper presented to the European Business History Association conference, Rotterdam, September 1999.

[8] D.R. Headrick, *The Invisible Weapon: telecommunications and international politics, 1851–1945* (Oxford University Press, Oxford, 1991), p. 16.

[9] Charles Bright, *The Story of the Atlantic Cable* (Hodder and Stoughton, London, 1903), p. 147.

Unlike other such contracts where support had been conditional upon the cable working, this agreement required only that each section tested successfully. No telegram ever passed the length of the line, but because the separate parts had worked during testing, the government was bound to pay Red Sea investors £36,000 a year for fifty years, an eventual bill to the Exchequer of £1.8 million.[10] After this disaster there was no further question of direct public finance for private cable schemes, but Galton's investigations, following soon afterwards, were carried out with strong government support and proved to be much more valuable to the industry than any number of financial guarantees.

Business Organization

Cable and telegraph companies experienced a series of mergers and take-overs during these first twenty years. There is no evidence that vertical integration was consciously intended at the outset; it evolved quickly, apparently through force of financial circumstances, and was eased by changes in company law, most notably the introduction of limited liability in 1856.[11] The unsuccessful attempts on the Atlantic in the 1850s marked the start of a decade of rapid change in company structures. In 1856 there had been no single cable-maker with the capacity to manufacture a core and armour an entire transatlantic cable in time to meet the tight deadline of a short summer cable-laying season. After a series of commercial battles, mergers and alliances, a single company, the Telegraph Construction and Maintenance Co. Ltd (known as Telcon), came to dominate the industry, manufacturing, arming and laying entire deep-sea cables, and with a number of prominent directors in common with the main submarine telegraph companies.[12]

Long deep-sea cable schemes required extraordinary amounts of capital. The Atlantic Telegraph Company's original capital, lost on the 1857 attempt, was £350,000.[13] The 1865 transatlantic telegraph, abandoned during laying, had cost £600,000. The same amount had to be raised for the 1866 expedition, which successfully installed a new cable and retrieved and completed the 1865 line.[14] Most of the money went immediately on making and laying the cable, as subsequent running costs were low even if the cable later suffered damage and needed repair. The main financial problem came if a breakdown could not be mended quickly, as there was then a total loss of income for the duration of the

---

[10] Headrick, *Invisible Weapon*, pp. 19–20.
[11] See Bishop Carleton Hunt, *The Development of the Business Corporation in England, 1800–1867* (Harvard University Press, Cambridge, Mass., 1936), ch. 6.
[12] See G.L. Lawford and L.R. Nicholson, *The Telcon Story 1850–1950* (Telegraph Construction and Maintenance Co. Ltd, London, 1950).
[13] Public Record Office, BT 31/62, number 237.
[14] Webb, *Public Telegraph Companies*, pp. 5–8.

break. This equation of high capital and low running costs, combined with the possibility of unpredictable interruptions to the service, was a matter of concern to directors and shareholders who were becoming accustomed to large dividends. In business terms, two main outcomes followed from this anxiety. First, it was necessary to find different approaches to raising capital which did not rely upon the capricious nature of small investors. After several well publicized-failures, the public in general had lost the confidence to invest in long cables – ironical as the knowledge and skill of electrical engineers was increasing fast, especially after Galton. Initial widespread enthusiasm, bolstered by the early cross-Channel success and by articles in journals such as the *Illustrated London News* whose pictures and accounts of submarine telegraph expeditions presented them as glorious adventures, had turned to scepticism and a suspicion that undersea cables were no more than a scheme to fleece the public. The second change in business strategy among submarine telegraph companies followed from the special nature of the technology and relates to attitudes towards competitors. The industry acted to maintain large returns on investment in a manner which could be defined as anti-competitive behaviour.

From the beginning, production of submarine cables rested with a very few companies. The most prominent were the Gutta Percha Company, which supplied cable cores to the rest; and two firms which armoured and laid cables, Glass, Elliot and Company, and R.S. Newall and Company of Gateshead. Newall, a litigious character and an outsider in more than just his northern origins, contrived to fall out with his main customers as well as his rivals. His reputation had also been damaged by his involvement with the Red Sea cable. From the early 1860s Newall diversified into other activities including chemicals and later asbestos processing, although he continued to produce and lay submarine cables for the Danish Great Northern Telegraph Company and others outside the mainstream British industry.[15] Glass and Elliot meanwhile cultivated their relationship with the major British-based telegraph companies. In 1864, as preparations were under way to relaunch the Atlantic scheme, Glass and Elliot amalgamated with the Gutta Percha Company to form the limited company known as Telcon.[16] Telcon went on to bolster its position through increasingly close links with the main submarine telegraph companies. A few smaller cable-making companies managed to survive alongside Telcon, either by seeking work abroad, like Newall, or by acting essentially as subcontractors to Telcon, as W.T. Henley did between 1864 and 1876. Significantly, Henley

---

[15] Kurt Jacobsen, 'Diplomacy and International Business: the Great Northern Telegraph Company, 1869–1921', paper presented at the European Business History Conference, Rotterdam, September 1999.
[16] Lawford and Nicholson, *Telcon Story*, p. 54.

went into liquidation within two years of the end of his working agreement with Telcon.[17]

Ultimately, the power of Telcon extended far beyond its near monopoly on making and laying submarine cables. The original investors who had supported telegraph schemes in the 1850s, many of them English or Scottish provincial merchants, were in the main reluctant or unable to risk further losses on the revived project. With the exception of one or two notable investors, there had been relatively little support from the United States for the Atlantic cables of the 1850s, and there was even greater reluctance when the scheme was revived during the American Civil War. When the Atlantic company could not raise the capital it needed, its future seemed uncertain. A London merchant bank came forward with some of the necessary funding, but it was Telcon itself which enabled the scheme to be renewed. The cable company put up a substantial proportion of capital for the Atlantic expeditions of 1865 and 1866, and deferred some of the debt, accepting telegraph company shares in part payment. Richard Glass, an accountant and Managing Director of Telcon, John Pender, the Chairman of Telcon whose fortune had come from the Manchester cotton trade, and Sir Daniel Gooch, previously saviour of the Great Western Railway, were instrumental in devising these arrangements. The model which they established, of Telcon accepting a sizeable stake in the contractor telegraph company, was repeated with the French Atlantic cable of 1869, which was only nominally French and involved some of the same group concerned in the 1866 Atlantic line. By following a pattern of receiving telegraph company shares in part payment, the cable company acquired a strong interest in the success of the enterprise. The convergence of the cable and submarine telegraph industries is indicated also by a crossing over of business leaders: Glass, Gooch, and especially Pender, became pivotal figures in international telegraphy. Glass accepted the chairmanship of the Anglo-American Telegraph Company, successor to the Atlantic Company; Gooch became Chairman of Telcon, was a director of the Anglo-American, and also of the company which owned Brunel's *Great Eastern*, the largest cable-laying ship in the world. Pender, a longstanding director of the Atlantic company, turned his attention to new projects and consolidated them into a still greater plan.

Launched in 1868 with a capital of £1.2 million, the French cable was one of a number of major schemes to follow closely upon the success of the 1866 transatlantic cable. Another was the long-awaited line to India, launched by John Pender as the British-Indian Submarine Telegraph Company.[18] British inland telegraphs had been nationalized in 1868, with inflated rates of compensation paid which made funds available for diversion into new

---

[17] Science Museum Library, HEN 29, 'The Early Life of W.T. Henley, by himself'.
[18] Barty-King, *Girdle Round the Earth*, p. 27.

submarine projects.[19] In 1869 seven major new submarine telegraph companies raised in total funds of £5.4 million.[20]

The general public may have been excited by international telegraphs, but most would never be able to afford to use them. Prices were exceptionally high, and the companies wanted to keep them so. Representatives of the submarine telegraph industry and investment advisors argued forcibly that telegrams must remain expensive if future schemes were to attract finance: 'There is no exception to the rule that a reduction of tariff leads to a diminution of the net product'.[21] *The Economist* pressed the same point, based on the limited capacity of lines: 'Reduce the price of messages and no amount that could be carried would pay'.[22] On the other hand, the rewards were huge. Investors were still nervous about the costs of repairs to cables, but were assured that although these could be great, 'the cost of effecting these repairs is a mere bagatelle when compared with the enormous receipts'.[23] The Anglo-American Telegraph Company Ltd was paying dividends of over 20 per cent per annum in 1868 and 1869, even when one of its two transatlantic lines was broken and telegram prices were falling.[24] Running costs were less than 14 per cent of the receipts, £30,000 against earnings of £220,000 a year. But *The Economist* foresaw problems if cable-making capacity expanded sufficiently to enable competing lines to be laid, or if the cost of cables fell: 'The business cannot flourish except where there is a monopoly ... If more companies are established than can do the work on given routes, they will cut each other's throats'.[25] Of course, the high cost of telegrams was a factor in encouraging competitors, like the promoters of the French Atlantic cable of 1869, some of them heavy users of the cable such as the news entrepreneur Julius Reuter, appalled by the Atlantic company's charges.[26] Landing concessions guaranteed companies a monopoly for a fixed number of years and provided some insurance of future profits, but *The Economist* predicted that the industry's uniform product would not engender customer loyalty so that once similarly routed lines were operating – 'There are so many shores that hardly any number of concessions would protect a company' – competition could prove fatal to the industry.

---

[19] James Foreman-Peck, 'Competition, Cooperation and Nationalisation in the Nineteenth Century Telegraph System' *Business History*, 31 (1989), pp. 81–102; Donald Read, *The Power of News: the History of Reuters* (Oxford University Press, Oxford, 1999), pp. 50–51.
[20] *The Economist*, 4 September 1869, p. 1042.
[21] Sir James Anderson, 'On the Statistics of Telegraphy' *Journal of the Royal Statistical Society*, 35:3 (1872), p. 297.
[22] *The Economist*, 4 September 1869, p. 1042.
[23] Wagstaff Blundell, *Telegraph Companies*, p. 7.
[24] Webb, *Public Telegraph Companies*, pp. 5–7.
[25] *The Economist*, 4 September 1869, p. 1043.
[26] Read, *Power of News*, p. 54.

At the time of the *Economist*'s gloomy prognosis telegram prices were tumbling. In the year to August 1869, which started with the launch of the French company and ended with the opening of their Brest–New York cable, the cost of sending a message on the Atlantic company's line fell by over 70 per cent, from 5 guineas (£5.25) to £1 10s (£1.50) for a minimum ten words.[27] The French company matched this price, and considered further reductions. Although all the lines were busy and returns remained healthy, both sets of directors believed that higher tariffs would be more profitable. Most telegrams were commercial, and price had little effect upon traffic levels. Following a number of interruptions to the 1866 line, there remained anxiety about loss of revenue if any of the three cables was broken and could not be repaired immediately.

In January 1870, less than six months after the opening of the French cable, a joint purse agreement was negotiated between the two Atlantic companies. Revenues would be pooled and shared, giving some security against cable breakage. The standard telegram tariff doubled to £3 later in 1870, and although it would subsequently fall back to £2, further damaging cuts below the level of late 1869 were avoided.[28] The joint purse agreement was the first step towards amalgamation of the Atlantic and French companies, which came to pass in 1873. By then, Gooch and Pender had been instrumental in setting up an early version of a unit trust, the Submarine Cables' Trust, which spread risk by holding a range of telegraph investments. Though much diminished, there was still some uncertainty attached to individual cable schemes, so Pender developed a technique of launching each cable as a free-standing enterprise, to be merged into his main companies once the line was safely established. For example, his four Mediterranean and Indian companies were formed into the Eastern Telegraph Company in 1872. This was the beginning of a virtual monopoly on international telecommunications by Pender and his associates which lasted until the century's end, and was the foundation of the business which was later to become Cable and Wireless.[29]

In business terms, the 1860s marked a shift of motivation in submarine telegraphy. As telegraph companies multiplied and amalgamated into big businesses, they became remote from the idealism and sense of opportunity which had, at least to a point, driven the transatlantic telegraph projectors of the 1850s. Some of the early participants had been spurred by a desire to improve international peace and understanding, others were prepared to give freely of their time and money in order to develop the technology and gain an opportunity to experiment in a way which was impossible in the laboratory.

---

[27] Anderson, 'Statistics of Telegraphy', p. 297.
[28] Gillian Cookson, '"Ruinous Competition": the French Atlantic Telegraph of 1869', *Entreprises et Histoire* (forthcoming).
[29] See Barty-King, *Girdle Round the Earth*.

The drive to fully-blown commercialism had a direct impact upon this last group, the electrical engineers, who found that in order to flourish in this brave new world a measure of business acumen was essential. The big adventure was over, any sense of joint enterprise with the promoters was lost, and the unsentimental men in charge of the industry after 1865 were reluctant to compensate engineers for any earlier sacrifices, even those which had been fundamental to the industry's success. Thomson, Jenkin and Cromwell Fleetwood Varley, the partners who for years had spent their own time and money developing instruments which made long distance telegraphy possible, were compelled to take legal action against the Atlantic company in order to extract any payment for use of their patents once the scheme had moved into profit.[30]

The Impact of Global Telegraphs

While most people could not afford to use the international telegraph, its effects were quickly felt across the whole of society. Daily newspapers and journals such as the *Illustrated London News* had brought a steady narrative of the progress of submarine telegraphy, an entertaining tale of adventure, of daring, of disaster and sometimes triumph. Once long cables were working, they were themselves the means of transmitting news and introduced a new immediacy which revolutionized the impact of international news reports. This is illustrated by reactions in Britain to the deaths of two American leaders, before and after the transatlantic telegraph. News of Lincoln's assassination took twelve days to reach London in April 1865 by the fastest combination of land telegraph and steamer, and responses were relatively muted. When President Garfield was assassinated in 1881, first news of the shooting was published in British newspapers within twenty-four hours, and after following Garfield's long struggle for life, the public mourned his loss with much greater feeling. The telegraph added to a sense of shared experience between Britain and the United States, and the excitement of fresh news boosted sales of newspapers.[31] Julius Reuter built his news agency into a formidable power largely through his talent for correctly anticipating developments in overland and submarine cables.[32]

Through the Atlantic cables, the near instantaneous transmission of share and commodity prices and other commercial information changed the nature of world markets. Telegraph lines to the east were inextricably bound to the politics of Empire. Pender, for all his success running a new scale of private

---

[30] Cookson and Hempstead, *Victorian Scientist*, ch. 4.
[31] Read, *Power of News*, p. 97.
[32] Read, *Power of News*, chs 1 and 2; Cookson, 'Ruinous Competition'.

business empire, and his protestation that 'Telegraphs know no politics', was firmly wedded to British government interests.[33] The empire provided his most important market, and he provided Britain with a tool which offered a means of controlling the farthest corners of its territories.[34] News and commercial information were as much a part of this as political and military direction.

Above all, early submarine telegraphs were a significant British industrial success story. Government had played an important part in this, not only as a user and supporter of the international telegraph, but in establishing the technological foundations of electrical engineering through sponsoring the Galton committee. This early advantage was enough to ensure that expertise in laying and running long deep sea cables remained concentrated in Britain for the remainder of the nineteenth century. But even as submarine telegraphy transformed itself from a great adventure to a serious business, its initial dynamism stalled, unable to co-exist with the new set of business institutions and relationships essential to the industry's continuance.

---

[33] Bruce J. Hunt, 'Doing Science in a Global Empire', in Bernard Lightman (ed.), *Victorian Science in Context* (University of Chicago Press, Chicago, 1997), pp. 320–22.

[34] See Headrick, *Invisible Weapon*.

Part II

Technology

Introduction to Part II: Technology

Ian Inkster

Other metals in fact dominated the British Golden Age, metals produced and worked by skills and in locations which were very far removed from formal educational or training establishments. The sites and agencies which together converted the baseness of iron or copper into the precious materials of industry, were not ancient universities on one hand, nor state-instigated polytechniques on the other. In our 'Events of the Golden Age', which acts as an extended preface to these essays, the pre-eminence of Britain as the engineering workshop of the world is suggested well enough. It is difficult to depict the economic returns on any one instance of technological advancement, or to array cases in order of their creative significance. But without any doubt Britain was replete with material, machine, chemical and product innovations and these emanated in the main from either centres of engineering excellence in London and the provincial urban centres, or from applications of more industry-specific trade skills in countless sites and across the complete range of British industries.

The three essays of the present section do combine to expose something of this technological base, the strength of which had been tested in the earlier years of industrial and technological progress. Frank James' paper is important because it reminds us of the scientific and institutional environs within which particular technologies were adopted, refined and diffused, and of the dependence of an industrial and commercial system upon key infrastructural developments. But it is also of interest because it very properly emphasizes the continuum of scientific and technical practices during the nineteenth century, particularly as embodied in such individuals as the Sandemanian scientist Michael Faraday (1791–1867), men whose very careers and incomes depended on such an eclecticism. In this sense at least, the social and intellectual resources of the Industrial Revolution continued on well past the supposed watershed of the Great Exhibition – Faraday was apprenticed to a London bookseller but was pensioned by Lord Melbourne and died in a house at Hampton Court given him by Queen Victoria.

The 1851 Exhibition – aspects of which are considered in two other essays in this volume and in the Introduction – did not mark any sharp break in the trajectory of technological change. Of course, the 'interests' could always utilize its metropolitan hugeness to illustrate their own problems or needs. The best way to succeed in this, then as now, was to identify such pressure-group concerns as 'in reality' those of the 'nation' at large. In Graeme Gooday's paper below we meet Sir Lyon Playfair (1818–98) doing this towards the end of our Golden Age, in 1867, as his considered response to the Paris Exhibition of that year. Graeme Gooday thus draws into this section another great figure of the century, a Scots chemist, MP, Postmaster-General, and lord-in-waiting to Queen Victoria herself. Perhaps the best-known of those who sought to highlight overseas industrial competition as a stimulant to more public investment in British science, technology and education, Playfair utilized the alleged advance of other industrial nations at the Paris Universal Exhibition, claiming that a fall-away in 'inventiveness' was undermining British industrial competitiveness, and thus he claimed his place amongst a long line of 'declinist' commentators that stretches back to the years before the industrial revolution really took hold of the British commercial economy.

But at the beginning of our period, in 1851, that same Playfair was quite as adamant. In a very lengthy commentary on the supposed significance of the Great Exhibition to industrial workers, Playfair astutely noted the earlier claims of scientists such as Babbage and Herschel concerning a decline of British (at least, English) science, and went on to persuade his readers that the nation was merely advanced in practice, but falling behind in theory, a juxtaposition that could not last for ever. In such a claim he was joined by other semi-official commentators, such as the Jacksonian Professor of Experimental Philosophy at Cambridge, Robert Willis. But although Cambridge made out a more thorough case than did London, with Willis proclaiming the importance of the application of formal mathematics to mechanical pursuits, the best argument that such a supporter could make was that the Exhibition would incite a greater interest in the sciences, and that educators might be so prepared. Little was presented to demonstrate any new relative technological backwardness. And Playfair was at the very least wrong on the main count – such a seemingly contrary combination, of advanced technique with unreformed science organization, could last for many a year after 1852. It certainly outlasted the Golden Age. Nor did Charles Babbage himself, in his *Exposition of 1851, Or Views of the Industry, the Science and the Government of England*, do more than reiterate his claims about a lack of science in England – although he exposed some inadequacies of the Exhibition at work, he did not demonstrate any relative decline of British technology as evidenced in displays and prizes. In this he reflected the attitudes of many skilled inventors and artizans. Thus the London journal, the *Artisan*, much nearer to the surface of things, reviewed the machinery displays in detail throughout the year, and complained repeatedly of their inability to represent the truth of British industry, but rarely spotted any technical backwardness when compared to other nations. In 1851, as at many other times, the demands of the special interests were forged from the cries of opportunism.

In the same place and time, but less famously and certainly far less fatuously than Playfair, Henry Hensman quietly got it right. Oxford, Cambridge and the old professions may or may not have lain dormant for some time – this was more properly a matter for the world of Playfair – but Hensman asserted correctly that the 'hidden mechanical talents' of British workers were what mattered, and these mostly lay well beyond the light of exhibition, curtailed if anything not by lack of science but more likely by insecurity of intellectual property. Indeed, in London itself, the Golden Age opened with the Exhibition, but also with the many metropolitan panopticons, polytechnics, panoramas, institutions, museums and galleries which displayed and, to an extent, elucidated, the underlying energies of British technology, skills and industry. The third paper of this section uses the abundant patent statistics to describe the geographic and social distances between the ordinary inventiveness of a successful industrial economy and the culture of its scientific and intellectual elites. It was just this distance which ensured that any cultural devaluation of the industrial undergirding of the Golden Age as exhibited by a vocal and highly literate elite was likely to have little or no effect on the actual course of technological change. Away from London, in new buildings and on old street corners, industrial Britain moved on.

Chapter 7

Michael Faraday and Lighthouses

*Frank A.J.L. James**

A prevalent interpretation of Michael Faraday (1791–1867) is that of a lone scientist working in his basement laboratory in the Royal Institution, laying the foundations of electrical technology and yet being one of the most gifted scientific communicators of the day. This interpretation was developed by his early biographers and has been fostered and developed by scientists and engineers as well as by some historians ever since. While I do not want to dismiss this interpretation out of hand, I do want to show that Faraday was actually far more interesting than it suggests. Faraday did not just do 'pure' science, which was then technologically applied by others (at a much later date), but was himself highly involved in bringing together science and technology into a single continuous spectrum of practice during the middle third of the nineteenth century within various contexts, mostly relating to the State and its agencies.

On 4 February 1836, Faraday was appointed 'Scientific adviser to the Corporation [of Trinity House] in experiments on Light'.[1] For his advice, which he gave until 1865, Faraday was paid an annual salary of £200. He then handed the job over to his successor at the Royal Institution, John Tyndall (1820–1893).[2] The Elder Brethren of Trinity House have been responsible for safe navigation round the shores of England and Wales since 1514 and are thus in charge of lighthouse, lightships, buoys and pilotage, as well as having some charitable functions towards retired seamen. So why in 1836 did they create this new post of scientific adviser and how did Faraday come to fill it? To answer these questions we must consider Faraday's own career and the political context in which he was appointed.

---

[*] I wish to thank the Royal Institution (RI) and the Royal Society (RS) for permission to work on manuscripts in their possession.

[1] Herbert to Faraday, 4 February 1836, in Frank A.J.L. James (ed.) *The Correspondence of Michael Faraday*, 4 vols (Institution of electrical Engineers, London, 1991–), vol. 2, letter 885. Hereafter cited as Faraday, *Correspondence* followed by volume and letter number.

[2] On Tyndall's lighthouse work see Roy M. MacLeod, 'Science and Government in Victorian England; Lighthouse Illumination and the Board of Trade, 1866–1886', *ISIS*, 60 (1969), pp. 4–38.

Faraday

Faraday was born on 22 September 1791 in Newington Butts, Southwark. His father, a blacksmith by trade, belonged to a sect of literal Christianity known in Scotland as the Glasites and in England as the Sandemanians. Throughout his life Faraday was fully committed to this sect. He made his confession of faith in 1821 (the same year that he married Sarah Barnard (1800–1879), a Sandemanian silversmith's daughter); he was a deacon in the church from 1830 to 1840; and an elder between 1840 and 1844 and again between 1860 and 1864.[3]

In 1805, at the age of fourteen, and just a few days before the Battle of Trafalgar, Faraday was indentured for seven years as an apprentice bookbinder to George Riebau in Blandford Street. Although Faraday completed his seven-year apprenticeship, he developed an overwhelming desire to become a natural philosopher and, in particular, a chemist. This was not an obvious career move at this time since there was no career structure in English science aside from that provided by medicine. He put this desire into effect in a number of ways. First, he was allowed to perform chemical experiments on Riebau's premises.[4] Second, he attended lectures given at a number of places in London, most notably at the City Philosophical Society in Fleet Street.[5] Third, he read texts such as *Conversations on Chemistry* by Jane Marcet (1769–1858) and the electrical entries in the *Encyclopaedia Britannica*.[6] Finally, in April 1812, the last year of his apprenticeship, a customer of Riebau's, William Dance (1755–1840), one of the founders of the Royal Philharmonic Society, gave him tickets to attend the last four lectures to be given by Humphry Davy (1778–1829) in the Royal Institution, which had been founded in 1799. Davy had just married a rich widow, Jane Apreece (1780–1855), and was thus able to retire from being professor at the age of thirty-four.

Faraday attended these lectures in which Davy dealt with a contemporary problem at the cutting edge of chemistry, the definition of acidity. In his letters to his close friend Benjamin Abbott (1793–1870), Faraday agreed entirely with

---

[3] On Faraday's Sandemanianism and its background see Geoffrey Cantor, *Michael Faraday: Sandemanian and Scientist. A Study of Science and Religion in the Nineteenth Century* (Macmillan, London, 1991).

[4] Benjamin Abbott, 'Jottings from Memory in reference to my dear and deceased Friend M. Faraday', in Frank A.J.L. James, 'The Tales of Benjamin Abbott: A Source for the Early Life of Michael Faraday', *British Journal for the History of Science*, 25 (1992), pp. 236–40, on p. 236.

[5] Frank A.J.L. James, 'Michael Faraday, The City Philosophical Society and the Society of Arts', *Royal Society of Arts Journal*, 140 (1992), pp. 192–9.

[6] Henry Bence Jones, *The Life and Letters of Faraday*, 2 vols (London, 1870), vol. 1, p. 11.

Davy's views.[7] Faraday wrote up his notes neatly[8] and sent them to Davy asking him for a job in science – Faraday was asking Davy to be his patron, a normal way of obtaining employment in early nineteenth-century Britain. Davy had no vacancy at the time but said he would keep him in mind. Faraday then commenced his career as a journeyman bookbinder. On 19 February 1813 there was a fight in the lecture theatre between the instrument-maker John Newman (c. 1783–1860) and the Chemical Assistant at the Royal Institution, William Payne. The managers (the committee that used to run the Royal Institution) placed the blame for the fight on Payne and sacked him.[9] They asked Davy to find a replacement; he remembered Faraday and called him for an interview. According to Faraday's later recollection of the meeting, Davy

> advised me not to give up the prospects I had before me, telling me that Science was a harsh mistress; and, in a pecuniary point of view, but poorly rewarding those who devoted themselves to her service. He smiled at my notion of the superior moral feelings of philosophic men, and said he would leave me to the experience of a few years to set me right on that matter.[10]

This recollection was written after Davy's death in 1829, and it seems to me unlikely that Davy would have held such a jaundiced view of the scientific profession, which had served him so well until that time; perhaps after his later disasters, particularly as President of the Royal Society, Davy may have expressed such views. Despite Davy's warnings, Faraday was undeterred and Davy recommended his appointment to the managers on 1 March 1813.[11] Faraday thus embarked, at the age of twenty-one, on what was effectively a second apprenticeship, this time in chemistry. Six months after Faraday's appointment, Davy was given a special passport by Napoleon to visit chemical laboratories and volcanoes in the French Empire. From October 1813 to April 1815 Davy, Lady Davy, her maid, and Faraday, acting as Davy's assistant, secretary and reluctant valet, toured the Continent.[12] On their return to England

---

[7] Faraday to Abbott, 19 August 1812; 1, 2 and 4 September 1812 and 9 and 11 September 1812, *Correspondence*, 1: 8, 9 and 10 respectively are Faraday's commentary on Davy's lectures.
[8] Faraday, 'Notes of Davy's 1812 Lectures', RI MS F4A.
[9] Royal Institution Managers' Minutes (hereafter RI MM), 22 February 1813, 5, p. 353.
[10] Faraday to Paris, 23 December 1829, *Correspondence*, 1, p. 419.
[11] RI MM, 1 March 1813, 5: p. 355.
[12] For Faraday's journal of this tour see Brian Bowers and Lenore Symons, *Curiosity Perfectly Satisfyed: Faraday's travels in Europe 1813–1815* (Peter Peregrinus in association with Science Museums, London, 1991). For a good account of Davy's and Faraday's relationship see David Knight, 'Davy and Faraday: Fathers and Sons' in David Gooding and Frank A.J.L. James (eds), *Faraday Rediscovered: Essays on the Life and Work of Michael Faraday, 1791–1867* (Macmillan, London 1985), pp. 33–49.

Faraday was reappointed Chemical Assistant at the Royal Institution,[13] where he worked mostly under the new Professor of Chemistry, William Thomas Brande (1788–1866).

It could be said that Britain is an island of coal surrounded by water. Both these geofacts have played hugely important roles in British history. Following 1815, Britain was the most powerful nation in the world. This was due partly to her navy, but it was claimed that it was also due to technological developments in the field of steam power and mining. As Sadi Carnot (1796–1832) put it in 1824 in his *Reflexions on the Motive Power of Fire*:

> If you were ... to deprive England of her steam engines, you would deprive her of both coal and iron; you would cut off the sources of all her wealth, totally destroy her means of prosperity, and reduce this nation of huge power to insignificance. The destruction of her navy, which she regards as the main source of her strength, would probably be less disastrous.[14]

With the importance of these core areas of British life, the navy and coal production, it should not be found surprising that much scientific activity of the day was centred around them. The Royal Institution, which had by far the best-equipped laboratories in the country, was ideally placed to undertake scientific work connected with mining and with the sea. It was therefore an obvious move to turn to Davy, and later Brande and Faraday, when it was found that practical problems needed an experimental approach to their resolution. There was really nowhere else where such work could be conducted. In many senses it was in the laboratory, and specifically, though not exclusively, the laboratory of the Royal Institution, where science and technology started to become so closely connected and in the process irrevocably change the nature and practice of both science and technology.[15]

Almost the first task that Davy undertook, with Faraday's help, following their return from the Continent was the invention of a lamp which could be used safely in mines;[16] the manuscripts of Davy's papers on the lamp are largely

---

[13] RI MM, 15 May 1815, 6: p. 58.

[14] Sadi Carnot, *Reflexions on the Motive Power of Fire: A critical edition with the surviving scientific manuscripts*, translated and edited by Robert Fox (Manchester University Press, Manchester, 1986), p. 62.

[15] Frank A.J.L. James, "Faraday in the Pits, Faraday at Sea: The Role of the Royal Institution in Changing the Practice of Science and Technology in Nineteenth Century Britain", *Proceedings of the Royal Institution*, 68 (1997), pp. 277–301.

[16] On this see David Knight, *Humphry Davy: Science and Power* (Blackwell, Oxford, 1992), pp. 105–20.

in Faraday's hand.[17] With his work for Davy on the safety lamp, Faraday, from the start of his career, was involved with technology and its relations with science. The development of the Davy lamp also demonstrated the crucial importance of the laboratory in this kind of work. Without access to the resources of the laboratory in the Royal Institution, Davy would not have been able to perform the experiments necessary for the invention of his safety lamp.

The crucial position which the Royal Institution occupied in British science meant that from the 1820s Faraday started giving expert advice himself. For instance, he analysed large numbers of gunpowders for the East India Company[18] and gave expert testimony at trials.[19] But there was a problem. He and Davy, by now President of the Royal Society, had fallen out seriously in 1823 and 1824 over priority in Faraday's discovery of the liquefaction of gases and about his election to fellowship of the Royal Society. to which Davy was opposed. There is a fine line between patronage where both sides gain something from the relationship and exploitation of the other person's abilities, and at this point Davy crossed that line.

Davy got Faraday to carry out experimental work for the Admiralty to electro-chemically protect from corrosion the copper bottoms of ships in the mid- 1820s. Davy discovered that if a piece of zinc was attached to the copper, the electro-chemical action engendered prevented the sheets corroding when they were placed in sea water. Faraday conducted many of the follow-up experiments for Davy. The method was then tested on ships moored in Portsmouth Harbour, before the entire fleet was ordered to be equipped with Davy protectors, as they were called. What the tests in Portsmouth had not shown was that one of the consequences of the corroding copper was that the salts produced in the process killed off the marine life which otherwise would attach itself to the bottom of ships. When the copper no longer produced these poisonous salts, because of the protectors, barnacles were able to adhere to a ship and thus drastically reduce its steerage. This disaster to the whole fleet did not put Davy, or initially science, in a good light with the Admiralty. This story is an excellent example of what can happen if it is not clearly understood how to transfer scientific knowledge from the laboratory to a practical situation.[20]

---

[17] For instance the manuscript of Humphry Davy, 'On the fire-damp of coal mines, and on methods of lighting the mines so as to prevent its explosion', *Philosophical Transactions* (1816), 106: pp. 1–22 is in Faraday's hand, RS MS PT 10.1.

[18] The laboratory notebook of the Royal Institution just for January 1820 contains details of twenty-four such analyses carried out by Faraday. RI MS HD 7A, ff. 92–5.

[19] See for example the sugar-refining insurance case of Severn and King discussed in June Z. Fullmer, 'Technology, Chemistry, and the Law in Early 19th-Century England', *Technology Culture*, 21 (1980), pp. 1–28.

[20] For a detailed account of this episode see Frank A.J.L. James, 'Davy in the Dockyard: Humphry Davy, the Royal Society and the Electro-chemical Protection of the Copper Sheeting of His Majesty's Ships in the mid 1820s', *Physis*, 29 (1992), pp. 205–25.

Faraday must have learnt a good deal from this episode. Where Davy's expert advice was short and enthusiastic, Faraday's was lengthy and cautious, full of caveats that everything must be tested properly, before large-scale installation. Furthermore, Faraday went to great lengths to explain the technical basis of his advice. Since most of those to whom he gave advice were civil servants or military men or suchlike, with a limited grasp of the technicalities of science but with an appreciation of its potential power, Faraday's capacity as a communicator was tested to the limit. That his advice was acted on and asked for again and again, is testimony to Faraday's success in communicating science. This is the more impressive since, in some cases, Faraday was dealing with science at the limits of knowledge at his time. In the hands of an able communicator like Faraday, science of even the most difficult kind can became transparent and understandable to a lay audience.[21] Arguably, it was this ability that made Faraday so valuable to the State. Even the State's experts acknowledged Faraday's value. As the Astronomer Royal, George Biddell Airy (1801–92), writing to Faraday about the standard yard in 1847, commented: 'We trouble you as a universal referee or character-counsel on all matters of science'.[22]

But the most time-consuming piece of work for which Davy was responsible was Faraday's project to improve optical glass on which, I have calculated, Faraday spent two-thirds of his available research time in the late 1820s.[23] Following Davy's death in Geneva in 1829, Faraday was free from the glass work, and that he regretted that he had ever taken it up is made plain in a letter to the new President of the Royal Society, Davies Gilbert (1767–1839),[24] a point he repeated the following year in a letter to one of the secretaries of the Royal Society, Peter Mark Roget (1779–1869).[25] Exactly eight weeks after this letter Faraday discovered electro-magnetic induction and how to produce electricity from magnetism. Faraday had no objection to undertaking practical work for the State, but he did not wish his time to be wasted as he considered it had been done in the glass work.

---

[21] A striking example of Faraday's approach in explaining science is his advice on lightning conductors to the East India Company. See Faraday to Melvill, 5 September 1839, 9 June 1841, 7 June 1845, *Correspondence*, 2, p. 1206; 3, p. 1351, p. 1745 respectively.

[22] Airy to Faraday, 19 May 1847, Faraday, *Correspondence*, 3 (1990).

[23] See Frank A.J.L. James, 'Michael Faraday's Work on Optical Glass', *Phys.Ed.*, 26 (1991), pp. 26, 296–300; Frank A.J.L. James, 'The Military Context of Chemistry: The Case of Michael Faraday', *Bull.Hist.Chem.* (1991), pp. 11, 36–40.

[24] Faraday to Gilbert, 13 May 1830, Faraday, *Correspondence*, 1, p. 446.

[25] Faraday to Roget, 4 July 1831, Faraday, *Correspondence*, 1, p. 501.

Faraday and the State

The outcomes of both the glass work and the electro-chemical protection of the copper sheeting of ships provoked a grave crisis in how the Admiralty viewed scientific advice. One of the immediate consequences was the abolition of the Board of Longitude in July 1828.[26] However, despite the Admiralty's far from favourable view of Davy, they realized that scientific understanding of practical issues was necessary for the Royal Navy to sustain its dominant position. Thus in late 1828 the Admiralty appointed a Scientific Committee, with Faraday as one of its members, to replace the Board of Longitude.[27] Faraday continued to advise the Admiralty on scientific matters until at least the Anglo-French war against Russia in the mid-1850s.

By the mid-1830s Faraday was recognized as one of the leading savants of the day and as certainly the best experimentalist in Britain. Furthermore, he was also seen as the leading scientific expert to advise the State on issues which required scientific advice. It was not just his work on electro-chemical protection of ships bottoms and in making optical glass that I have already alluded to, but in many other areas as well. He taught chemistry to the cadets at the Royal Military Academy in Woolwich,[28] he analysed adulterated oatmeal for the Admiralty, he worked on gunpowder and lightning rods for the East India Company and a whole range of other matters besides. One can almost hear the cry 'send for Faraday' go up in the corridors of Whitehall and Westminster when a technical problem required solution.[29] It was Faraday's work for the State that was stressed in 1835 when it was proposed to the Tory Prime Minister Robert Peel (1788–1850) that Faraday should be awarded a Civil List Pension. This could not be awarded immediately for Peel went out of office and was replaced by the Whig William Lamb, 2nd Viscount Melbourne (1779–1848), for his second term, but that is a long complex story which I shall not pursue here.[30]

Five years earlier, on 22 November 1830, the Whig politician Charles, 2nd Earl Grey (1764–1845) had become Prime Minister ending more than twenty years of continuous Tory rule. Under the electoral rules then applying the Tories could have continued in office almost indefinitely, but they had lost

---

[26] 'An Act for Repealing the Laws Now in Force Relating to the Discovery of Longitude at Sea', 9 Geo IV, c.66.

[27] Admiralty to Young, Sabine and Faraday, 31 December 1828, *Correspondence*, 1, p. 384.

[28] James, 'Military Context'.

[29] For a good example of this see Frank A.J.L. James and Margaret Ray, 'Science in the Pits: Michael Faraday, Charles Lyell and the Home Office Enquiry into the Explosion at Haswell Colliery, County Durham, in 1844', *History and Technology*, 15 (1999), pp. 213–31.

[30] See the introduction to volume 2 of Faraday's *Correspondence*, p. xxxiv for a discussion.

confidence in themselves and had become a split party over the issue of Catholic emancipation. Earl Grey, with Lord John Russell (1792–1878), then piloted through Parliament the Reform Act which did away with most rotten boroughs and gave the seats to the rapidly industrializing towns of the Midlands and the north, and, for the first time, gave the vote to members of the non-landowning middle class. This bill meant that the Whigs could never again be excluded from power for so long.

The general election of 1832 returned the Whigs to power with a landslide majority of 300. It was always intended that the reform of Parliament would be the first necessary step in the reform of Britain which was seen in various quarters as having been stagnant at the place where the reaction to the French Revolution of 1789 and the French wars had left it in 1815. In the ensuing years Grey and his successor, Melbourne, instituted a series of reforms which had far-reaching consequences. Bank notes became legal tender, the first Companies Act was passed, the East India Company was made a government agency, local government was introduced as was the civil registration of births, marriages and death. It was widely believed that these and other reforms prevented revolution in Britain. One must remember that the early 1830s saw the successful revolution in France and the failed one in the Russian province of Poland.

Faraday and Trinity House

It is in this political context of root and branch reform of the State that the reform of Trinity House must be viewed. Although the 1514 charter theoretically gave the corporation full responsibility for the safety of navigation, in practice they controlled very little and even by the 1830s many lighthouses were still in the hands of private owners. Following parliamentary inquiry, an act was passed in 1836 giving Trinity House control of all lighthouses in England and Wales and secure funding from harbour dues. Furthermore, the government gave the corporation £1,182,546 to purchase all private lighthouses,[31] a figure which puts Faraday's salary of £200 in perspective. The reform also formalized the payment of harbour and light dues to the corporation.

With secure funding and the knowledge that it could gain economy of scale with any innovation it introduced (since it now controlled all English and Welsh lighthouses), Trinity House embarked on a major programme of lighthouse development, largely, though not exclusively, in imitation of the

---

[31] Douglas B. Hague and Rosemary Christie, *Lighthouses: the architecture, history and archaeology* (Gomer Press, Llandysul, 1975), p. 59.

French lighthouse service. New lighthouses were built where none had previously existed and out-of-date lights such as those at St Catherine's on the Isle of Wight and South Foreland in Kent were replaced. Until this time many English lighthouse establishments comprised two or more lighthouses, each emitting a constant beam, produced by an oil lamp of some sort. When the lights were in line, sailors would have a guide as to their position. Such was the system used at St Catherine's and South Foreland. The middle third of the nineteenth century saw a changeover to rotating beams of fixed duration and time intervals. These used both mirrors and Fresnel lenses which, though invented by Augustin Jean Fresnel (1788–1827), were put into practice by his brother Lénore Fresnel (1790–1869) who was secretary and director of the French lighthouse service.

Implementing and overseeing this programme of technological improvement was John Henry Pelly (1777–1852) who had been appointed Deputy Master of Trinity House in 1834. He decided that the corporation, to undertake its programme, needed a scientific adviser and on 28 January 1836 sounded out Faraday to see if he would be interested.[32] Faraday signified that he would be willing to accept the position. He was then appointed and given considerable resources to do his work. He had access to facilities at Trinity House and use of the test bed for lights at Trinity Wharf, Blackwall. In total more than 500 letters between Faraday and Trinity House have been found which is more than 10 per cent of his surviving correspondence.[33]

In 1860 Faraday described his work for Trinity House to the Royal Commission on Lighthouses thus:

> a large part of my attention has been given to the lighthouses in respect of their ventilation, their lightning conductors and arrangements, the impurity and cure of waters, the provision of domestic water, the examination of optical apparatus, & c., the results of which may be seen in various reports to Trinity House. A very large part also of my consideration has been given to the numerous propositions of all kinds which have been and are presented continually to the Corporation; few of these present any reasonable prospect of practical and useful application, and I have been obliged to use my judgment chiefly in checking imperfect and unsafe propositions, rather than in forwarding any which could be advanced to a practical result.[34]

It is clear both from this account and from many of his letters that much of Faraday's work for the corporation was rather mundane in nature. What this

---

[32] Pelly to Faraday, 2 February 1836, *Correspondence*, 2, p. 882.
[33] For a discussion of the archive see Faraday's Correspondence, 3, pp. xxv–xxvi.
[34] Faraday's written evidence to the Royal Commission on Lighthouses, 25 February 1860, *Parl. Papers*, 1861 [2793] 25, pp. 591–2.

passage hides is that Faraday undertook a vast amount of travelling to lighthouses and to lighthouse manufactories both in England and, on occasion, in France. He also proactively innovated lighthouse technology in two areas, in ventilation and in the application of electric power to lighthouses.

Faraday invented a chimney to improve the ventilation of lighthouse lanterns which he made over to his brother Robert Faraday (1788–1846), a brass-founder and gas engineer, who patented it;[35] the only invention of Faraday's to be patented. This chimney was quite successful and it was installed in buildings other than lighthouses including the Athenaeum[36] and Buckingham Palace where, as *The Times* noted in 1846, Faraday's lamp illuminated the christening of Princess Helena (1846–1923).[37]

But perhaps the most innovative work that Faraday did for the corporation was on the electrification of lighthouse illumination. As early as 1847 it was proposed to electrify buoys. Faraday commented on this scheme that the buoys had to work under all weather conditions 'the light must be a *certain* light & in no respect liable to cease: for an uncertain indication is worse than none'.[38] This was to be a constant theme in the ensuing years both in lectures that he delivered[39] and in reports that he wrote on the merits or otherwise of various systems of electrification proposed to Trinity House.[40]

Faraday believed that Trinity House should not deploy the new science-based technology of electric light just because it existed. If such a new system was to fail then lives would be at risk. As one would expect in a Sandemanian, such as Faraday, moral and social considerations took priority over his deep scientific interest when reaching a technological conclusion. Science and technology for Faraday thus had a moral and social dimension and could not, *ipso facto*, be viewed as value free. This seems to repudiate Morris Berman's rather simplistic analysis of Faraday's work. Faraday was not simply 'London's leading technological "fixer"'[41] as Berman put it, but viewed the technical component of a problem within its overall context which generally took precedence when technical decisions were made.

---

[35] Patent 9679, 25 March 1843, 'Ventilating gas-burners, and burners for consuming oil, tallow, and other matters'. See also Faraday to Robert Faraday, 10 December 1842 and 10 January 1843, *Correspondence*, 3, pp. 1452 and 1460; Faraday, 'On the Ventilation of Lighthouse Lamps; the points necessary to be observed, and the manner in which these have been or may be attained', *Ministry of the Institution of Civil Engineers*, 2 (1843), pp. 206–9.

[36] Frank Richard Cowell, *The Athenaeum: Club and Social Life in London 1824–1974*, (Heinemann, London, 1975), pp. 24–5.

[37] *The Times*, 27 July 1846, p. 5.

[38] Faraday to Brown and Brown, 17 December 1847, *Correspondence*, 3, p. 2038.

[39] M. Faraday, 'On Lighthouse Illumination–the Electric Light', *Proceedings of the Royal Institution*, 3 (1860), pp. 220–23.

[40] Faraday report to Trinity House, 15 August 1854, Faraday, *Correspondence*, 4: 2878.

[41] Morris Berman, *Social Change and Scientific Organization: The Royal Institution, 1799–1844* (Heinemann, London, 1978), p. 156.

Thus, in his report on testing the system of illuminating lighthouses by using batteries and carbon arcs which had been proposed by Joseph John William Watson,[42] Faraday wrote:

> Much, therefore, as I desire to see the Electric light made available in lighthouses, I cannot recommend its adoption under present circumstances. There is no human arrangement that requires more regularity and certainty of service than a lighthouse. It is trusted by the Mariner as if it were a law of nature; and as the Sun sets so he expects that, with the same certainty, the lights will appear.[43]

This rather poetic passage, making the striking comparison between the laws of nature (which God had written into nature at the Creation)[44] and a human technology, emphasizes the moral seriousness with which Faraday worked on lighthouses. As with all his work, in whatever context, in the final analysis it was underpinned by his religious beliefs. He concluded a Friday Evening Discourse at the Royal Institution in 1858 with the view that by 'enabling the mind to apply the natural power through law, it conveys the gifts of God to man';[45] humankind can, indeed is obliged, to use knowledge of the Creation for practical purposes – in this case recommending the rejection of a technology (which Trinity House proceeded to do).

The second system of electrical illumination, was proposed in the late 1850s, by Frederick Hale Holmes,[46] used a magneto-electrical generator which powered a carbon arc. Faraday subjected this system to the same thorough testing as he had applied to Watson's proposal, but Holmes passed and it went into practical operation at the South Foreland lighthouse. There Faraday continued to monitor its performance during the first half of the 1860s, sometimes to the detriment of his health. Over the next fifteen to twenty years electric light was installed in several lighthouses.[47] However, in the end this system was deemed a failure due mainly to its expense, which Faraday had

---

[42] Joseph John William Watson, *A Few Remarks on the Present State and Prospects of Electrical Illumination* (London, 1853). For a detailed discussion of Watson's Work see Frank A.J.L. James, 'The civil-engineer's talent: Michael Faraday, science, engineering and the English lighthouse service, 1836-1865', *Trans. Newcomen Soc.*, 70 (1999), pp. 153–60.

[43] Faraday report to Trinity House, 15 August 1854, Faraday, *Correspondence*, 4, p. 2878

[44] Cantor, *Michael Faraday*, pp. 201–5.

[45] Faraday, 'On Wheatstone's Electric Telegraph in relation to Science (being an argument in favour of the full recognition of Science as a branch of Education)', *Proceedings of the Royal Institute*, 2 (1858), p. 560.

[46] Frederick Hale Holmes, *Magneto-Electric Light as Applicable to Lighthouses* (London, 1861).

[47] On this programme see James Nicholas Douglas, 'The Electric Light applied to Lighthouse Illumination' *Ministry of the Proceeding Institute of Civil Engineers*, 57 (1879), pp. 77–165.

pointed out would be an issue.[48] In 1880 a reverse programme of replacing the electrified lights with gas and oil lamps was undertaken.[49] It was not until 1922 before electricity was again used for lighthouses when the South Foreland was fitted with a filament lamp.[50]

Conclusion

Faraday's work for Trinity House was part of a process which by the end of the nineteenth century had brought science and technology into one common practice. But the process was not a straightforward one. What Faraday emphatically did not do was simply to take scientific knowledge out of the laboratory and apply it in a practical situation. Scientific knowledge was only one aspect of his work for Trinity House. Economic, moral and religious considerations played their part in the recommendations that Faraday made. Some of these Faraday was quite explicit about in his reports; others he did not mention. In the career of Faraday we have a case study of how the practice of science and technology became embodied in a single individual and his practice within the context of working for the State and its agencies.

Faraday was thus a key figure in bringing science and technology together during his lifetime (and beyond, I might add, in the iconic image that was created for him after his death[51] but which, ironically, had little to do with the work discussed in this essay). But the purpose of bringing together knowledge of the world with power over that world was, in Faraday's words, to convey the 'gifts of God to man'. In addition to Faraday's overt theological meaning, there was also a social meaning since, as Faraday was well aware, the interaction of science and technology with society was a highly intricate one. This he made clear in the opening passage of his Friday Evening Discourse on the electrification of lighthouses:

> The use of light to guide the mariner as he approaches land, or passes through intricate channels, has, with the advance of society and its ever increasing interests, caused such a necessity for means more and more perfect, as to tax to the utmost the powers of both the philosopher and the practical man, in the development of the principles concerned, and their efficient application.[52]

---

[48] Faraday, 'Lighthouse Illumination', p. 223.
[49] Kenneth Sutton-Jones, *Pharos: The Lighthouse Yesterday, Today and Tomorrow*, (Michael Russell, Salisbury, 1985), p. 117
[50] Hague and Christie, *Lighthouses*, p. 160.
[51] Graeme Gooday, 'Faraday Reinvented: Moral Imagery and Institutional Icons in Victorian Electrical Engineering', *History of Technology*, 15 (1993), pp. 190–205.
[52] Faraday, 'Lighthouse Illumination', p. 220.

In this passage, Faraday encapsulated his belief in the moral responsibility that the practitioners of science and technology should have towards the demands of society and the state. It is little wonder that Faraday became such an icon for the Golden Age.

Chapter 8

Lies, Damned Lies and Declinism: Lyon Playfair, the Paris 1867 Exhibition and the Contested Rhetorics of Scientific Education and Industrial Performance

Graeme Gooday

I have just returned from Paris, where I acted as a juror in one of the classes of the Exhibition ... I am sorry to say that, with very few exceptions, a singular accordance of opinion prevailed that our country has shown little inventiveness and made little progress in the peaceful arts of industry since 1851 when I found some of our chief mechanical and civil engineers lamenting the want of progress in their industries, and pointing to the wonderful progress which other nations are making [and] our chymical[sic] and even textile manufacturers uttering similar complaints, I naturally devoted attention to elicit their views as to the causes.[1]

Referring to the Exhibition, I was somewhat disappointed to find very few new things in the English Department of machinery. I am quite sure we could have made a much better show in this respect if we had wished to ... Concerning the want of novelty among the English engineering [sic], I have been informed that this may partly be accounted for by the disinclination which is felt by our firms to show their newest arrangements ... very many men of eminence in my business have lost faith in Exhibitions altogether; they do not approve of them, and if they are compelled, from various causes, to exhibit at all, they send the least, instead of the most[,] which they feel will suffice to keep their name before the public.[2]

---

[1] Lyon Playfair, Letter to Right Hon. Lord Taunton, London, 15 May 1867, reproduced on 7 June, *Journal of the Society of Arts* [*JSA*], 15 (1866–67), pp. 477–8. The original letter appeared first in *The Times*, 29 May 1867, and two letters critical of Playfair's conclusion that subsequently appeared in *The Times* were republished in the same issue of the *JSA*, pp. 478–9.

[2] John Evans, *Reports of Artisans selected by a Committee appointed by the Council of the Society of Arts to visit the Paris Universal Exhibition 1867* (London, 1867), part 1, p. 422.

> Among the most remarkable omissions is that of the Transatlantic Telegraph cable and its accessories. Since the last international exhibition no nation [other than Britain] has achieved an act of higher practical difficulty, of larger results, or one which has obtained the same amount of admiration and applause; and it is certainly one of the strangest facts connected with this exhibition that the union of the old and new worlds by two electric cables should be totally without record on this occasion.[3]

The spectacular and controversial Paris Universal Exhibition ran from May to October 1867. Considerably larger than both its predecessor in 1855, and counterparts held in London in 1851 and 1862, it displayed wares from over eighty different countries or colonies, in ninety categories ranging from oil paintings to surgical equipment, perfumes to machine tools, and from porcelain to prize pigs. Approximately 12 million visits were made to the huge elliptical palace that housed the exhibition on the Champ du Mars, and like so many of the international extravaganzas that followed the Great Exhibition in London (1851), it had been organized with the nationalistic messages directed at both domestic audiences and competitor countries.[4] Although three years later these messages proved to have been in vain as France collapsed under the Prussian invasion, the 1867 'Exposition Universelle' was the cue for more immediate self-examination in Britain. Some voluble commentators considered that the exhibition had furnished evidence of severe complacency in British manufacturing industry and each prescribed one of a variety of cures for this. In this chapter, I shall devote particular attention to the testimony of the juror to Class 51 'apparatus used in chemistry, pharmacy and tanning yards', the Professor of Chemistry at Edinburgh University, Lyon Playfair, who used alarmist allegations about the nation's lack of 'inventiveness' manifested at the exhibition as a springboard to relaunch his previously unheeded campaign to promote scientific education in Britain.

Some historians of British science and technical education have followed Playfair's judgement as reliable and disinterested testimony on the actual state of affairs revealed by the Paris Exhibition of 1867. For Donald Cardwell and Michael Sanderson, this episode represented the end of sixteen years of unrivalled British industrial supremacy that had followed the remarkable

---

[3] [Anon] Editorial 'The Paris Exhibition', *JSA*, 15 (1866–67), p. 476.

[4] As Victor Hugo wrote in the *Guide de l'exposant et du visiteur avec les documents officiels, un plan et une vue de l'Exposition* (1867): 'O France adieu. You are too great to be merely a country. People are becoming separated from their mother and she is becoming a goddess. A little while more, and you will vanish in the transfiguration. You are so great that you will no longer be. You will cease to be France, you will become humanity; you will cease to be a nation, you will be ubiquity' cited in Bob Brain, *Going to the Fair: Readings in the Culture of Nineteenth Century Exhibitions* (Whipple Museum of the History of Science, Cambridge, 1993), p. 152.

success at all international exhibitions held since 1851. The 1867 exhibition in Paris supposedly shattered British (or alternatively English) delusions of international invincibility by showing that other nations, especially those on the continent of Europe, eclipsed the United Kingdom's performance in almost all areas of industrial endeavour. Within the terms of this book this might be translated into a claim that if such international exhibitions are to be taken as reliable indicators of relative industrial performance, the Golden Age of British (or English) manufacturing was over by the last third of the nineteenth century. I challenge the plausibility of such an account and suggest that a rather different interpretation might be placed on the significance of the events in Paris in 1867, especially in their relation to the history of science education in Britain.

Three questions lie at the heart of this paper. Did the Paris Universal Exhibition of 1867 unequivocally signal the (relative) decline of 'English' global industrial dominance? I suggest that this was not the case: by examining the testimony of a range of visitors, I show that there were many diverse views on this subject, and not a few who claimed that the exhibition did not provide a sufficient evidential base for such claims to be made. Given that the exhibition did not self-evidently furnish evidence for a relative British industrial decline, it is important to ask, then, why certain characters, notably Lyon Playfair, were so forthright in their declinist conclusions and why Playfair in particular diagnosed the cause of this alleged decline to be an inadequate system of scientific education. In a later section of the paper I consider what connection can reliably be drawn between the outcome of the exhibition and the subsequent development of British science education – which ironically seems to have begun its own Golden Age just as the industrial campaign was fading.

The Declinist Literature on the Paris Exhibition

> It has been remarked that exhibitions, national or international, are not necessarily indices of a nation's technical and industrial progress. Indeed, to form a comprehensive judgement from an exhibition would demand near omniscience. Only perhaps in the case of specific industries can fairly objective conclusions be drawn by competent, specialized experts, and even in these cases it is difficult to establish whether or not the exhibits are really representative; the choicest fruit is not necessarily put in the front of the stall.[5]

The theme of British 'decline' is usually conceived as a relative decline in world industrial standing and or economic dominance and has been burdened

---

[5] Donald Cardwell, *The Organization of Science in England* (Heinemann, London, 1972), p. 111.

with nearly as many chronologies as it has putative historical explanations. Where Martin Wiener saw the rot setting in immediately after the 1851 exhibition, historians of science education pinpoint the *annus horribilis* as 1867, economic historians more generally point towards the early 1870s,[6] Corelli Barnett suggests the relevant threshold was the Second World War, and David Edgerton holds that important forms of relative decline only really began in the 1970s.[7] As Edgerton himself notes, however, most accounts of decline have involved self-interested attempts to nurture a sense of national crisis by those who wish to peddle a lucrative solution to it. As long ago as 1938, Sir John Clapham noted with admirable scepticism of declinist claims, that an engineer with great 'faith' in the application of electric trams for cities or in milling cutters for all kinds of metal-working was apt to 'overdo the backwardness of his countrymen and the deadly universal efficiency of someone else'.[8]

The critical historian should indeed always be wary of taking at face value any pronouncements of decline: rather than accepting them as unproblematic evidential claims, a more useful strategy is to interrogate the purposes to which such claims are directed. Since nobody in the late nineteenth century seems to have undertaken a dispassionate quantitative analysis of the decline issue, I suggest all claims about declinism were driven by some form of vested interest that was served by the nurturing of a widespread impression of industrial 'decline'. Edgerton's deconstruction of accounts of declinism unequivocally commences in 1870, and he wastes no time in considering 1867 as a plausible candidate for the beginning of a national malaise. Not all historians, though, have been as sceptical as they might have been about the impartiality of Lyon Playfair's open letter to the chairman of the Schools' Inquiry Commission, Lord Taunton, in May 1867, in which he heralded the commencement of such a decline. Quite why Playfair addressed this letter to the chairman of the Commission into the financing and management of Endowed Schools is a point to which I shall return later, but it was pointedly also published in *The Times* of 29 May and reproduced shortly afterwards in many other places, including the *Journal of the Society of Arts.*[9]

In this letter Playfair claimed that in the wake of the Paris Exhibition an almost universal opinion prevailed in Britain and elsewhere that the award of

---

[6] Michael Dintenfass, *The Decline of Industrial Britain, 1870–1980* (Routledge, London, 1992).

[7] Martin Wiener, *English Culture and the Decline of the Industrial Spirit* (Cambridge University Press, Cambridge, 1981); Corelli Barnett, *The Audit of War: the illusion and reality of Britain as a great nation* (Macmillan, London, 1986); David Edgerton, *Science, technology and the British Industrial 'Decline' 1870–1970* (Cambridge University Press, Cambridge, 1996).

[8] Cited in Edgerton, *Science*, p. 12.

[9] See citations in note 1.

'scarcely' a dozen top prizes in the ninety exhibition categories – implicitly a dreadful contrast with previous exhibitions – meant that the nation had lost its pre-eminence in international industry. He further claimed that there was 'unanimity of conviction' concerning the explanation for the cause of this apparent calamity: although he conceded Britain's poorer record of industrial relations might partly be responsible, he flatly denied that 'deficient representation' could have been a relevant factor in explaining this outcome. Playfair contended instead that the reason for Continental incursions on hitherto exclusively British prize territory was that France, Prussia, Austria, Belgium and Switzerland possessed good systems of industrial education for the masters and managers of their factories and workshops. To support this claim, Playfair recounted strolling round the exhibition alongside the eminent French chemist and senator, J.B. Dumas. Dumas had apparently 'assured' Playfair that the development of technical education in France had given a 'great impulse' to the industry of France. Indeed, whenever anything of particular excellence in a French exhibit grabbed Dumas's attention, he had, Playfair averred, quickly ascertained that most of the managers responsible had once studied at the Ecole Centrale des Arts et Manufactures. In addition, the Director of the Conservatoire des Arts et Metiers had apparently told Playfair that he had recently sat on a commission which had established the finest technical education for workmen was to be found in Austria, whereas 'higher instruction' of masters was best in France, Prussia and Switzerland.[10]

In his classic work *The Organization of Science in England*, the late D.S.L. Cardwell opens his chapter on 'The Age of Enquiries' (1868–90) with the sceptical remark about the evidential significance of exhibitions cited in the epigraph above. Nevertheless, he then relates the contents of Playfair's letter to the head of the Schools Enquiry Commission without any critical scrutiny of the latter's selective citation from witnesses. Noting that Playfair's letter moved the commission to circularize British jurors at Paris with a request for their views on the matter, Cardwell then contends that 'without exception' those who replied concurred with Playfair's views. As we shall see, however, by no means all these jurors were prepared to acquiesce in Playfair's claims. Some refused to concede the relevance of education to the exhibition's outcome, others ironically prefigured Cardwell's scepticism about what could be inferred from exhibitions; still others noted that British presence had not accurately represented the state of industry in the UK – as may be seen in the discussion below.

Nevertheless, Cardwell infers from Playfair's testimony that the exhibition was widely believed to have revealed a state of affairs 'highly discreditable' to Britain, and that the subsequent alarm naturally generated a movement for

---

[10] Playfair, Letter 1867.

technical education. This final point is the key to understanding why Cardwell takes Playfair's judgements at face value: Cardwell's historiographical purpose in his book is to account for the rapid and sudden emergence of the technical education movement in the years immediately following the Paris Exhibition. Accordingly, the stark evidential significance of the exhibition – as interpreted at least by Playfair – has great explanatory power for Cardwell in accounting for the relatively sudden outbreak of interest in scientific and technical education in the United Kingdom in the late 1860s.[11]

Subsequent historiographies of science education have followed Cardwell's interpretation, centring on the putative impact of the exhibition. Not all have so directly acquiesced, however, in Playfair's alarmist evidence. This is apparent, for example, in several essays in Roy McLeod's collected volume on the development of science examinations in later Victorian England. Whereas Frank Foden claims that the Paris Exhibition by itself eroded British 'complacency' about its technical development, others claim less contentiously that the relationship between industrial success and scientific education merely became more widely debated in Britain in the aftermath of the exhibition. Christine Heward suggests that such debates were conducted with a view to ensuring that Britain maintained its existing supremacy in industry – hardly the view that Playfair supported; Harry Butterworth notes, moreover, that the connection between the outcome of the Paris Exhibition and the subsequent growth of science teaching was somewhat indirect, and mediated by other indigenous factors in the contemporary British education scene, such as the abolition of certain highly-selective teachers' examinations, and Joseph Whitworth's philanthropic offer of scholarships to trainee teachers in the Department of Science and Art.[12]

Notwithstanding such subtle revisionist claims, and irrespective of Edgerton's forceful critiques of the declinist literature, Michael Sanderson has recently reiterated the Playfair position on the awful significance of the Paris Exhibition as a herald of the decline of British industry. A main purpose of this chapter is to counter Sanderson's claims on this score, especially the assertion that there was a consensus in favour of Playfair's apocalyptic vision of relative industrial decline.[13]

---

[11] Cardwell, *Organization of Science*, pp. 111–12.

[12] See Roy Macleod (ed.), *Days of Judgement* (Nafferton, Driffield, 1982), especially Harry Butterworth, 'The Science and Art Department Examinations: Origins and Achievements', pp. 32, 41; Christine Heward, 'Education, Examinations and the Artisans: The development of Science and Art in Birmingham, 1853–1902', p. 3; Frank Foden 'The Technology Examinations of the City and Guilds', p. 66. For a more recent reiteration of theses theme see W.H. Brock, 'Science Education', in Robert Olby et al., *Companion to the History of Modern Science* (Routledge, London, 1990), pp. 946–59, especially p. 957.

[13] Michael Sanderson, 'French Influences on Technical and Managerial Education in England, 1870–1940' in Y. Cassis, F. Crouzet, and T.R. Gourvish (eds), *Management and*

Reassessing the Paris Universal Exhibition of 1867

> I hardly think that an Exhibition in Paris furnishes the means of accurately testing the comparative merits of English and French education. The simple inconvenience of transport tends to render England worse represented than France.
> John Tyndall, letter to Taunton Commission, 3 June 1867.

In his widely republished open letter to the Taunton Commission in May 1867, Professor Playfair cited three possible causes for the apparently poor showing of the British exhibitors in the Parisian prize competitions in comparison to that in 1862: deficient representation, inadequate scientific education and poor industrial relations. He quickly dismissed the first of these accounts, arguing that this could not explain why British exhibitors had won top prizes in only ten of the ninety categories: Playfair evidently found it inconceivable that the nation's best manufacturers might have such little faith or interest in international exhibitions that they could boycott it to this unprecedented extent. He gave somewhat greater credence to the third explanation, noting that it was widely accepted that Britain had indeed suffered from a 'want of cordiality' between employers and workmen, arguing even that trade unions had actively inhibited workmen from 'giving free scope' to their skills; such concerns were noted by several other jurors as relevant to explaining the international distribution of prizes. Strikingly, however, Playfair did not follow up this observation with any recommendation that more be done to alleviate Britain's industrial relations problems, and chose instead in a somewhat partisan fashion to stigmatize Britain's system of science education as the chief source of the problem. From conversations in Paris, he argued that the explanation about which there was most 'unanimity' was that whereas France, Prussia, Austria, Belgium and Switzerland possessed 'good systems of industrial education' for the masters and managers of factories, Britain had none.[14]

Looking at the correspondence elicited from Paris jurors by the secretary of the Taunton Commission, H.J. Roby, it is clear that Playfair's interpretation of conversations he had held in Paris was somewhat selective and that he had reinterpreted evidence on the educational issue to suit his particular polemical purposes. Whilst several shared Playfair's concern about the deleterious effects of problematic industrial relations,[15] a significant number disagreed with him

---

Business in France and Britain, 1996; *Education and Economic Decline in Britain, 1870 to the 1990s* (Cambridge University Press, Cambridge), p. 14

[14] Playfair, Letter 1867.

[15] See testimony of carpet manufacturer Peter Graham that the influences of trades unions tended to reduce the intelligent and industrious workmen to the 'same level as the stupid and lazy'. Unlike Playfair, A.J. Mundella noted that such impediments to 'industrial progress' could be properly dealt with by local boards of arbitration and conciliation, and he had himself

that the exhibition had fairly represented British manufacturing; they thus did not share Playfair's basic premiss that there was (yet) any British decline to be explained on the basis of exhibition evidence. In so far as many agreed with Playfair that educational issues mattered, their concern on this point was not necessarily about science education, and tended to refer to ensuring that the future industrial performance of Britain did not go into decline. Commentaries, moreover, written by working men sent by the Society of Arts to Paris show further evidence of the complexity of the response of British visitors to the outcomes of the Paris Exhibition.

The possibility that British manufacturing was not fairly represented at the Paris Exhibition was dismissed by Playfair with striking rapidity. Since he clearly wished to formulate a strong case for the significance of educational matters in the number of prizes won, Playfair needed to play down all doubt on the international representativeness of the exhibition, lest his alarmist conclusions about its outcome were dismissed with equal rapidity. Although Playfair's letter in May notes that British companies had only been awarded top honours in less than a dozen classes, Playfair was relying only on provisional announcements, since the official jury prizes were not awarded until early July. Moreover, although Playfair pinpointed the industries of engineering, chemicals and textiles as key areas of apparently lost British prestige, these covered only about ten of the ninety-five classes exhibited. Looking, by contrast, at the competition from the point of view of overall number of prizes awarded it is difficult to see a dramatic decrease in the numbers awarded to the United Kingdom between 1862 and 1867. Of the ninety classes represented in 1867, there were only eighteen in which Britain won no prizes at all, and indeed sixteen of those related to perishables such as agriculture and foodstuffs – where the French could so much more easily display fresh produce. Indeed, it was pointed out by a commentator for the *Journal of the Society of Arts* that not a single British cheese-maker had bothered to send any cheeses over to the exhibition, despite the fact that they had won many prizes in previous years.[16] As noted above, John Tyndall, Professor of Physics at the Royal Institution in London, believed that the complex and expensive logistics of transporting goods to Paris had significantly deterred British exhibitors from accomplishing full representation.[17]

There is a broader interpretive problem for the historian of what can or cannot be proven by the award of prizes at an international exhibition in the nineteenth century. One need not be a hardened cliometrician to suspect that

---

been involved in such enterprises. See their accounts in letters to the Schools Enquiry Commission, 6 June 1867, reproduced in *JSA*, 15 (1866–67), pp. 602–4, 612–15, for example.

[16] [Anon], 'The Paris Exhibition,' *JSA*, 15 (1866–67), p. 477.

[17] John Tyndall, letter to Schools Enquiry Commission, 6 June 1867, reproduced in *JSA*, 15 (1866–67), p. 602.

prize awards at such exhibitions hardly offer rigorous indices to measure the comparative industrial expertise of the exhibiting countries: even Cardwell himself notes that the choicest fruit was not necessarily put in the 'front of the stall'.[18] Quite a number of British jurors in Paris and other commentators noted such problems of representativeness in ways that were completely occluded in Playfair's letter to the Taunton Commission. A former HM Schools Inspector, Rev. Canon Morris, replied to the commission's circular letter on Playfair's themes that international exhibitions had 'already outgrown those conditions under which the justice of awards could be considered at all certain'; he contended that an injustice had been done to the nation's merits in the prize categories since 'we were not at all fairly represented'.[19] The engineer Robert Mallett observed that for many reasons 'England' was in various industries 'either ill-represented or not represented in Paris at all'.[20] Similar comments were made by the mining juror, Professor Warrington Smyth, of the Royal School of Mines, and John Fowler, President of the Institution of Civil Engineers.[21]

More tellingly, perhaps, we can observe that the French exhibition organizers had, in the nationalistic traditions of exhibition hosts, not taken great trouble to ensure that other countries' manufacturers were equally represented. Such nationalistic chauvinism had been apparent at the 1862 London Exhibition at which British dominance among prize-winners had been underscored by the proportionately large amount of display space – approximately half – granted to the host nation; accordingly it is not surprising that the organizers of the Paris Exhibition had arranged for French exhibitors to have more display space than any other country, occupying nearly half of the entire site.[22] Moreover, as Henry Cole noted in his official report on the exhibition, the French exhibitors were visible on an extensive panorama along a long edge of the oval display, whereas the British display was on a somewhat tighter curve at one end of the exhibition. According to Cole, this entailed two things: viewers could get a broad view of the French displays, but not of the British, and the British display areas were not rectangular or even necessarily symmetrical, which entailed particular difficulties with mounting objects in an effective or convenient manner.[23]

---

[18] Cardwell, *Organization of Science*, p. 111

[19] Rev, Canon Norris, Letter to Schools Enquiry Commission, 3 June 1867, reproduced in *JSA*, 15 (1866–67), pp. 601–2.

[20] Robert Mallett, letter to Schools Enquiry Commission, 17 June 1867, reproduced in *JSA*, 15 (1866–67), pp. 612–13.

[21] See letters by Fowler Smyth, reproduced in *JSA*, 15 (1866–67), pp. 603; 604 respectively.

[22] See plans in frontispiece of *Reports on the Paris Exhibition (presented to both Houses of Parliament), vol.1 containing the Report by the Executive Commissioner* (London, 1867).

[23] Henry Cole, *Reports on the Paris Exhibition*, ibid., p. ix.

Then again, even if the British exhibitors had been offered display space comparable to that in London in 1862, it is by no means obvious that it would have been occupied by wares from the nation's most prestigious manufacturers. Although Playfair had pinpointed the textiles industry as a key area of apparently lost British prestige, Peter Graham of Messrs Jackson and Graham (furniture and carpet makers) offered the Taunton Commission some conflicting testimony. Graham argued that the Paris Exhibition did not 'fairly represent' progress made since 1862 or the existing state of 'several of our most important manufactures', and more specifically that of all 'our great textile manufacturers' the only ones 'fully and fairly represented' were Scots tweeds and Irish linens. In response to Graham's comments the historian may well ask why the relevant English makers should have stayed away from Paris. One obvious reason is that successful British companies with secure domestic and overseas markets hardly needed to exhibit their wares in order to prove their merits and, indeed, then also to have to pay substantially for the privilege of so doing. More crucially, such successful companies probably did not wish to have their designs scrutinized at such close quarters by competitors. A number of exhibition commentators argued that if there had been any improvement in 'foreign' goods between 1862 and 1867 it was because French and Prussian companies had just copied English designs,[24] and thus by implication it was prudent for the best British manufacturers to withhold their most advanced products from scrutiny by competitor countries.

The mechanical engineer John Evans wrote on this theme in the *Reports of Artisans selected by the Society of Arts to visit the Paris Universal Exhibition.* He was sure that many of the 'English' contingent could have displayed many more new items if they had so 'wished', but having 'lost faith' in exhibitions altogether they were disinclined to do so; if they sent anything for display at all, it was the least that would 'suffice to keep their name before the public'. Certain foreign observers, he contended, were always on the look out for a 'successful English invention' so that it could be copied and sent abroad and not infrequently reimported to Britain as a 'foreign discovery'. Alas, he noted, inventors did not consider that English patent law provided adequate protection against infringements of this sort. Accordingly, it would be a gross commercial blunder for a British manufacturer to send highly innovative machinery to a foreign exhibition where it could more easily be inspected at close quarters by a potentially much larger corps of non-British plagiarists. As it turned out, though, Evans considered that in terms of 'finish and sound workmanship' the British displays demonstrated great superiority of a degree that could not so easily be appropriated by Continental rivals. Whilst some American

---

[24] Capt. R.E. Beaumont, Letter to Schools Enquiry Commission, 10 June 1867, reproduced in *JSA*, 15 (1866–67), p. 604.

productions approached the faultless sophistication and exactness of the Whitworth company's machine tools, the work of French producers showed 'much waste of metal' and evidence of 'want of skill' in design. Unsurprisingly, Evans drew no inferences whatsoever about Britain's need for more advanced technical education to attain French standards or workmanship.[25]

If prestigious British mechanical engineering companies sent only their less recent products to the 1867 Exhibition, the most famous concern in the new field of submarine telecommunications engineering, the Atlantic Telegraph Company, sent no exhibits to Paris at all. In the previous year, this company had quite spectacularly succeeded in providing direct telegraphic communication between the British Isles and North America on a scale much more permanent than the problematic and ultimately failed cable of 1858. The official guide to the London Exhibition in 1862 recorded that there had been 'no great discoveries in electrical science, nor any important practical application of principles' since the 1851 exhibition, yet the catalogue for the British Section of the Paris Universal Exhibition noted that by far the 'greatest achievement' since 1862 was the establishment in 1866 of a working cable 1,670 nautical miles long between Ireland and Newfoundland.[26] It was with some irony, then, that in the same issue that it published Playfair's letter to Lord Taunton, the *Journal of the Society of Arts* noted how extraordinary it was that the Atlantic Telegraph Company was not represented in display class 64 'Telegraphic apparatus and processes' at Paris. As documented in the epigraph at the head of this paper, an editorial commented that one of the 'strangest facts' about the exhibition was that the union of the Old and New Worlds by two electric cables 'should be totally without record'.[27]

It is difficult to establish precisely why the Transatlantic Telegraph Company boycotted the Paris meeting. It is perfectly reasonable to surmise that it not only wished to limit the gaze of Continental copycats, but needed no further publicity than it had received among the paeans of praise heaped upon it in 1866 for both laying a new cable and reviving the broken 1865 cable.[28] Playfair's neglect of this point in his radically declinist account is all the more telling since it has been uncritically accepted by practically all historians of science education since then. As Daniel Headrick has pertinently noted, the British dominance of the submarine telegraph industry – in the manufacture, laying and maintenance of cables – was as unrivalled as it was lucrative up to

---

[25] John Evans, *Reports of Artisans*, 1867, p. 422.
[26] *Catalogue for the British Section for the Paris Universal Exhibition* (London, 1867), p. 102.
[27] [Anon] 'The Paris Exhibition', *JSA*, 15 (1866–67), p. 476
[28] Crosbie Smith and Norton Wise, *Energy and Empire: A biographical study of Lord Kelvin* (Cambridge University Press, Cambridge, 1989), pp. 682–3.

the end of the nineteenth century.[29] And yet although this field (along with electric light engineering from the 1880s) was a major area of innovation and economic success in the British economy for the rest of the century, it has been rather egregiously omitted from all major declinist accounts of British industry. Playfair's own remarkable lack of reference to the phenomenon of telegraphy should itself be sufficient indication of the pains he took to avoid any reference to British industrial successes that were manifestly not represented at the Paris exhibition. Quite why he should do so will be the subject of the next section.

Reassessing the Educational Implications of the Paris Exhibition

> The facilities for scientific education are far greater on the Continent that in England, and where such differences exist, England is sure to fall behind as regards those industries into which the scientific element enters.
> (John Tyndall, Testimony to Taunton Commission)

In previous sections I have questioned the historically 'received view' that there was a consensus in agreement with Playfair's interpretation of the 1867 exhibition. Not a few jurors and visitors denied that its displays could be credibly interpreted as showing evidence of stagnation or relative regress in British industrial performance, and thus saw no direct link between the number of prizes awarded and the state of the nation's education. However, to be fair to those historians of scientific education who have adopted Playfair's declinist rhetoric, it must be admitted that at least some of the jurors to whom his letter was sent did concur with his critical diagnoses; notable among these were the London chemist Edward Frankland, the locomotive expert James McConnell, the Leeds Woollens manufacturer Edward Huth, and the naval architect John Scott Russell.[30]

Even those who did not accept Playfair's gloomy pronouncements, or his faith in the impartial justice of the exhibition award system, still acknowledged that the performance of Continental rivals certainly held important implications for British scientific education. An important distinction to note is that those who followed Playfair interpreted the exhibition as revealing major flaws in existing schemes of industrially relevant training, whereas others considered only that Britain would at some future stage need to improve its educational facilities in order to prevent rival nations from overtaking its industrial

---

[29] Daniel Headrick, *The Invisible Weapon: Telecommunications and International Politics, 1861–1945* (Oxford University Press, Oxford, 1991), especially pp. 11–111.

[30] See their respective Letters to Schools Enquiry Commission, reproduced in *JSA*, 15 (1866–67), pp. 602–3.

performance. Moreover, two further subtle differentiations that the historian must observe are the training of the 'artisan' versus that of managers, and that of primary versus higher education since several commentators in both groups tended to pick out one rather than the other as bearing most closely on standards of industrial performance. These distinctions are important to highlight since they are unhelpfully glossed over in the claims of Cardwell and Sanderson.

For example, on the first point, John Tyndall noted with Playfair that facilities for scientific education were much greater on the Continent than in the United Kingdom, but warned in virtue of this only that England must 'one day' soon 'find herself outstripped' by such nations – not that it had already been thus outstripped.[31] Similarly, the Paris juror on iron, David Price, PhD, replied to the Taunton Commission's circular that 'extended scientific education' was of the 'highest consequence' if Britain wished to retain its present position in the 'scale of nations'.[32] On the second point, we can note that the Rev. Canon Morris, until recently one of Her Majesty's School Inspectors, focused attention on the working classes, claiming that whilst Britain's facilities for primary education were 'well abreast' of France, Austria and Prussia, these nations were more proficient at providing the higher education that converted the 'mere *workmen*' into the skilled 'artizan'. By contrast, the Lecturer in Mining and Mineralogy at the School of Mines in London, Warrington Smyth, contended that the considerable advancement in mining techniques made by France, Prussia and Belgium since 1862 was not due to the efforts of workmen but owed much to the 'superior training' of the managers and sub-officers of works.

It might be *prima facie* perplexing to find such a diversity of judgements about the relationship between educational provision and industrial performance in the aftermath of the Paris Exhibition. Yet those who are familiar with recent historical relationship on this subject by Fox and Guagnini, and Kees Gispen, will know that most authors now see no universal connection between these two parameters. There are plenty of examples from the last two centuries, of nations which had sophisticated educational systems which did not render them global industrial leaders, whilst others nations were industrially successful for reasons quite independent of their educational resources.[33] Why, then, have some historians of education been so keen to draw

---

[31] Tyndall, Letter to Schools Enquiry Commission, p. 602

[32] David Price, letter to Schools Enquiry Commission, 6 June 1867, reproduced in *JSA*, 15 (1866–67), p. 614.

[33] See essays in Robert Fox and Anna Guagnini (eds), *Education, Technology and Industrial Performance in Europe, 1850–1939* (Cambridge University Press, 1993), and discussion of the relations between school culture and 'shop culture' in Kees Gispen, *New Profession, Old Order: Engineers and German Society, 1815–1941* (Cambridge University Press, Cambridge, 1989). For an alternative view see Göran Ahlström, *Engineers and*

a tight deterministic correlation between industrial training and industrial performance? The answer is probably not merely that of wishful thinking, uncritical educational functionalism or historiographical laziness. It is arguable that the positions adopted by Donald Cardwell and Michael Sanderson were closely related to their personal convictions about the social value of science education for industrial welfare, and perhaps their defence of highly expensive forms of science education in financially inclement conditions.

As far as Lyon Playfair's case is concerned, it is not difficult to see what motivated him to interpret the outcome of the Paris Exhibition so reductively in terms of British educational deficiency. In the last two paragraphs of his letter to Lord Taunton in May 1867, Playfair alluded to a booklet that he wrote in the aftermath of the Great Exhibition two years before, entitled 'Industrial Education on the Continent', in which he argued that the rate of progress in manufacturing abroad would inevitably be greater than in Britain if the latter failed to develop a comparable system of training. Giving the overwhelming dominance of British success at the 1851 exhibition, Playfair's document had not been taken especially seriously. Having been personal adviser to Prince Albert for the Great Exhibition and having managed the prize-awarding juries in 1851, Playfair became Inspector-General of Government Museums and Schools of Science in 1854, and tried to launch a national scheme of science schools coordinated through the central administration of the Department of Science and Art (DSA) in South Kensington. In his autobiography, Playfair unhappily described this episode as 'preaching in the wilderness' to an unsympathetic population, so complacent at their nation's success at the 1851 exhibition that they were not persuaded that much needed to be done to maintain industrial supremacy.[34]

Bitterly disillusioned with the DSA and science education and bereaved at the loss of his first wife, in 1858, Playfair accepted the Chair of Chemistry at the University of Edinburgh where he stayed for ten years. Only when the jury awards at the 1867 exhibition seemed to Playfair to vindicate his earlier jeremiads about the loss of British industrial dominance did he return to the fray, and once again argue the case for promoting science education at a national level. The specific timing of this had a further significance: debates about extending the electoral franchise stimulated renewed consideration in Britain of the need for widespread primary education; Playfair shrewdly realized that science educators such as himself needed to raise the profile of science in the plans for new primary education if they were to increase the

---

industrial growth: higher technical education and the engineering profession during the nineteenth and early twentieth centuries: France, Germany, Sweden and England (Croom Helm, London, 1982).

[34] Thomas Wemyss Reid (ed.), *Memoirs and Correspondence of Lyon Playfair, First Lord Playfair of St. Andrews* (Cassell, London, 1899), p. 152.

scope of their discipline, and in the longer term to build up a new cadre of researchers. Having learnt from his failure in 1853, however, that personal appeals were of little avail in such important matters, Playfair pleaded in his letter to Lord Taunton that either his commission or an equivalent successor would investigate the matter as authoritatively as possible. And Playfair managed to persuade enough people who agreed with something like his interpretation of events in Paris, to ensure that a Select Committee was set up in early 1868 under Bernard Samuelson to investigate technical instruction in the United Kingdom and on the Continent, and its relation to industrial progress.

Playfair was one of dozens interviewed by the Samuelson Committee, but the majority of the industrialists and workers interrogated did not sympathize with the specific educational focus of Playfair's claims. The Select Committee later concluded that much of Britain's progress had hitherto been linked to its advantageous stores of coal and ores, geographical position, climate, and the 'unrivalled energy' of its population. In addition to this, many witnesses had maintained that the 'acquisition of scientific knowledge' was only one among several such important factors in the nation's industrial success and, indeed, was only relevant at all in certain specific trades (such as dying and electroplating). For all other trades the 'indispensable' skills were not to be drawn from 'science' but from practical experience and manipulative skill acquired uniquely in the workplace; and witnesses to the committee had testified that such skill was widespread 'in a pre-eminent degree' throughout all the social grades of the industrial population:

> Although the pressure of foreign competition, where it exists, is considered by some witnesses to be partly owing to the superior scientific attainments of foreign manufacturers, yet the general result of the evidence proves that it is to be attributed mainly to their artistic taste, to fashion, to lower wages, and to the absence of trade disputes abroad, and to the greater readiness with which handicraftsmen abroad, in some trades, adapt themselves to new requirements.

The Committee concluded that it was because of these characteristics, rather than scientific attainment, that the lacemakers of Calais, the locomotive manufacturers of Creuzot and Esslingen, were now effectively competing with Britain in both 'neutral' and home markets. One witness who ran collieries in Northumberland and County Durham had explicitly stated that the only reason he had sought supplies of machinery in Anzin, France, was the low rate of wages and the high-quality workmanship in executing his own designs. Only three out of the dozens of witnesses attributed any loss of markets to the superior scientific skills of German, Belgian and American manufacturers, although the Krupps steel works in Westphalia was the only establishment named as superior to any counterpart in the United Kingdom. Nevertheless, many had reported that the scientific skills of managers had brought a palpable improvement in the quality of overseas manufacturing. A substantial number of witnesses thus argued that to maintain its still dominant industrial position, Britain would need to develop a more comprehensive system of scientific education; the viability of which would clearly depend on the growth of numeracy and literacy that could only be established by means of a major expansion of the system of primary education.[35] So to explain the subsequent growth of primary education and industrially-related higher education in Britain during the next two decades it is to that sort of consensus that the historian must appeal – not to the self-serving claims about a consensus on industrial declinism concocted by Lyon Playfair and his allies in Paris during May and June 1867.

Conclusion

Historians of science education have lessons to learn from revisionist historians of declinism. Only by recovering the contextually-located and often subtle politics of past claims about relations between industrial performance and education, can we judge how to interpret the historiographical significance of such events as the Paris Exhibition of 1867. Given that the Paris Exhibition did not unequivocally prove what Lyon Playfair claimed of its evidential import concerning the decline in British industrial prestige, historians of science education and economic historians alike should also reconsider the relationship between the latter phases of the 'Golden Age' of industrial performance and the genesis of flourishing higher scientific and technical education in the United Kingdom.

---

[35] Report of the Select Committee appointed to enquire into the Provisions for giving instruction in theoretical and applied science to the Industrial Classes, *Parl Papers (Reports of Committees)*, 15 (1867–68), pp. iii–ix, especially pp. vii–viii.

Chapter 9

Machinofacture and Technical Change: The Patent Evidence

Ian Inkster

The present chapter is part of a larger project designed to utilize patenting systems in order to measure the siting and agency of incremental technological change, the international patterns of technological transfer and information diffusion, and the extent to which institutions may both serve and hinder the process of industrialization through technological modernization. In this chapter we focus on the geographical and social location of inventive activity in Britain during the Golden Age by utilizing the data produced by the Patent Office and its commissioners between c. 1840 and 1881.[1] In doing so we also provide some measure of the value of the stock of technological knowledge, some insight into the nature of the information system, estimates of the openness of British technology and its conduits, and a hopefully useful perspective on the vexed questions of rise, decline and fall. The core of the approach is analysis of some 30,000 patent applications drawn from the entire 1840s, the select years of the Golden Age, 1852–53, 1855, 1860, 1865 and 1870, with an additional inclusion of 1881 in a passing effort to capture a slightly longer trend.

Town and Country: Patenting as a System

Because they clearly do not directly capture equal items of technological change, and because of the great differences between systems and periods, patents have long been neglected by economists, usually following summary

---

[1] The geographical and occupational data for the sample years 1855–70 are entirely derived from the original patent applications, each one of which has been sighted and summarized in an EXCEL spreadsheet system. All other data unless mentioned otherwise and including the Belgian material in section 8, is derived from data in *The Commissioner of Patents Journal,* annually from 1852 to 1883.

Table 9.1: Regional Distribution of Patenting, 1855–70

	1855	1860	1865	1870	Total	% System
London	1152	1348	1472	1532	5504	37
Industrial Counties	1042	1073	1204	1295	4614	31
Home Counties	47	38	89	99	273	2
Other England	470	511	583	547	2111	14
Scotland	163	165	184	302	814	5
Ireland	73	65	58	42	238	2
Wales	17	42	47	53	159	1
Foreign	492	343	301	214	1350	9
Total	**3456**	**3585**	**3938**	**4084**	**15063**	

Chart 9.1: Direct Patentees in Britain, 1855–70

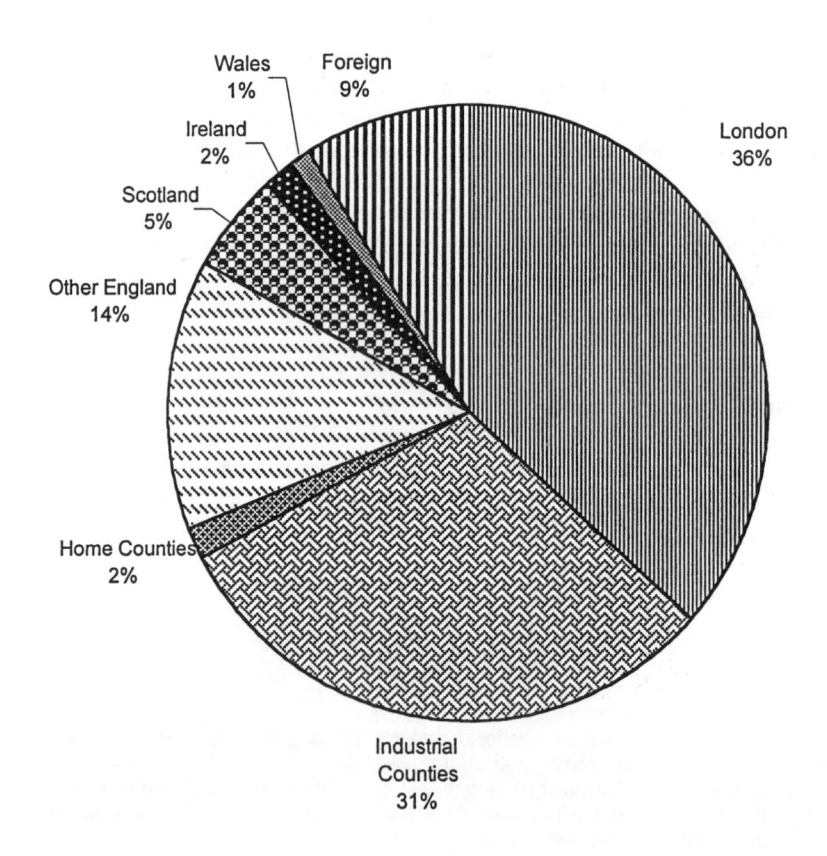

- Wales 1%
- Foreign 9%
- Ireland 2%
- Scotland 5%
- Other England 14%
- Home Counties 2%
- Industrial Counties 31%
- London 36%

dismissal.[2] A small group of economic historians have opened up some aspects of the British and other patent systems, but their analyses have generally been based on relatively accessible, processed data, such as that provided in the mid-nineteenth century by that arch-systemizer and defender of the patent faith, Bennett Woodcroft.[3] Until now, no one has undertaken any serious consideration of the British patent system in the years after the Patent Law Amendment Act of 1852, which reduced fees, tightened up the sealing, registration and maintenance systems and established one Patent Office in which all stages of patenting were placed under the control of the Commissioner of Patents.[4] The system in Britain remained mainly unaltered until the British accession to the International Convention for the Protection of Industrial Property, Paris 1883, in March of 1884, which permitted easier and cheaper international patent lodgement. So, in our period, the British patent system represented a large, powerful, increasingly efficient, open technological system, the analysis of which allows the historian to identify instances and patterns of technological transfer to and from the country.

Table 9.1 and chart 9.1 show the regional distribution of British patenting, the totals excluding foreign patenting by communications.[5] The predominance of London at 37 per cent of the total reflects exactly the weight of the metropolis in the system throughout the 1840s. The Home Counties are represented by Hertfordshire, Surrey, Buckinghamshire, Bedfordshire, Essex, Berkshire, Oxfordshire and Hampshire, the industrial counties by Lancashire, Yorkshire, Cheshire, Nottinghamshire, Staffordshire and Warwickshire, and London includes the metropolitan areas of Middlesex, Kent and Surrey. These figures clearly dispel the claim that patenting was common in London because of proximity only – patenting was relatively uncommon in the most closely surrounding counties but disproportionately represented in the more distant

---

[2] For an excellent summary in the context of a quite brilliant essay on technological change in history, see J.D. Gould, *Economic Growth in History*, Methuen, London, 1972, pp. 322–6, 397–8. See also, Ian Inkster, 'The Ambivalent Role of Patents in Technology Development', *Bulletin of Science, Technology and Society*, 2 (1982), pp. 181–90.

[3] Dirk van zyl Smit, *The Social Creation of a Legal Reality: A Study of the Emergence and Acceptance of the British Patent System as a Legal Instrument for the* Control of New Technology, PhD thesis, University of Edinburgh, 1980; Harry Dutton, *The Patent System and Inventive Activity during the Industrial Revolution*, University Press, Manchester, 1984, the best book on the subject and one which takes up the theme of internal industry expertise in the patenting system; Christine Macleod, *Inventing the Industrial Revolution. The English Patent System 1660–1800*, Cambridge, Cambridge University Press, 1988.

[4] For some beginnings, however, see Fritz Machlup and Edith Penrose, 'The Patent Controversy in the Nineteenth Century', *Journal of Economic History*, 10 (1950), pp. 1–29; Barbara Smith, 'Patents for Invention: The National and Local Picture', *Business History*, 4 (1962), pp. 109–19.

[5] After 1852 foreigners could lodge patents in Britain as direct patentees from a British address, or by communications, normally through a London-based patent agent. See details below.

industrial counties. Patent incidence was related to industrial and occupational distributions. Table 9.2 confirms this by removing the effect of population densities. Intensity of patenting was five times higher in the industrial than in the Home Counties during the Golden Age. The case of London will be examined more closely at later points.

Table 9.2 Patentees per 1,000 Population, 1855–70

	Population	Per Capita	Per 1,000 Population
London	2,362,236	0.00232999	2.33
Home Counties	1,765,560	0.000154625	0.15
Industrial Counties	5,684,919	0.000811663	0.81

Table 9.3 Urban Ranking of British Patents, 1700–1881

	1700–1858	1852–53	1855–70	1881
London	1	1	1	1
Manchester	2	2	2	2
Birmingham	3	3	3	3
Glasgow	4	4	4	4
Liverpool	5	5	5	5
Leeds	6	7	6	6
Sheffield	7	8	7	7
Bristol	8	14	16	13
Bradford	9	6	8	8
Nottingham	10	11	7	9
Edinburgh	11	10	11	10
Newcastle	12	13	9	14
Dublin	13	16	12	18
Rochdale	14	8	13	26
Leicester	15	17	15	21
Bolton	16	15	10	11
Halifax	17	12	15	16
Oldham	18	-	14	17
Huddersfield	19	-	17	15
Derby	20	-	18	-

MACHINOFACTURE AND TECHNICAL CHANGE 125

Chart 9.2: Lancashire Patentees per 1,000 Population

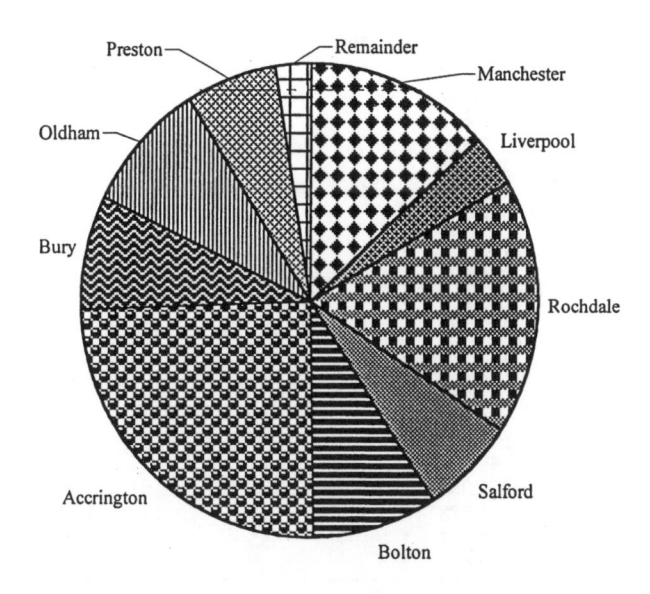

Chart 9.3: Yorkshire Patentees per 1,000 Population

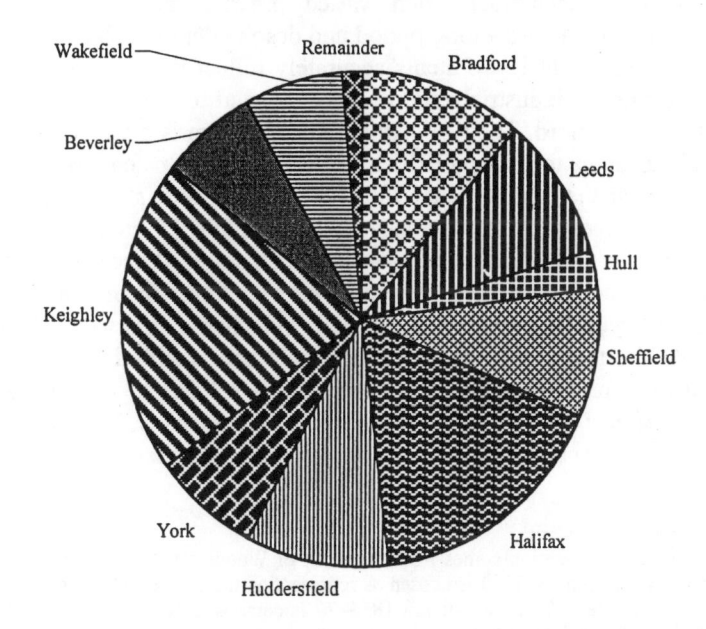

Table 9.4 Number of Patents per City or Town and Ranking Per Capita, 1855–70

	Population	Number Patents all Sample Years	Patents per 1,000 Population	Rank
Rochdale	29,195	71	2.43	1
Nottingham	57,407	133	2.32	2
Birmingham	232,841	526	2.26	3
London	2,362,236	5,133	2.17	4
Manchester	303,382	616	2.03	5
Halifax	33,582	61	1.82	6
Huddersfield	30,880	43	1.39	7
Bradford	103,778	135	1.30	8
Bolton	61,171	79	1.29	9
Oldham	52,820	65	1.23	10
Leeds	172,270	187	1.09	11
Sheffield	135,310	139	1.03	12
Leicester	60,584	60	1.00	13
Glasgow	329,097	312	0.95	14
Newcastle	87,784	77	0.88	15
Derby	40,609	27	0.66	16
Liverpool	375,955	221	0.59	17
Dublin	146,778	78	0.53	18
Bristol	137,328	55	0.40	19
Edinburgh	222,015	77	0.35	20

Table 9.3 summarizes the urban bias of patenting over the period from 1700 to 1881.[6] Although magnitudes often varied greatly, rankings show a tremendous consistency across a long period and despite demographic changes and the alteration in patent law. Patents accurately reflect the relative fall of towns such as Bristol as industrialization proceeded, and the rise during these years of Bolton or Bradford. The rise of the industrial town is made clearer in table 9.4 which ranks the top twenty British towns in terms of patents per 1,000 of the population for the sample years 1855–70. London now slips down the ranks and the newer industrial towns appear in force. Table 9.5 provides urban breakdowns of the industrial counties of Yorkshire, Lancashire and Warwickshire. Of the large cities, in crude terms, Birmingham massively dominated its county at 92 per cent of the total, Manchester less so at 36 per cent, with Yorkshire towns such as Leeds, Sheffield and Bradford sharing the incidence of invention more equally. However, charts 9.2 and 9.3 demonstrate dramatically the effects of sheer size. In terms of patent intensity, Hull was the least active of towns, Accrington the most active, with Manchester reduced to modest status overall.

---

[6] The 1700–1858 summation comes from the work of Woodcroft at the Patent Office during the 1850s. The column 1852–53 is chosen as representing the first 4,297 patents under the new system. As in other tables, the column 1855–70 is composed of all patenting in the sample years 1855, 1860, 1865 and 1870.

Table 9.5 Patentees in Yorkshire, Lancashire and Warwickshire

Yorkshire

	1855	1860	1865	1870	Total	%
Bradford	31	33	36	76	176	15.90
Leeds	28	37	57	128	250	22.58
Hull	12	8	0	13	33	2.98
Sheffield	22	36	65	49	172	15.53
Halifax	23	19	22	15	79	7.14
Huddersfield	13	10	14	17	54	4.88
York	5	7	2	0	14	1.26
Keighley	1	21	10	9	41	3.70
Beverley	5	4	0	0	9	0.81
Wakefield	4	5	11	2	22	1.99
Remainder	59	42	65	91	257	23.22
Total	203	222	282	400	1107	

Lancashire

	1855	1860	1865	1870	Total	%
Manchester	199	157	193	219	768	36.42
Liverpool	64	54	89	52	259	12.28
Rochdale	26	26	20	27	99	4.69
Salford	27	26	14	14	81	3.84
Bolton	19	22	33	31	105	4.98
Accrington	8	8	11	9	36	1.71
Bury	18	9	7	13	47	2.23
Oldham	23	23	25	18	89	4.22
Preston	28	18	14	25	85	4.03
Remainder	122	116	141	161	540	25.60
Total	534	459	547	569	2109	

Warwickshire

	1855	1860	1865	1870	Total
Birmingham	169	153	180	142	644
Remainder	6	22	14	18	60
Total	175	175	194	160	704

Birmingham as % of Warwick = 91.48

The Information System

Elsewhere in this volume, chapters by Inkster and Brown and Barton touch upon aspects of a technical information culture. The patent system added to this very greatly. A principal source of exact information was the Free Public Library in the Office of the Commissioner of Patents, which was visited over 400,000 times between 1855 and 1879, an annual average of 16,082 which rose to 26,655 for 1875–79. Attendance at the museum of the Patent Office in South

Kensington was, of course, far greater, and totalled over 4.5 million in the period 1858–80, with the annual visits for 1879 totalling 262,909. If the metropolitan basis of the information system helped create a bias in earlier years, by the second half of the century this was mitigated in a tremendous effort of the Patent Office to itself redress the balance. By 1870 the patent commissioners had donated large bundles of patent material and journals free of charge to 75 provincial cities and towns, 19 public offices and 8 centres of learning, whilst 27 donations had been made to the British colonies and 29 to foreign patent offices on the basis of an approximate reciprocity. These donations varied, but by 1882 the system had been regularized. Free of all charge but at a cost per instance of £3,900, 129 locations were in receipt of 3,820 volumes of patent and technical literature. At a total cost to the British government of £503,100, some half a million volumes of material had been dispersed, including those to 56 provincial towns and cities. Such large donations went to public libraries, mechanics' institutes, reading societies and so on, and included in the standard arrangement 527 volumes of patent specifications under the old law, 1,775 volumes of new specifications, chronological, alphabetical and subject-matter indexes, legal literature relating to patents and technology, reports of commissioners, library catalogues and the journal of the patent commissioners in thirty-eight volumes. The latter was the most important of all, for it contained substantial consolidated indexes, the complete list and descriptions in English of all foreign patent systems, including the massive listings from France and America, and reliable information on foreign and domestic procedures and regulations. Provincial facilities and usage expanded quickly. By the early 1850s the Royal Museum and library of Salford held full patent specifications and related literature, and the number of separate references to this local stock was reported as 3,619 in 1855–56, rising to 7,194 in 1856–57.[7] The last four years of the decade witnessed no less than 115,000 separate cases of reference to patent specifications at the Free Library, Campfield, Manchester. Table 9.6 categorizes some 60 per cent of them under nine major headings. Textiles and the steam economy still loom very large, but worth noting is the quick emergence of sewing machines and chemical process technologies. It is very clear that there was a high degree of informed usage of technical literature, and that this was increasingly dispersed throughout the provincial towns. We might hazard that this should have exerted some accelerating effects on the geographical distribution of patenting, a point to which we return below.

---

[7] Report of the Commissioners on the Working of the Laws Relating to Letters Patent for Inventions, Both Houses, House of Commons, London, 1865, p. 57.

Table 9.6. References to Patent Specifications in Manchester, 1857–60

	1857	1858	1859	1860	Total
Total	20,877	27,856	36,972	29,241	114,946
Chemical	-	355	1,829	283	2,467
Dyeing &Colouring	257	260	692	625	1,834
Gas Manufacture	264	718	8,785	660	10,427
Motive Power	1,243	437	221	1,797	3,698
Railways	2,185	1,295	839	1,154	5,473
Steam Engines	728	3,169	2,717	2,834	9,448
Sewing Machines	578	626	1,100	2,514	4,818
Textiles	1,689	6,261	8,414	9,313	25,677
Telegraphs	503	120	100	1,865	2,588
Sub-Total	7,447	13,241	24,697	21,045	66,430

Table 9.7. Occupations of Patentees in Britain, 1855–70

	1855	%	1860	%	1865	%	1870	%	Total	%
Engineer	843	33	1,019	36.4	1,126	40 1	1,187	42 8	4,175	38 2
Manufacturer	508	19 9	532	19.0	486	17.3	535	19 3	2,061	18.8
Small Manufacturer	7	0.3	9	0.3	12	0 4	16	0.6	44	0 4
Artisan Tradesman (Industrial)	258	10 1	321	11.5	209	7 4	194	7 0	982	9 0
Artisan Tradesman (Agricult'l)	14	0.5	5	0 2	23	0.8	20	0 7	62	0 6
Merchant	102	4 0	121	4 3	104	3 7	110	4 0	437	4 0
Patent Agent	154	6 0	168	6.0	198	7.1	169	6 1	689	6 3
Retail Tradesman	53	2 1	53	1 9	71	2 5	43	1 5	220	2 0
Professional	117	4 6	125	4 5	152	5.4	127	4 6	521	4 8
Supervisor	28	1.1	30	1 1	26	0.9	24	0 9	108	1 0
Commercial/ Clerical/ Agent	88	3 4	104	3 7	102	3 6	101	3 6	395	3 6
Other	30	1 2	20	0 7	28	1 0	16	0 6	94	0 9
Farmer/ Yeoman	12	0 5	15	0 5	11	0 4	11	0 4	50	0 5
Instrument Maker	5	0 2	4	0 1	8	0 3	8	0 3	25	0 2
Gent/Esq	338	13 2	270	9 7	251	8 9	214	7 7	1,073	9 8
Total	2,557		2,796		2,807		2,776		10,936	

Table 9.8: Occupations of Patentees, 1855–70: Regional Distributions

	Industrial	% of Industrial	Home	% of Home	London	% of London
Engineer	1350	37.05	78	39.59	1653	41.60
Manufacturer	1080	29.64	19	9.64	353	8.88
Small Manufacturer	25	0.69	0	0.00	5	0.13
Artisan Trades(Ind)	426	11.69	20	10.15	234	5.89
Artisan Trades(Ag)	21	0.58	2	1.02	21	0.53
Merchant	140	3.84	1	0.51	129	3.25
Patent Agent	21	0.58	0	0.00	679	17.09
Retail Tradesman	80	2.20	11	5.58	62	1.56
Professional	72	1.98	15	7.61	202	5.08
Supervisor	83	2.28	2	1.02	11	0.28
Commercial/Clerk	193	5.30	3	1.52	59	1.48
Other	29	0.80	4	2.03	26	0.65
Farmer/Yeoman	7	0.19	8	4.06	0	0.00
Instrument Maker	4	0.11	0	0.00	16	0.40
Gent	113	3.10	34	17.26	524	13.19
Total	3644		197		3974	

Chart 9.4: Occupations of Patentees in Manchester 1855–70

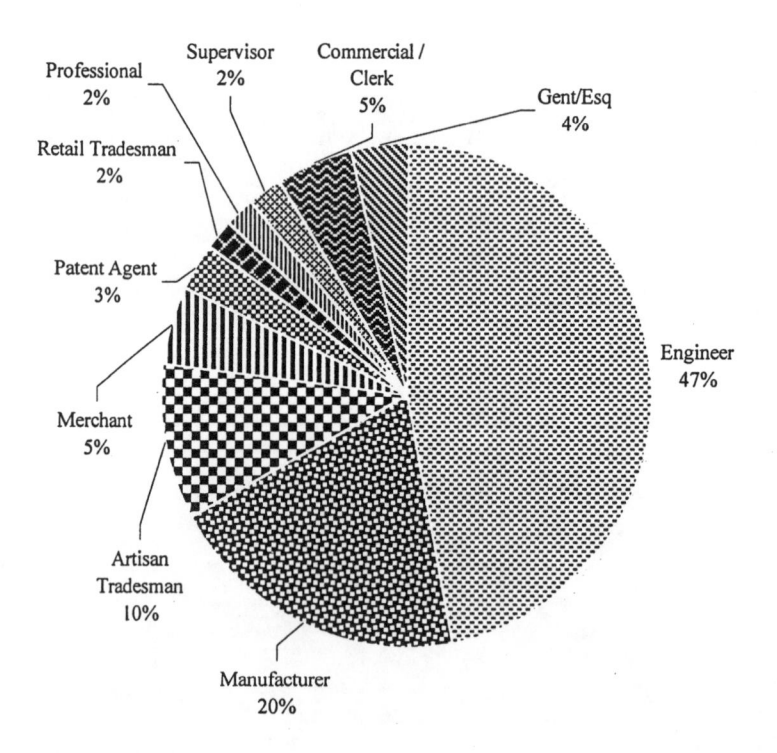

- Professional 2%
- Supervisor 2%
- Commercial/Clerk 5%
- Gent/Esq 4%
- Retail Tradesman 2%
- Patent Agent 3%
- Merchant 5%
- Artisan Tradesman 10%
- Manufacturer 20%
- Engineer 47%

Engineers and Tradesmen: The Patentees

Table 9.7 summarizes the occupational distribution of the patentees, covering over 10,000 cases for the sample years 1855–70. Approximately 40 per cent of patentees were engineers, this exactly consistent with the proportion for the entire 1840s decade. In contrast, where self-described gentlemen and esquires represented 34 per cent of all 1840s patentees in Britain, by our later years this had fallen to 10 per cent. During the Golden Age, the great bulk of patenting was either by engineers (44 per cent if we add patent agents) or by manufactures and artisan tradesmen (around 30 per cent), with strong activity also from merchants, professionals and various types of commercial agent and managers. Through these years the engineers were rising in importance. We may also analyse patentee occupations in terms of our designated regions, as in table 9.8. It is of importance that the engineers were evenly distributed, and of significance that manufactures had almost equivalent impact in the six manufacturing counties. With the artisan tradesmen the strictly industrial occupations represented 80 per cent of patentees in industrial locations, and innovation in such areas was disproportionately the result of applications of specific expertise from within industry itself. Gentlemen favoured London. Chart 9.4 illustrates the occupational distribution in Manchester, the largest centre of patenting outside London. Manchester appears to reflect the tendency within the industrial counties more generally, with perhaps a slightly larger role for general engineering skills.

Agency: Partnerships and Professionals

Most patenting took place as an individual activity. However, in some cases partnerships and the use of patent agents served to spread and share both information and costs. In the sample years 1855–70 there were 1,960 cases of patent partnerships, of which 88 per cent were of just two partners. Partnerships represented 15 per cent of all patenting, and engineers were represented in partnerships in exact proportion to the representation of engineers amongst patentees generally, that is, around 38 per cent. Table 9.9 provides the general picture. During the 1840s, partnerships in provincial cities such as Manchester tended to be between engineers, between tradesmen, between tradesmen and engineers, and between tradesmen and manufacturers and there is evidence of much cross-skilling, that is, as in partnerships between machine-makers and dyers or coal merchants and engineers. This was a self-help culture. For instance, in Manchester all cases of patenting involving cross-

Table 9.9 Partnerships, 1855–70
(a) Number of Patents

Number of Partners	1855	1860	1865	1870	Total
2	380	408	422	515	1725
3	47	7	80	76	210
4	7	1	6	7	21
5	0	0	0	1	1
6	0	0	2	1	3
Total Patents involving Partnerships	434	416	510	600	1960
Total all Patents	2,958	3,196	3,381	3,405	12,940
% all Patents Partnerships	14.67	13.02	15.08	17.62	15.15

(b) Partnerships involving Engineers

	1855	1860	1865	1870	Total
No. Partnerships involving Engineers	169	161	188	234	752
% of all Patents	5.71	5.04	5.56	6.87	5.81
% of all Partnerships involving an Engineer	38.94	38.70	36.86	39.00	38.37

skills led to skill-specific patenting. Inventions were arising from industrial-specific knowledge or experience at times combined with general engineering knowledge or commercial acumen. The sprinkling of gentleman–engineer partnerships in a city such as Manchester should not be dismissed. For instance, the gentleman John Spear combined with engineer Joseph Whitworth in Manchester in 1840 to produce cutting and shaping machinery, a patent epitomizing the mechanisms of machinofacture. Very few partnerships involved geographical spread – most partners were in the same city and certainly in the same county. In small numbers for the 1840s, nascent corporate patenting is identifiable. We might instance the iron-shipbuilding patents of Thomas Vernon and James Kennedy of the firm of Bury, Curtis and Kennedy of Liverpool, the various engineers of the Vulcan Foundry, Warrington patenting improvements in boilers, the combinations of managers and engineers from the Sutton Glassworks of Manchester, all relating to glass manufacture, or the managers and toolmakers of Messrs William Collier and

Co., Salford, all concerning turning, boring, and cutting machinery for metalworking.

Such 1840s trends were continued across the reform watershed, but were modified by some increase in the presence of the engineers and of overlookers, foremen and managers. There was also an increased tendency towards corporate patenting by firms which had already formed around shared technical expertise. Such partnerships were increasingly outside of the core machinofacture processes – hat manufacturers in Denton, Lancaster, gun and pistol manufacturers in London, carpet manufacturers of Durham. Partnerships by foreigners abroad remained rare apart from the ubiquitous array of gentlemen and with the more important exception of Parisian chemists and engineers.

Table 9.10 Patents Involving a Patent Agent, 1855–70

	1855	1860	1865	1870	Total	%
London	263	414	527	459	1663	84.2
Industrial Counties	4	10	16	34	64	3.2
Other Provincial	2	5	2	4	13	0.7
Paris	3	0	5	18	26	1.3
Other Foreign	9	0	2	0	11	0.6
London & Glasgow	82	7	0	19	108	5.5
London & Foreign	0	2	39	49	80	4.0
Total	363	438	591	583	1,975	
As % of all Patents	12.27	13.7	17.48	17.12	15.14	

Table 9.10 shows that professional patent agency for British inventors was almost entirely confined to London. During the 1855–70 period 85 per cent of the 1,880 patents issued via a patent agent were based on metropolitan agency. Our next section shows how this directly reflected the importance of foreign communications into the British system.

Metropolitan Mechanisms: London and the Foreigners

Table 9.11 provides details of all foreign patenting in Britain, combining patenting directly within Britain with communications from overseas via a British patent agent. Whereas at the beginning of the 1850s foreign patenting came to perhaps 10 per cent or less of the total, the average for the Golden Age was not much less than 30 per cent, and kept on rising during the 1870s. France was the largest foreign element, but was being overtaken by the United States. Whereas direct patenting was of greatest importance around 1855, by 1870 patenting by British agency had taken over. It was such agency work which provided the bulk of the work of London patentees acting as patent agents. Table 9.12 gives full details of the occupations of patentees in Holborn, the area containing the patent facilities and offices of Chancery Lane, Fleet Street and Lincoln's Inn, and representing around 20 per cent of all London patenting. Of over 1000 cases, around 80 per cent came from patent agents and engineers acting as agents. Here was a very specific institutional siting of the mechanisms of technology transfer and information dispersal.

Table 9.11 Foreign Patents in Britain, 1855–70
(a) Direct and by Communication to a Patent Agent

	1855	1860	1865	1870	Total
France	397	494	435	293	1,619
USA	140	298	389	448	1,275
Belgium	28	35	27	21	111
Other Foreign	90	93	142	156	481
Total	655	920	993	918	3,486
As % of all Patenting	22%	29%	29%	27%	27%

(b) As Percentage of Total Foreign Patents

	1855	1860	1865	1870	Total
France	61%	54%	44%	32%	46%
USA	21%	32%	39%	49%	37%
Belgium	4%	4%	3%	2%	3%
Other Foreign	14%	10%	14%	17%	14%

(c) Percentage of Foreign Patents Direct and Communicated

	1855	1860	1865	1870	Total
Communicated	25%	63%	68%	75%	58%
Direct	75%	37%	32%	25%	42%

Table 9.12 Occupations of Patentees in Holborn, London, 1855–70

	1855	1860	1865	1870	Total
Patent Agent	34	100	117	88	339
Eng/Patent Agent	46	189	250	249	734
Engineer	3	12	11	19	45
Manufacturer	12	8	5	6	31
Artisan Tradesman	4	6	3	2	15
Merchant	1	0	2	2	5

	1855	1860	1865	1870	Total
Retail Tradesman	0	3	2	0	5
Professional	2	0	2	4	8
Commercial/Clerk	0	0	0	1	1
Other	2	0	3	0	5
Gent/Esq	8	7	6	2	23
Unknown	25	21	32	64	142
Total	137	346	433	437	1,353
Holborn as % London	12	26	32	28	25
% Patentees in Holborn acting as Patent Agents	80	84	85	77	79

Evaluating Patents and the Elite

We may well wish to analyse the reasons why so many engineers, agents, tradesmen and manufacturers might wish to bear the costs (including extensive opportunity and organizational costs) of lodgement of patents, particularly those lodging from overseas. By 1870, the cost in US dollars of securing patents, calculated on the basis of price per annum covered, was considerably higher in Britain than elsewhere, by a factor of three compared to Belgium and France, of ten compared to Prussia, and of thirty compared to the United States. Yet the American system was almost empty of foreigners, the British system burgeoning with overseas communications and visitations. By 1881 foreign patenting represented perhaps 6 per cent of the United States total, but around 37 per cent of the British total. Of the foreign total in Britain at that time, United States patentees composed 35 per cent, French 26 per cent, and German 22 per cent. Of 38 major foreign cities whose residents lodged 2,057 patents in Britain during the three years 1867–69, only Parisian inventions outnumbered those of New York. Within Britain, only London and Manchester outranked New York, and only five British cities outranked Boston. Analysis of a sample of over 800 of the American patentees shows that the majority resided or worked in New York, Boston or Philadelphia, surely urban centres of extreme commercial and technical competition.[8] That is, more broadly, technicians and engineers from the more mature of the industrial nations struggled to patent within the most costly system of them all, a process requiring information search, expensive agencies and visitations and large blocks of time. Assuming their rationality, we might presume that British and foreign patentees lodged British patents because of high rates of return and because they thereby bought into a superior information system.

---

[8] Report of the Select Committee on Letters Patent, Proceedings of Committee, appendices, House of Commons, London, 1871: appendix 4.

Table 9.13 UK Patent System, 1853–77: Applications and Elite Patents

	Applications	% Sealed	% £50	% £100
1853–57	15,073	67.89	27.78	9.53
1858–62	15,969	64.07	28.17	9.78
1863–67	17,131	62.53	27.61	10.02
1868–72	18,681	65.44	32.17	11.94
1873–77	23,365	68.90	28.79	-
1853–77	90,219	65.84	29.00	10.39

We may refer to elite patents as those which were lodged by expert partnerships, or extended to seven and then fourteen years beyond initial granting (at a cost of £50 and £100 respectively), or were also lodged separately in foreign patent systems. Table 9.13 isolates elite patents by cost of maintenance of over 90,000 initial applications in the years 1853–77.[9] Nearly 66 per cent of this total were successfully sealed at a minimal direct cost of around £25. Of these some 30 per cent paid another £75 for maintenance to a seventh year, over 10 per cent another £125 for maintenance to the fourteenth year of coverage. If we assume rational expectations and a prevailing interest rate of 4 per cent, then we may estimate the valuations placed on these stocks of patents by their owners, in much the same way as we would any investments. That is, we may assume that an individual will pay out £50 for patent maintenance if he expects at least that much return upon it, that is, he values the patent (assuming 4 per cent rate of interest) at 50 x 25=£1250, the capital value of an investment at 4 per cent which would yield £50. From table 9.13 this yields stocks of patent value, of sealed patents voided at third year of £37,125,625 (£25 @ 4 per cent prevailing = £625 x 59,401 cases), of patents extended to seventh year of £30,495,000 (£75 @ 4 per cent prevailing = £1875 x 16,264 cases), of patents extended to fourteenth year of £14,087,500 (£125 @ 4 per cent prevailing = £3125 x 4,508). Such calculations may seem somewhat heroic, but are based merely on the assumption that the rational calculation of the individual is to only spend on patent costs up to the expected income return from that patent, in other words, that patenting costs were investment items, not consumption items. This approach yields very high values for the stock of technical knowledge captured within this system of intellectual property rights.

---

[9] The applications are surveyed from 1879, which means that from then the fourteenth year figures were unknown. The 10.39 per cent of highest elite patents for 1853–77 are based on 1853–72 only.

Table 9.14 The Belgian Patent System, 1860, 1865, 1870: Patents Granted

	Total	% French	% British	% All Foreign
1860	1,815	28.9	7.2	37.2
1865	1,685	34.0	8.8	43.8
1870	1,680	36.5	9.2	47.0
Total	5,180	33.0	8.4	42.6

Another way of isolating elite patents is to consider those British patents which were additionally registered at extra cost and effort in foreign systems. In order to consider the qualities of a relatively small group, table 9.14 presents data for three years in the Belgian patent system, covering some 5,000 cases of patents granted. This system was even more open than that of Britain, with most foreign patenting being French, British patentees taking a strong second position with over 8 per cent of the total. Counting each of these yielded a time lag between initial British registration and Belgian grant of 21.2 weeks for 1870, although this embraced the range of one week to seventy–eight weeks, with 50 per cent of the total at twenty weeks or less. This suggests a fairly astute and professional pursuit. The patents here show strong similarities with those for partnerships within Britain. Most British patents went through Brussels-based agents, many of whom, such as John Piddington and R. Culliford, were major British patentees in their own right. At least 75 per cent were based on British partnerships, especially of manufacturers and engineers, and several of these were from different British locations such as, Leeds and Bradford, York and Halifax. This relatively small sample contains the greater proportion of British breakthrough technologies of the Golden Age, including Bessemer's steel patents, Whitworth's armaments, Neilson's steam hammer, Ramsden's steam boilers, and Fairlie's locomotive engines, but also a goodly array of highly competitive new product technologies such as saccharine and collodion. Interestingly, the patents with the shortest time lag between British and Belgian registration were biased towards more traditional elements of the steam economy – steam valves and boilers and so on, but represented also were railway technologies, sewing machines, waterproofing fabrics, power looms and new fuels. There was a relatively high proportion of original American patentees who had secured British patents and then moved on to Belgium, these including George Westinghouse and Cyrus MacCormack of reaper fame.

A very large amount of time, effort and money was spent on British patenting during the Golden Age of machinofacture. Probably between 10 per cent and 20 per cent of all patenting may be regarded as an elite activity, involving real skills and experience in the pursuit of commercial gain. The relative openness of the system meant that it served as an information complex as well as a mechanism by which intellectual property rights were more or less secured. The evidence suggests that high valuations were placed on secured patents in very many cases, and that otherwise sophisticated individuals and partnerships in Britain and elsewhere were willing to pay relatively high prices for the opportunity of securing technical and market advantages and access to an extensive and improving information system.

Conclusions: Machinofacture in a Golden Age

This essay has shown that the gross patent data does tell us something about the geographical dispersal and social composition of technical innovation activity. Even brief discussion has thrown some light upon technology transfers, the size of the industrial culture, possible paths of mobility and achievement, knowledge diffusion and usage, and the comparative importance of general engineering versus specific industrial knowledge in the innovation process, and hints at the origin of many enterprises in patent partnerships. Further work will explore whether secured knowledge may have been as great an importance as original capital formation in the foundation of industrial enterprises during and beyond the Golden Age. The identification of elite patenting by cost and persistence (partnerships, price of maintenance and overseas registration) allows some brave estimations of the extent of knowledge capital stock and the importance of international patenting as a mechanism of technology transfers. Examination of agency, especially in the form of partnerships and patent agents, serves to identify the penetration of professionalism as one of the social resources required for the establishment of an effective culture of machinofacture.

Patent data allows us to capture the culture of '*machinofacture*' defined as (a) the engineering and legalistic organization of skills and property rights around (b) key institutions, understandings (implicit or tactic knowledge of how things might work) and information networks, through (c) specific conduits and agencies (from Paris to Southampton Buildings) which (d) is, because of its very informality, difficult to emulate or functionally substitute for by all those who lie outside it, that is, in relatively backward economic systems such as Europe, Russia, Japan or elsewhere. Machinofacture may have been a necessary factor explaining Britain's continued competitive position through and well beyond the Golden Age.

Weiner's now infamous 'cultural *cordon sanitaire* encircling the forces of economic development' was always a shaky conception.[10] No nation has ever been entirely 'industrial', and even fast-growing Japan boasted a strong

---

[10] Martin J. Wiener, *English Culture and the Decline of the Industrial Spirit 1850–1980*, Cambridge, Cambridge University Press, 1981, quote p. ix.

counter-culture and, very notably, a massive machinofacture sector, still visible in the disproportionate numbers employed in small businesses. Modern industry everywhere has unleashed a romantic backlash and ideologies of resistance or poetries of dismissal. Yet there was very little that was Toynbeean, or Priestleyist or even Dickensian (despite his wonderful "Poor Man's Tale of a Patent") in the culture of technique and industry as revealed by the patent data. It is difficult to imagine that it was ever much affected by the cultural imperatives of a portion of the British intelligentsia in the years prior to 1914. After which date things did get more complicated. Nor does the material of this essay reflect the very public statements of contemporary politicians, intellectuals and industrialists concerning an insufficiency of British technical education, polytechnical training and apprenticeship in the later nineteenth century.[11] We have been reminded that the material of the polemics of the public sphere, especially when produced by expertise, is no substitute for the activities of thousands who work at the coalface of any system.[12] Again, that distant Japanese example has shown that industrial productivity arising over a considerable historical period is not any close function of investments in formal education, and far more an outcome of enterprise culture itself and the manner in which the social system distributes its rewards to work. The patent data reveals a strong pro-technology culture during the Golden Age. As Edgerton has recently suggested, such an element is not in itself equivalent to a strong industrial or capitalist culture *per se*.[13] The innovative activities of engineers, tradesmen and a miscellany of partnerships can always fall upon a stony ground prepared over time by short-sighted management or unimaginative bureaucracy.[14]

---

[11] For the alternative view see Pollard n.14 which shows the advantages of the British educational and training systems over those of much-vaunted Germany; Michael Sanderson, *The Universities and British Industry 1850–1970*, London, 1972, the best early counter-blast; and P.L. Robertson, 'Technical Education in the British Shipbuilding and Marine Engineering Industries 1863–1914', *Economic History Review*, 27 (1974), an excellent industry study. For apprenticeship, which is still neglected in the analysis of industrial culture and its productivities, see J.B. Jefferys, *The Story of the Engineers*, Letchworth, Garden CityPress, 1945; Charles More, *Skill and the English Working Class 1870–1914*, Croom Helm, London, 1980; Joan Lane, *Apprenticeship in England 1600–1914*, Boulder, Westview Press, 1996; and comments by Inkster and Nicholas in Inkster ed., *The Steam Intellect Societies*, Nottingham University Press, Nottingham, 1985, pp. 80–96, pp. 185–92.

[12] Frank M. Turner, 'Public Science in Britain, 1880–1919', *Isis*, 71 (1980), pp. 589–608.

[13] David Edgerton, *Science, Technology and the British Industrial 'Decline' 1870–1970* (Cambridge, Cambridge University Press, 1996), pp. 7–8, 67–9.

[14] For the latter see the brilliance of Sidney Pollard, *Britain's Prime and Britain's Decline*, Edward Arnold, London, 1989, ch. 4. For example, commenting on government policy, industry was 'something to be watched, limited, controlled: never anything to be fostered' (p. 253).

Part III

Social Institutions

Introduction to Part III: Social Institutions

Jeff Hill

Economic historians have probably been foremost in the discussion of a 'Golden Age'. The period has long attracted them, certainly ever since Sir John Clapham's description of it as one in which Britain enjoyed 'undisputed superiority in commerce and in money-lending', combined with 'a superiority in manufacturing which was the more marked because it was shown most in the making of those things which the world chiefly needs'.[1] With such measurable attributes the period has for many appeared to contain the mysteries that might explain the attainment of the elusive goal of 'economic growth'. But historians of a social and cultural orientation, for whom qualitative measures are perhaps more important, have equally been fascinated by the apparent combination in this period of social harmony with intellectual dynamism. It was this feature which led many to see in these years a 'cohesiveness' that was captured in W.L. Burn's famous description of 'the age of equipoise'.[2] Much of this rested upon a bourgeois ascendancy, articulated through Harold Perkin's idea of the 'entrepreneurial ideal', whose triumph in these years caused it to influence, if not permeate, a large part of British society.[3] Marxist historians also felt that there was something to be explained here, taking their cue from the sardonic observation of Engels: 'the English proletariat is becoming more and more bourgeois, so that this most bourgeois of all nations is apparently aiming ultimately at the possession of a bourgeois aristocracy and a bourgeois proletariat as well as a bourgeoisie'.[4]

Looking back in this way, a Golden Age seems clearly imbued in the historical fabric. To those who lived through the time itself, however, the golden qualities were often less clearly visible. The contributions to this section of the volume tease out some of the dualism implicit in the idea of the Golden Age: on the one hand, the sense of intellectual ferment that Geoffrey Best saw

---

[1] J.H. Clapham, *An Economic History of Modern Britain: Free Trade and Steel 1850–1886* (Cambridge University Press, Cambridge, 1963 edn), p. 12.
[2] W.L. Burn, *The Age of Equipoise* (Unwin University Books, London, 1964).
[3] Harold Perkin, *The Origins of Modern English Society 1780–1880* (Routledge and Kegan Paul, London 1972 edn), ch. 8.
[4] Quoted in David McLellan, *Engels* (Fontana/Collins, London, 1977), p. 53.

as part of a 'national self-congratulation', that the storms of the 1840s had abated and given way to the years of peace earlier perceived by G.M. Young;[5] but on the other, a feeling of doubt about what was being created in British society. Much of the former is revealed in Su Barton's discussion of popular participation in the Great Exhibition, an event that has long been seen as emblematic of the new harmony. Barton shows that there was a marked interest in the exhibition by many of those whose political agitation had contributed to the turbulence of earlier years. Some radicals and socialists saw the exhibition as an international forum for promoting their ideas, and for reviving the revolutionary urge that had subsided after 1848. But if the authorities had fears that the event would trigger conflict, such anxieties soon proved groundless. Instead, Hyde Park became the focus of genuine national interest, as millions of ordinary people travelled to London and enjoyed the exhibition. It was not only a symbol of national unity of purpose but a testament to the organizational and fund-raising capacity of innumerable local voluntary committees. Their activities are charted in Vicki Brown and Ian Inkster's short chapter, which reveals an extensive geographical spread, involving men and women, of the ideology of industrialism, and which gives credence to the idea of a populist infrastructure of 'steam intellect' ferment that explains enthusiasm for the Exhibition. If an image were needed of a peaceable and orderly nation, its collective mind fixed firmly on the industrial ethic, there could have been no better representation of it than that provided by the events in Hyde Park.

By contrast, the essays of Gary Moses and Sarah Wilson show the Golden Age as a time of tension. In what has been termed the 'Golden Age of agriculture' the development of advanced capitalist farming in the East Riding of Yorkshire actually strengthened an archaic form of agrarian labour – farm service – and the customary contracts struck at annual hiring fairs that regulated it. The continuing relevance of farm service in a district beset by shortage of agricultural labour contrasted with its disappearance in other parts of the country, and became the occasion for conflicts of interest within what might be termed the local 'ruling class'. For clergy of the Church of England like R.I. Wilberforce, Archdeacon of the East Riding, the 'Golden Age of agriculture' was equated with sins of omission on the part of the gentry. Farmers were seen as increasingly concerned with profits, to the detriment of the duty of care they had formerly exercised over their employees. The period was thus one of 'moral crisis' for Wilberforce and those like him, whose memory of a past moral order provided the impetus among Anglicans to seek a restoration of a proper balance in social relations. Sarah Wilson similarly exposes the contradictions inscribed in the emergence of a new form of financial

---

[5] Geoffrey Best, *Mid-Victorian Britain 1851–75* (Weidenfeld and Nicolson, London, 1971), p. 229.

irregularity occasioned by the boom in railways. 'High art' crime, as it was known, posed a particular difficulty for the social order, since it publicized illegalities by the 'respectable' members of society. Contemporaries were aware of the qualities needed for advancing economic prosperity, and saw that these existed in the dynamic, risk-taking ventures of businessmen. But at the same time these very qualities were likely to produce practices that not only jeopardized the economic structure, but which also showed the better-off sections of society as being capable of the same transgressions of the law as those found among the lower orders. The 'criminal classes' were now a truly classless group.

Finally, Jeff Hill's piece looks at the continuing presence of Golden Age thinking in that part of the economy whose peak was experienced at a time when other parts were in relative decline. In north-east Lancashire the late flourishing of cotton weaving allowed people to create their own local, late nineteenth-century version of the ethos of progress and dynamism that was elsewhere associated with the mid-Victorian years. It was worked into an ideology of local identity and town patriotism that celebrated 'good times', and which served to remind weavers and their families that 'better days' were no longer a memory of the past; they could be experienced in the present and anticipated in the future. 'Golden Age-ism' thus remained as a narrative of industrial capitalism, until the decline of the cotton industry after the Second World War finally revealed it to be a myth.

Chapter 10

'Why Should Working Men Visit the Exhibition?': Workers and the Great Exhibition and the Ethos of Industrialism

Su Barton

> Friends and fellow countrymen – We live in strange times – in times when the different nations of the earth are called upon to wage war against each other; but not in deadly array – not in deluging our fertile fields and plains with the gore of our favourite sons – not in making widows, and orphans, and childless parents – not in creating famine, pestilence and disease; but in multiplying, in a thousand degrees, every source and avenue of human enjoyment, happiness and social ties. A war – who shall be the best Samaritan, and who shall excel in acts of charity, munificence, benevolence, and humanity, – war of love, of concord and affection.
>
> There are a number of you who ask 'Of what use and benefit would be a visit for us (the working classes)?' I ask, in return, of what use are the rains and fertilising dews of spring to all the vegetable kingdom? Do they not make all animated nature to bud forth and themselves in all the splendour of their foliage and verdure? Do they not impart fecundity and fertility to the lap of nature, and cause the earth to bring forth her fruit in abundance?[1]

This is an extract from an essay entitled 'Why Should Workingmen Visit the Exhibition' published anonymously in Thomas Cook's *Exhibition Herald* in 1851. The article from which it is taken, described as the first part of the essay, eulogizes the benefit to be gained by skilled craftsmen from observation of the work of others. This essay, probably the prize-winning text referred to in the Bolton Local Committee for the Great Exhibition Minute Book,[2]

---

[1] *Exhibition Herald and Excursion Advertiser*, 1, 31 May 1851, p. 2. Cook told his readers that this essay was the work of a Bolton weaver and that it had won a £5 prize donated by the Mechanics' Institute of Bolton for writing by a working man on the benefits of the Great Exhibition.

[2] The Bolton Local Committee for the Great Exhibition of the Works of Industry of All Nations' Minute Book records J.R. Bridson as the sponsor of an essay-writing contest for

provides an illustration of the interest shown by working people in the Great Exhibition. In fact, the Exhibition provided, both intellectually and physically, a focus of national proportion for people of all regions and classes, which explains why it came to be regarded as a symbol of unity and harmony.

The Exhibition provided a national focus and a common destination and motivation for visitors. The distances and time involved in travel to London made staying away from home essential for those who came from outside the metropolis. The railway network made the journey possible, in a reasonable length of time, from all around Britain. The emerging culture of the commodity and consumerism,[3] together with preparations and arrangements for large numbers of working people to travel to London, to stay there for several days and visit the Exhibition and other places of interest, are all of vital importance in illustrating the ethos of industrialization that was emerging in mid-Victorian Britain.

The nineteenth century's mid-point, as well as being the time when the population of England became mostly urban, was also of significance in the history of working-class politics. Only a short time earlier massive Chartist demonstrations called for the extension of democratic rights, indicative of the European-wide surge of nationalist and democratic demands that had surfaced in the revolutions of 1848. In Britain, this movement had involved artizans, independent tradesmen and skilled workers. From the peak of Chartist activity in 1842 and its resurgence in 1848, there was an apparent dramatic decline in working-class political activity. A report that the Financial and Parliamentary Reform Movement could not raise enough subscriptions to carry on its campaign was regarded by the mainstream press as a sign of the 'return of common sense to those classes that had hitherto been the dupes of a set of selfish demagogues'.[4] Organization and contact between groups and individuals probably survived but their campaigning took on a different focus. Traditional forms of struggle often seemed like attempts to turn the tide of industrial development. It now appeared a more likely proposition that reform could be won by accepting the existing social order and attempting to change it by constitutional means. This emerging new outlook is demonstrated by attitudes towards the Great Exhibition.

---

working men with £5 of books as the prize, to be judged by the committee. The title of this essay was to be 'The Advantages to the Working Man from Visiting the Great Exhibition of 1851'. Mill-wright Thomas Briggs was described by Audrey Short in 'Workers Under Glass' *Victorian Studies* (1966), pp. 193–202, as the author of a prize-winning essay with the similar title 'The Advantages to be Derived by the Working Man from Visiting the Great Exhibition of 1851', published in Bolton in 1850.

[3] Thomas Richards, *The Commodity Culture of Victorian England – Advertising and Spectacle, 1851–1914* (Verso, London 1990), pp. 1–72.

[4] *Stockport Advertiser*, 31 January 1851.

Events in Europe and in England, and the arrival of *émigrés* post-1848, aroused a growing sense of internationalism among the English people. Socialist and radical newspapers devoted large sections to international issues and reports from revolutionary movements overseas. The very idea of the World Exhibition may have been suggested by the sentiments advocated and popularized by Julian Harney and other advanced proletarian spokesmen. Writing in the 1920s, the labour movement historian and trade unionist Theodore Rothstein asserted that it was 'tolerably certain that the internationalist movement which arose among the liberal bourgeoisie about that time was greatly influenced by this proletarian propaganda and must be considered as a semi-conscious attempt at competition with it'.[5] The bourgeoisie of that period also began to court trade unions and the cooperatives, endeavouring by fostering Mechanics' Institutes and popular libraries, as well as the publication of cheap literature, to wrest the working people from the intellectual influence of the still active 'agitators'. Other inspirations were the Mechanics' Institute exhibitions held in several provincial towns between 1838 and 1840[6] and the 'National Bazaar' of the Anti-Corn Law League held in Covent Garden in 1845 where products were shown to celebrate free trade.

Plans for a Great Exhibition of the Works of Industry of All Nations to be held in London were announced in 1850. The Royal Commission organizing the event wished to emphasize social harmony. It had initially recommended the establishment of a Working Classes Committee. The proposal was soon retracted. The short-lived Working Classes Committee had comprised two Members of Parliament, the Bishop of Oxford and the Chartists Lovett, Place and Vincent who had remarked that 'the working-class regarded the Exhibition as a movement to wean them from politics'.[7] However, the idea had caught on and working people participated in local groups that met to discuss regional contributions to the Exhibition. Every major town established a committee to organize local arrangements. Committee functions included collecting financial contributions towards the costs, which were met by public subscription with very little central government financial support, soliciting locally, manufactured exhibits, promoting the event, estimating the numbers likely to travel and liaising with the national organizers in London. Working-men's committees were often set up as subcommittees, often involving many former Chartists, Mechanics' Institute members and workplace

---

[5] Theodore Rothstein, *From Chartism to Labourism* (Martin Lawrence, London, 1929), p. 158.

[6] Susan Barton, 'Mechanics Institutes – Pioneers of Excursion Travel' *Transactions of the Leicestershire Archaeological and Historical Society* (1993), p. 48; J. Pudney, *The Thomas Cook Story* (Michael Joseph, London, 1953), p. 53.

[7] Short, 'Workers Under Glass', p. 195.

representatives. However, of the leading Chartists who were involved it was usually those associated with the right-wing of the movement or 'moral force' section such as Place and Lovett.[8]

Some Chartists welcomed the Exhibition for the opportunity it presented to them for advancing their own ideas and the aims of socialism. At a large meeting of trades delegates in Glasgow, a number of men identified as Chartists, although disagreeing on other issues, urged 'the necessity of union and energetic action amongst all shades of democrats and the importance of ... the Great Exhibition of spreading their principles, and helping forward the Great European struggle for liberty'. For them, the Exhibition presented a means of reviving the Chartist organization through providing a focus around which to agitate.[9]

A Central Committee of Social Propaganda was formed in London with a number of local committees supporting it through raising funds. This body thought the congregation of so many visitors, including large numbers from overseas, offered a wonderful opportunity of spreading the socialist message. 'What moment more opportune for promulgating these views so well calculated to make the world happy, than the time when the world is there to listen to you?'[10] The ageing Robert Owen agreed to write a series of tracts with French and German translations and, with other socialists, to give public lectures during the exhibition season.[11] As pointed out in the *Friend of the People*, 'as many come from countries where freedom of speech and press are almost unknown, such an opportunity for getting political and social information may be to them of double value'.[12]

International unity was not just the aspiration of the radical movement in Britain. The New York Industrial Congress resolved to send a delegation to London, 'to meet in convention the delegates of trade societies and labour associations from other parts of the world, during the Fair of 1851, for the purpose of interchanging opinions with each other in relation to the state of labour, and the condition of the labouring classes in the various countries they represent'. In February 1851, Parsons E. Day was appointed delegate to London where he met with inventors' clubs, trade societies and labour associations in order to arrange a convention of mechanics and working men during the exhibition period.[13]

The aim of class unity and educational improvement was attractive to members of Mechanics Institutes. These working-class campaigners had their

---

[8] Short, 'Workers Under Glass', p. 194.
[9] *Friend of the People*, 8 March 1851, p. 98.
[10] *Friend of the People*, January 1851, p. 59.
[11] *Friend of the People*, 19 April 1851, p. 168.
[12] *Friend of the People*, January 1851, p. 59.
[13] *Friend of the People*, March 1851, p. 136.

own political agenda: the fight for the franchise and for the consolidation and extension of the Ten Hour Day. Commentators predicted the end of the contempt shown for tradesmen and mechanics once the world could witness the skill involved in the production of artefacts for display. Many believed that the event would provide the ideal opportunity for working men to demonstrate not just their skills and intellectual capabilities in the design and making of exhibits, but also their respectability and responsibility through their behaviour. It was not just the middle class who despised drunkenness amongst the poor; class-conscious workers did too. They promoted the idea of 'rational recreation', not as a means of social control but for self-improvement, a better quality of life and to use as propaganda to get support for shorter working hours. Education was a political necessity, a reform to be fought for in opposition to the ruling classes who 'well know that Knowledge and Freedom go hand in hand, and therefore do they attempt to stem her liberty-bringing torrents fearful that they will sweep away the pillars of Ignorance and Prejudice on which the oligarchical power is based'.[14]

However, many other working-class militants were totally opposed to the Exhibition and the open class collaboration it seemed to involve. Julian Harney, writing as 'l'Ami de Peuple' in the Chartist paper the *Friend of the People*, described the opening pageant which attracted crowds to watch the Royal Family's cavalcade pass by as being inspired by 'the spirit of flunkeyism'.[15] He went on to describe the 'works of art and plunder wrung from the people of all lands, by their conquerors, the men of blood, privilege, and capital'. A truly worthy industrial exhibition could only happen when workers had renounced flunkeyism and substituted for the rule of masters and a degenerated monarchy 'the Supremacy of Labour, and the Sovereignty of the Nation'.[16]

There were public meetings in towns all around Britain. Local committees were formed in 297 different localities,[17] many of which had working-men's subcommittees. In Leicester a meeting addressed particularly to the working classes was held in the New Hall, a building used by the Mechanics' Institute.[18] On the platform were Commissioner Highmore Rosser, the Mayor and William Biggs, a former mayor and renowned radical having campaigned in favour of the extension of the franchise.[19] The highly- attended meeting

---

[14] *Friend of the People*, January 1851, p. 51
[15] *Friend of the People*, 10 May 1851, p. 189.
[16] Ibid.
[17] *Journal of Design and Manufacture* (1851), pp. 155–6.
[18] *Payne's Leicestershire and Midlands Advertiser. Leicestershire Mercury*, 20 July 1850.
[19] Biggs was a supporter of the Leicester Mechanics Institute Exhibition in 1840 and the rail excursions associated with it and a concurrent exhibition in Nottingham. Susan Barton, 'Mechanics Institutes'.

elected John Matts as Chairman of the Leicester Working Men's Committee. At a follow-up meeting in the Town Hall, Biggs criticized the Royal Commission's demand for local collections to finance the exhibition; he thought the government ought to pay, not the people.[20] Francis Warner moved that a society be formed called the 'Working Men's Provident Association' to enable the working classes of the town to visit the exhibition which would be a symbol of peace and harmony between nations and classes. This man might have been the same Warner who was active in the Working Men's Chartist Association in Leicester.[21] Formerly a physical-force Chartist, in 1848 he had split from the main Chartist group in the town.[22] Seconding the resolution proposing the formation of the Working Men's Provident Association, Mr Parker hoped that an equal number of the fair sex would go to London. Parker may have been Mr W. Parker, the Leicester Anti-Corn Law Association's working-men's secretary, listed in *Thomas Cook's Guide to Leicester*.[23] He had also been an active Chartist campaigner.[24] The rules drawn up by the committee were read; subscriptions were 6d, 9d or a 1s a week. John Matts had been in touch with other committees and the rules were therefore similar. Writing to other secretaries for advice was a common initial action.[25] Matts calculated that if a person saved 6d a week for a year it would come to 26s; of this 5s would be needed for the train fare. He hoped that 5 or 6,000 working-class people from Leicester would visit London and the Exhibition the following summer. Later in the same meeting Joseph Dare, a Unitarian minister, another supporter of reform, said there was no reason why every working man and woman should not go. Apart from Biggs and Dare, the occupations of the speakers are unknown, but such confidence in public speaking to a very large audience in surroundings like the Town Hall[26] probably indicated previous political experience.

Proposers of resolutions to Manchester's Working Men's Committee were themselves working men.[27] Their resolutions are worth quoting in full as they

---

[20] *Leicestershire Mercury*, 10 August 1850.
[21] *Leicestershire Chronicle*, 17 June 1848; J.F.C. Harrison, 'Chartism in Leicester, Asa Briggs (ed.), *Chartist Studies* (Macmillan, London, 1959), pp. 99–146; 118.
[22] *Leicestershire Chronicle*, 17 June 1848.
[23] *Thomas Cook's Guide to Leicester* (Leicester, 1843), (pages not numbered).
[24] *Leicestershire Chronicle*, 17 June 1848
[25] Several examples of letters between local committee secretaries have been found from places as diverse as Bath, Bristol, Oldham, Bolton and Northampton in Manchester Reference Library.
[26] The former Town Hall in Leicester is now known as the Guild Hall since a replacement was built in 1867.
[27] A handbill reporting a meeting held at the Mechanics' Institute records that resolutions passed were proposed or seconded by William Mellor and Thomas Greenhaugh, mechanics at Sharp Brothers; Robert Crichton and Jonathan Ogilvie employed by Messrs Fairburn; James Haughton and Charles Howarts in the employ of Messrs W. and D. Morriss

encompass the main objects of most other committees which involved workers:

> That we, as working men, feel gratified to find ourselves consulted upon a matter of such importance to the industrious classes of the whole world; and, if the co-operation of the employers can be obtained, we therefore pledge our exertions in furtherance of the object, so as to prove that the confidence of the commissioners is not misplaced.
>
> That a committee of two men from each principal workshop and manufactory in Manchester be formed, to assist in carrying out the objects of the Great National Exhibition of 1851; such committee to meet on the first Friday of each month, in this Institution.
>
> That the committee be requested to originate an active canvass amongst the artizans in our different machine shops and manufactories, to ascertain how many individuals, or associated bodies, will prepare specimens of their skill for exhibition, and to make a list of such articles, to be reported if possible at the second monthly meeting.
>
> That it be an object with the committee to arrange for a cheap trip on a series of days, so as to allow all interested to visit the exhibition at the lowest possible cost.[28]

Judging by their letters, leaflets and quotations in newspapers, the committee members and those most active in the working men's groups were highly literate, and enthusiastic about an event that promised self-improvement. A chance to use their skills, to be taken seriously and to be given responsibility must have been welcomed, especially if political activity was no longer a viable interest.

For political reasons, then, many working-class activists and spokesmen joined, or indeed chose not to join, the local committees planning for the exhibition. As well as committee involvement, the main participation of working people was in the workers' travel and savings clubs formed to enable those 'of moderate means' to visit Hyde Park's Crystal Palace on the cheaper 1s days. Many of these clubs were organized at community and workplace level.[29] Employers sometimes gave support and granted financial contributions or time off to make the visit to London.

---

and Jonathan Ryder and John Wrigley, mechanics at Sharp Brothers. Manchester Reference Library. Handbill Manchester Working Men's Committee, 1850.

[28] Ibid.

[29] *The Journal of Design and Manufacture* (1851), pp. 155–6, gives details of exhibition savings clubs in about thirty towns reported to the commissioners, but this is obviously only a small proportion of the final total

Not everyone was in favour of this involvement by workers. There was political argument as to whether the entry fee of 1s would be enough to restrict entry to only the respectable working classes;[30] some middle-and upper-class people were worried that the poorer and less desirable elements might be able to afford admission. There were fears of sedition and uprising, as well as of increased crime and unruly behaviour. The middle classes feared that the descent on London by large numbers of working people would attract the seditious elements not just from Britain but from abroad too, who would use the opportunity to stir up class hatred and rebellion amongst the congregated masses. The press started to campaign against foreign 'agitators' alleged to be planning a revolution for the occasion. Even Feargus O'Connor thought it necessary to join in the outcry and, in the *Northern Star*, warned Chartists to beware of the foreign revolutionary crowd and spies,[31] to the indignation of other Chartist groups such as Harney's Fraternal Democrats.[32] The French and other European revolutions of 1848 were still fresh in middle-class minds and the change in political tactics by many radicals was not yet acknowledged. Fears of social revolution were linked to general worries of working-class lawlessness. This itself, some believed, was a result of the moral degradation inherent in industrial life. Lord Ashley was convinced that everywhere he looked there was a 'wild and satanic spirit' abroad.[33] He was particularly concerned with the ill-behaviour of young people in the industrial areas, such as Manchester, Sheffield and the Potteries, and their involvement in violence associated with Chartist demonstrations.

The Barrack Master at Canterbury, Captain James Thomas, wrote to the commissioners outlining a complicated set of proposals for organizing workers staying in London. The commissioners, however, were 'indisposed to throw any restraint on individual plans or wishes' of visitors who after all would be coming at their own expense.[34] Unperturbed, Captain Thomas communicated his suggestions for the guidance of visitors to the Manchester Committee.[35] He proposed a committee for working-class visitors which could divide London up into neighbourhoods where, so far as possible, accommodation for the industrious classes should be in contiguous vicinities. These neighbourhoods were to be subdivided into districts corresponding with the counties and those districts be further divided into sections corresponding

---

[30] Short, 'Workers Under Glass', p. 199.
[31] *Northern Star*, 5 April 1851.
[32] Rothstein, *From Chartism to Labourism*, p. 159.
[33] Geoffrey Pearson, *Hooligan, a History of Respectable Fears* (Macmillan, London, 1983), p. 160.
[34] Manchester Reference Library, Letter, Manchester Local Committee to Captain James Thomas, February 1851.
[35] Manchester Reference Library, Letter, Captain James Thomas, to Manchester Local Committee, 31 January 1851.

with the chief towns of each county. The actual number of beds provided in each section and district was to have been carefully ascertained. Local committees could judge the proposed number of working-class visitors and these could be divided by the total number of beds in each section so that an appropriate area could be allocated. A person appointed as a leader in each district would meet visitors from the trains and conduct them to their lodgings after explaining rules of behaviour. As far as possible, apart from time at the Crystal Palace, working-class visitors should stay in their own allocated area of London. Captain Thomas also thought it best if no carpets were laid in houses for working-class occupancy, nor any woollen furniture used except for bedding. Floors and stairs were to be kept clean by daily rubbing with sand and a dry scrubber. He calculated that London could accommodate 55,000 working-class visitors staying for two nights with a day allowed for turnover. He was convinced that the working classes would cheerfully fall in with these well-intentioned plans for their comfort.

Despite such middle-class anxiety about large groups of workers gathered together, the event passed off peacefully. Even the public opening ceremony, performed by Queen Victoria, was without incident, after initial plans to hold it behind closed doors because of the worry of an assassination attempt. An additional 150 policemen were recruited and extra troops were garrisoned to guard London, but their presence for counter-revolutionary purposes proved to be unnecessary.[36] To the authorities' surprise, according to the First Report of the Committee for the Exhibition of 1851, crime in London actually appeared to decrease during the Exhibition. More than half of the arrests were made for drunkenness. Incredibly, there were only eleven reported thefts despite a total of approximately six million visitors.[37] Good behaviour was also in evidence at other sites in London, such as the National Gallery and British Museum, to which visitors also thronged. Elaborate preparations and precautions to ensure good behaviour proved to be either unnecessary or very successful. Polite London society was pleasantly surprised by the good behaviour of the crowds, having feared the worst in drunken brawls from the invasion by the rough provincial masses.

Accounts kept in Bolton give a gauge of the enthusiasm for the exhibition in individual workplaces as each contribution to both the General and the Operatives Funds is recorded.[38] The General Fund provided subscriptions to be sent to London towards the costs of the exhibition, after local expenses had been deducted. The Operatives' Fund was a local collection to enable people

---

[36] *Northern Star*, 3 May 1851, p. 3.
[37] Short, 'Workers Under Glass', p. 202.
[38] The Bolton account of subscribers is useful as it gives a list of names and the place of employment of those workers likely to have been the most probable visitors to the Exhibition. Bolton Records Office.

to work on preparing items for display by granting them the equivalent of wages for the duration of their projects, and to pay for someone to go to London and assist in finding and inspecting lodgings and supervizing the installation of exhibits.[39] These factory collections symbolize a commitment by workers to the idea of the Exhibition; at the time of the initial subscriptions there was no guarantee that the Exhibition would even take place or that contributors would be able to go and see it for themselves.

In Sunderland, on the advice of the circular from the Royal Commissioners on behalf of Prince Albert, a public meeting was convened to which the industrial classes were 'explicitly invited'. Circulars distributed in large factories and advertisements in local newspapers publicized the meeting.[40] A few workmen were invited to work with the committee and some employer committee members were asked to promote this among their workmen. William Armstrong and George Rochester were added to the Sunderland Executive Committee very early on in its existence because of their connection with the association for collecting subscriptions to allow workmen to see the Exhibition. Like Bolton's, Sunderland's committee was concerned with enabling working people to become involved in all aspects of the Exhibition, including financing the production of exhibits. As early as April 1850, at a committee meeting, Francis Gray Ross volunteered to make any medals which might be needed, his object being to earn enough to visit the exhibition.[41] The same meeting also heard a letter from Mr G.R. Taylor, who needed financial help to complete models of the bridge and a lifeboat. Later a fund was established to provide models of shipbuilding, Sunderland's staple industry.[42]

---

[39] Bolton Records Office. Bolton Local Committee Minute Book, 1850–1851. The most usual donation by operatives seems to have been 6d to each fund, although in the account book there are beside some firm's names a note saying 'Men declined to give' and in some cases blank collecting cards were returned. Altogether £725 11s 8d was collected in Bolton.

[40] Sunderland Reference Library, Sunderland Local Committee Minute Book, 1850–1851.

[41] Ibid.

[42] A Mr Hedley was one of those to benefit from this, receiving the equivalent of three weeks' wages for the completion of a model; he received a payment of £6 so presumably a skilled worker's wages in Sunderland at that time were £2 a week. This is interesting, as when the committee members received a letter from Mr Harrison in December 1850 advertising his Mechanics' Home lodgings for artisans, they resolved to write to him to express that in their opinion the charges were too high for working men. Posters and bills for the Mechanics' Home advertised charges of 1s 3d a night, not a huge sum compared with a wage of £2, only about 3.5 per cent of a week's pay. This seems comparatively less than the modern equivalent of cheap bed and breakfast at about £20 related to £200 as an example of a normal worker's pay; the cost has increased to 10 per cent of income. The comments of the Bolton Local Committee show they seemed to expect about 1s 2d to be the normal rate to pay for lodgings, not much less than the Mechanics' Home charges. Maybe Hedley commanded a higher rate of pay than the usual rate in the locality because of his skills.

The collection and saving of money was successfully organized by workers' travel clubs. Some of them also arranged travel and accommodation for their members. In some cases the clubs were organized under the auspices of the official local committees; some committee members, as prominent citizens and employers, were able to push the idea of the exhibition amongst their employees. Others seem to have been independent working-class organizations, established spontaneously by enthusiasts. The many clubs operating from public houses would have been outside the sponsorship and control of employers.[43] A network developed of club and committee secretaries dealing with the working classes.[44] From the collection of letters in Manchester, the usual topics of correspondence related to the desirability of having working men on the committees[45] and guidelines on the running of travel clubs.

The People's Club assisted Bolton workers and their families to go to London. A meeting between the Local Committee of Bolton and a deputation from the People's Club, represented by Mr J. Swift and Mr Kirkman, was held in 14 May 1851. From Bolton's records it is possible to deduce something about those People's Club representatives. Swift was probably the James Swift who donated 6d to the Operatives' Fund and 1s to the General Fund and who was employed by Benjamin Hicks and Son at the Soho Ironworks, where four other Swift family members also worked. If this was so, it was likely that his companion was Joseph Kirkman, also employed at the Soho Ironworks,

---

[43] Examples found in Manchester Reference Library include the Feathers Club and the Albert Exhibition Club in Manchester, and the Chapel Inn, Stalybridge. The Salford Working People's Association was run from the Albert Hotel (Salford Records Office).

[44] Other evidence of workers' involvement on the official local committees comes from the 'Miscellaneous' columns of the *Journal of Design and Manufacture* (1851) where short reports of information given to the commissioners are given. In Aberdeen it was resolved at a public meeting that a committee of twelve to be named by the working classes be added to the local committee; at Darlington it was resolved that a number of practical working men be invited to join the committee; at Woolwich the foremen from the Royal Dockyard and the Royal Arsenal were added to the committee.

[45] Manchester Reference Library, Letter Bristol Local Committee to Manchester's Local Committee. On the issue of workers on committees, the Bristol secretary wrote to Manchester's to ascertain what the effect of such a proceeding had been. Letters asking for information about the rules for travel clubs came from such diverse sources as Oldham, Runcorn, Bristol, Bath, Dundee, Salford, Sheffield and from the Chapel Inn, Stalybridge. One man even wrote in on behalf of men in his singing classes. It is possible that the committees wrote to all the others but this seems unlikely as the letters are all handwritten; copying them out 297 times would have been an onerous task for even the most committed secretary. Bolton Committee had been in touch with Northampton's, which supplied a copy of the rules upon which Bolton's own were based.

who contributed 6d to each of the funds.[46] The company showed a high commitment to the Exhibition with two medals won for products displayed.[47]

Southampton Local Committee formed auxiliary committees in the surrounding villages.[48] At the first such meeting for Shirley, Millbrook and the surrounding area, crowded by an audience made up almost entirely of the working classes, a collection of £10 was raised in the first few minutes by small sums. A travelling fund was eagerly established in connection with Southampton's 700-member Great Club.[49] Both men and women were reported being present, 'a large number of both sexes enrolled their names and paid the weekly subscription of one penny'. The next week there was another meeting at Romsey, where the Mayor hoped to enrol hundreds more to assist in 'This noble contest of mind over matter'.[50] As well as the Exhibition Travelling Fund, Southampton's Admission and Provision Fund enabled members 'to go to the door of the Exhibition with a silver key in the shape of a shilling'.[51]

Not all the travel clubs had enough members to meet the railway companies' requirements for groups to travel in parties of 250 or 200. Letters were sent to Manchester from Warrington where only twenty people wished to travel and from Glossop where the Working Men's Club of the district did not amount to more than eighteen.[52] Those clubs with fewer members approached their local committee secretaries to enquire whether they could combine with similar larger clubs of Manchester excursionists.

Money clubs were often established independently of the local committees, frequently on the premises of public houses. Landlords often looked after trade union accounts as well as having other social functions, such as organizing the goose clubs. Newspaper readings took place in pubs, which would have generated interest in the Exhibition. Information about the exhibition would have been learnt from newspapers or through conversation so it is therefore quite natural that the pub should have been the focus of local interest for many regarding the Great Exhibition, especially those working in small workshops rather than large factories.

---

[46] Bolton Records Office, Bolton Local Committee Accounts from the Minute Book; the *Bolton Chronicle*, 6 April 1850, advertises a concert by Benjamin Hicks's works' brass band.

[47] From entries in the Bolton Local Committee's accounts it can be seen that the firm itself gave £20 in January 1851 and from collections among the operatives, £10 8s 3d in February followed by a further £2 given in June. To the Operatives Fund an even more generous donation from operatives of £18 17s 3d was given on 20 February of that year.

[48] Manchester Reference Library, Handbill, Southampton Local Committee, May 1850.

[49] The club had 700 members in May 1850.

[50] *Journal of Design and Manufacture* (1851), p. 156.

[51] Ibid.

[52] From letters held at Manchester Reference Library.

A letter requesting information about clubs was sent to Manchester's Local Committee from the Chapel Inn, Stalybridge.[53] In Bradford a club was established at the Hope and Anchor; similar clubs were formed at various inns in the town and neighbourhood. The Feathers Exhibition Club was based at the Feathers Inn, Deansgate in Manchester.[54] William Hancock, the landlord, was listed as treasurer of the club whose shares were 1s a week. Copies of the rules were available at the bar of the house for 1d. Weekly meetings were held at the Feathers to collect subscriptions. An additional weekly charge of 1d for expenses was levied. A fine of 1d a week was imposed on shares not duly paid up. If a member wanted to withdraw his share a fine of 2s 6d was imposed or a 6d fee to transfer the shares to someone else.

Another Manchester organization was the Albert Exhibition Club.[55] Full shares of 5s a fortnight covered the cost of the journey, bed and breakfast for six nights, exhibition catalogue and conveyance to the lodgings, in effect an inclusive tour; half shares included only the cost of the rail journey, members to make their own accommodation arrangements. Membership of this club cost 1s which included a copy of the rules and, interestingly, a guide to all the free exhibitions and places of interest in London.[56] The headquarters of this club was 25 Abraham's Court, Market Street, the home of Richard Stanley who styled himself 'Manager and Conductor'.[57]

The Bristol Association in Connexion (sic) with the Great Exhibition of 1851 allowed members to pay in instalments of 6d at any convenient interval.[58] This club's organizers appear to have been shopkeepers and include a salt-store keeper, stationer, two booksellers, three grocers, eight druggists, a linen-draper, brushmaker and an oil and colourman. They aimed to provide comfortable and economical accommodation for members during their stay.

Salford Working People's Association again met in a public house, the Albert Hotel in New Bailey Street. Members here paid 2s 6d per week and at the appointed time would receive £2 9s together with a railway ticket allowing seven days in London, from the first Saturday in July. If membership exceeded 250, a second train would be arranged for the following Saturday.

---

[53] Manchester Reference Library, Letter Landlord Chapel Inn, Stalybridge to Manchester Local Committee.

[54] Manchester Reference Library, Handbill, advertising the Feathers Exhibition Club, Manchester, 1851.

[55] Manchester Reference Library, Handbill, advertising the Albert Exhibition Club, Manchester, 1851.

[56] Ibid.

[57] Ibid.

[58] Manchester Reference Library, Handbill, advertising the Bristol Association in Connexion with the Great Exhibition of 1851.

The People's Exhibition Club of Bolton charged £1 14s in June and £1 9s afterwards for transport and accommodation. Membership cost 1s 6d.[59] Bolton Local Committee encouraged workers to go through a newspaper notice; they claimed that if a person were to 'Neglect to avail himself of the advantages offered, he will be in a worse position than his fellow workmen who embrace them. The Exhibition would thus injure him if he refuse to benefit from it'.[60] The Bolton Committee estimated £4 as the sum required for a weeklong visit to London: £1 for the journey; eight breakfasts (at 6d), eight dinners (at 1s), eight suppers (9d) totalling £1 for food; six nights bed (1s 2d) and malt liquor or tea (1s a day), making 16s; admission to exhibitions and a steamer trip on the Thames another £1; plus 4s extra for spending. In order to enable savings of this amount to be made, the committee recommended an early start, saving 1s $6^{1}/_{2}$d a week for a year. Blank savings cards to be adopted by any club were printed. People of the same trade and of a similar age were advised to form groups up to twenty people to open savings accounts, the interest to be shared.

Travel clubs offered members a large range of facilities that differed from place to place. Some simply offered a savings bank facility; others bulk purchase of cheap excursion train fares; some secured both transport and accommodation; others offered a complete package of transport, accommodation, exhibition ticket and catalogue, guide books and the help of an appointed person in London. In effect some offered a fully inclusive tour or package, such as the Albert Exhibition Club in Manchester which arranged train travel, six nights' bed and breakfast, an exhibition catalogue and conveyance to the lodgings.[61] Some savings might even have included an amount to compensate for loss of income while off work or perhaps some spending money.

So imperative was the need to provide suitable accommodation for working people that the Royal Commission set up a special agency in London to correspond with the local committees.[62] This agency provided information about lodgings, their quality and price. It regulated the dates for the various cheap trips from the large manufacturing districts, to subdivide as much as possible the immense masses converging on London. Providers of accommodation had to register with the Central Agency. As well as assisting visitors find comfort and security, this system would 'prevent respectable persons locating themselves in houses of doubtful reputation'.[63] Only persons of good character could register and references were required. Regulations

---

[59] *Bolton Chronicle*, 17 August 1851.
[60] Ibid.
[61] Manchester Reference Library, Handbill, Albert Exhibition Club, Manchester, 1851.
[62] *Journal of Design and Manufacture* (1851), p. 192.
[63] Manchester Reference Library, Circular London Central Registry, London, May 1851.

forbade families and couples, and single men and women to be lodged in the same house.

A range of accommodation was on offer, even within the category designated for the working classes. One of the cheapest was Castle's on Baltic Wharf. For 1s a night, 200 men could sleep in apartments in the style of emigrant ships.[64] The most likely number conveyed at one time from any one establishment was expected to be 200. This would reduce the likelihood of disagreement, which might result if several large bodies of men were accommodated together.[65]

Samuel Herapath accommodated people of moderate means in two houses with eighty to one hundred beds in Holborn.[66] John Parker, a carpenter and builder of Commercial Road East, lodged 150 artizans in a fine, open and healthy part of London for half a guinea a week, each person having a bed and washing convenience to himself, boots cleaned and hot water.[67]

The Clarence Club House for Artizans and Others provided distinct apartments with bed, bedding, basin, soap and towels and box with lock and key for 2s a night.[68] Thomas Harrison ran the Mechanics' Home, Ranelagh Road, Pimlico. A letter handwritten by Harrison to the Manchester Committee included a prospectus for his establishment saying he would be taking bookings from 1 March.[69] Another lodging, on similar lines, was Jones's of Rochester Row where for 2s a night one hundred artizans could sleep in

---

[64] Prospectus produced by H. Castle and Company, Baltic Wharf, Westminster, 1851. For only 1s a night, each man had a berth to himself, a flock bed, pillow, blanket, two sheets and a coverlet – all clean. Breakfast, though, cost an additional 9d. Guests could bring their own cheese, biscuits, coffee and the like for which a storeroom was available. No smoking was allowed in the building or on the wharf; to smoke a man had to go on to a ship lying beside the wharf, the decks of which made a good promenade Manchester Reference Library, Handbill Castle and Co., Manchester 1851).

[65] *Circular London Central Registry Office for House Accommodation for Visitors to the Exhibition of the Works of All Nations* (London, 1851), p. 5.

[66] Manchester Reference Library, Handbill advertising Samuel Herapath's lodging house, 1851.

[67] Manchester Reference Library, Letter, John Parker to Manchester Local Committee.

[68] Manchester Reference Library, Advertising handbill produced for Clarence Clubhouse, 1851.

[69] Manchester Reference Library, Letter from Thomas Harrison, to Manchester Local Committee 6 December 1850 enclosing prospectus. The poster format prospectus describes the accommodation as occupying 2 acres in a perfectly airy, well-ventilated situation. For 1s 3d, up to 1,000 guests a night slept in dormitories, arranged so as to give privacy. A smoking room where ale and porter could be purchased also had a band playing in it every evening. Mr Harrison made 'effectual provision for the comfort, convenience and discipline' of the large body of men resident there. The premises on the engraving on the publicity bills show what appears to be a riverside warehouse, next door to Cubitt's works on the accompanying map, Manchester Reference Library. Thomas Harrison's Mechanics' Home Prospectus 1851.

dormitories on bedsteads and hair or wool mattresses with blankets. Bagatelle and smoking rooms provided recreation space.[70]

Eversfield and Horne's register listed rooms to rent in private houses, many of which were suitable for families and women. All price ranges were available, from mansions with rooms for servants to shared bedsits. Some were run like boarding houses, perhaps by genteel families fallen on hard times. The Bucklands of Euston Square promised cheerful and good society in their comfortable and elegant home. They offered terms from £1 11s 6d to £2 2s 0d a week in their highly respectable accommodation where French and German were spoken.[71]

Vast numbers of tents also provided a place to stay for the poorest or those who did not book early enough. All lodgings would have been at a premium with a massive influx not just from within Britain but from all over the globe.[72]

The exhibition presented a golden opportunity for Londoners with initiative to improve their finances, perhaps too easy an opportunity. Some commentators feared not for London with this influx of strangers of doubtful respectability but for the safety of the visitors at the mercy of worldly Londoners. That the capital's natives might corrupt or trick the unwary visitor not familiar with London ways was the fear of Reverend Morgan of Leeds. He spoke of the need for protection 'from the serious disadvantages which will at once occur to those who know what London is – what it is to be dropped from a railway train, a perfect stranger to the place and its ways, from a railway train in the middle of that vast wilderness'.[73] The story of the honest northern artizan wandering in London, pocket picked and baggage stolen, victim of the cunning Londoner, false friend, with his theatres, dens of vice and drink was a common theme of the almanac and chap book.[74] This presented an almost child-like image of workers, vulnerable to evil influences and ready to be led astray once away from the control of the middle classes.

Going to London and actually staying there for several days was quite an adventure and a completely new experience for working-class travellers, even those who had been on excursions which became common from 1840. The main obstacle to going away from home would have been not just the cost but the loss of wages. Time off had to be saved up for so that normal household

---

[70] An engraved picture on Jones's advertising poster shows a smart building with hanging gas lanterns outside, headed by the name 'Jones, Wine & Spirit Merchant'. The poster is bordered by a list of information about the attractions of London such as Madame Tussauds, galleries and museums.
[71] Manchester Reference Library, *Accommodation List* (Eversfield and Horne, London, 1851).
[72] *Journal of Design and Manufacture* (1851); *Northern Star*, 10 May 1851.
[73] Supplement, *Leeds Mercury*, 1 February 1851.
[74] Supplement, *Leeds Mercury*, 1 February 1851.

bills could be paid. An article in Thomas Cook's *Exhibition Herald* asked 'how shall the people be spared from their employment with the least injury to business and the future comfort of the workpeople themselves?'[75] In reply, a Leicester workman suggested that employers should allow workers wages for the time they would be absent, so that on their return their families would not be destitute of the week's supplies. Some of the clubs with higher subscriptions might have taken this into account. Some employers gave leave and encouraged their employees to go. Some gave money towards expenses, while others even went so far as to give paid time off in some of the earliest examples of paid leave.[76]

The Exhibition was the first time many had encountered people of other nationalities. All contributing nations were allocated their own Exhibition space.[77] This would have enhanced awareness of places and cultures beyond Britain and emphasised that industrialization was becoming a worldwide phenomenon.

Working-class involvement in all aspects of the Exhibition showed that the risk of insurrection was over and an era of incorporation, collaboration and reformist politics had commenced. Workers had come to terms with their status as wage labourers and adapted themselves culturally and politically to the ethos of industrialization. During unsuccessful campaigning for the 1852 Reform Bill, a demonstration in Leicester the day after the Exhibition closed was addressed by a man named Thompson. He cited the thousands who had visited the Exhibition in peace and behaved in a responsible manner. At 'yesterday's closing ceremony' he said, 'there were 107,000 people at the Crystal Palace and not one breach of the peace. This was in a country with only 850,000 voters and only 250,000 of these independent'.[78] Six million visits were made to the Exhibition.[79] It had attracted more visitors than comprised the entire electorate at the time. There was no justifiable reason why the majority of the people should be excluded from the franchise. They had proved themselves to be respectable, sensibly behaved and thrifty, posing no threat to the existing order. London had coped with the unprecedented

---

[75] *Exhibition Herald*, 1, 31 May 1851, p. 7

[76] There are examples of time off with pay from Leicester, which must have been repeated elsewhere around the country. Berridge and Macauley gave £5 to their clerks and leave to go to the exhibition as did the bankers Parsons and Dain; clerks at Adcock and Dalton were also given leave; these seem all to have been white-collar workers. But in what is undoubtedly one of the first examples of a paid holiday anywhere in the world for manual workers, Goodwin and Hobson, brewers, of Leicester gave each man in their employ four days' holiday with pay and the rail fare to go to London *Payne's Leicestershire Advertiser*, 5 July 1851.

[77] Ibid., pp. 26–33.

[78] *Leicestershire Mercury*, 11 October 1851.

[79] Short, 'Workers Under Glass, p. 202.

influx of visitors; the visitors had coped with London, thanks to the innovative and resourceful combination of activities by working people.

Chapter 11

Estimating a Public Sphere: Intellectual and Technical Associations at the Time of the Great Exhibition

Vicky Brown and Ian Inkster

The following short chapter is based on a census of voluntary associations undertaken by government in order to identify societies worthy of nil-rate assessments, which was published in a consolidated form within the 1851 Census as 'List of the Literary and Scientific Institutions from which Returns were Procured at the Census of 1851'. The list was arranged by registration districts or Poor Law unions, topographically ordered as in the Registrar-General's reports.[1] It embraced 1,020 institutions in England and Wales boasting some 165,000 members. The data has been referred to before only in a passing manner,[2] and in this chapter we shall analyse it in terms of geographical and demographical distribution, gender and accessibility, with particular reference to the distribution of 'steam intellect' (the interests and associations which serviced industrial culture) and to the possibility of extending the database by a combination of local research and statistical inference.

The purpose of the chapter is to shift an earlier research programme on the urban scientific-technical culture of industrializing Britain towards a study of the associations of civil culture and civic life as a means of identifying entry into and exclusion from a relatively new platform of the public sphere. Where earlier work, especially the standard paper by Arnold Thackray on Manchester, addressed matters of intellectual networking, technical innovation, and the relations between intellectual interests and religious and political beliefs in a very fulsome manner,[3] our intent in this very brief essay is to ask questions of

---

[1] Unless otherwise stated, basic data is calculated from 1851 *Population Census of Great Britain, Sessions 1852–54*, British Parliamentary Papers, Irish University Reprint Series, vol. 11

[2] D.S.L. Cardwell, *The Organisation of Science in England*, William Heinemann, London 1957, p. 57; Ian Inkster, 'Introduction: The Context of Steam Intellect in Britain', in Inkster ed. *The Steam Intellect Societies – Esssays on Culture, Education and Industry circa 1820–1914*, University of Nottingham Press, Nottingham, 1985, pp. 3–19

[3] Arnold Thackray, 'Natural Knowledge in Cultural Context: the Manchester Model', *American Historical Review*, 79 (1974), pp. 672–709

inclusion and gender as well as to hazard the relations between intellectual and practical pursuits and those between the intellectual associations and the development of a civil culture which may have acted as an underpinning of the constitutional changes of the last years of the Golden Age.

General

The great majority of associations boasted libraries, reading rooms and lecture meetings in their own accommodation, and those of the 'mechanics' institute' type frequently held regular classes, with some institutions having laboratory and other additional facilities. Scientific subjects of study were explicitly mentioned in some 500 instances, literature and the arts around 400 times, while specific exclusion of religion and politics was mentioned 37 times, as against overt inclusions of such matters, which were mentioned 88 times. Associations varied in size between those of a dozen or so members to institutions of well over 1,000 members.

Table 11.1 Membership of Institutions, 1851, per 1,000 of Population

County	Main Towns	Remainder of County	County	Main Towns	Remainder of County
Berks	23	5	Northampton	25	4
Cambridge	35	5	Northumberland	17	9
Chester	9	3	Nottingham	19	4
Cumberland	21	7	Oxford	0	2
Derby	16	5	Somerset	0	4
Devon	25	7	Stafford	5	6
Durham	16	8	Suffolk	20	4
Gloucester	17	3	Sussex	19	9
Kent	13	5	Warwick	3	2
Lancaster	18	4	Worcester	7	6
Leicester	16	5	York	35	10
Norfolk	9	2			

Table 11.1 lists the membership of associations in England by county per thousand of the population in order to neutralize simple demographic effects, and also shows the membership distribution in terms of main towns in each county. The urban membership is consistently higher than the county membership for each county location. Although the industrial county of Yorkshire contains an urban high, so too does Cambridgeshire, and there are high urban memberships in such non-industrial counties as Sussex, Suffolk and Cumberland, and medium participations in Leicester and Nottinghamshire. The

conclusion from such figures would suggest that some thirty years of effort had failed to skew the distribution of urban intellectual and technical associations towards the industrial locations.

A slightly more precise targeting is provided in chart 11.1, which shows membership distributions per thousand adults in the populations of London (including metropolitan Middlesex, Kent and Surrey), six industrial counties (Lancashire, Yorkshire, Nottinghamshire, Cheshire, Staffordshire and Warwickshire), eight Home Counties (Hertfordshire, Essex, Surrey, Berkshire, Buckinghamshire, Oxfordshire, Hampshire and Bedfordshire), and the remaining English counties. London fares very well, mostly because of the large number of national intellectual societies located there. When this effect is excluded, the industrial counties do generally just outmatch the rest of England on average, but to a relatively small extent.

Gender

A total of 13,205 females were members of the associations overall, or 7.9 per cent of the total known cases. In terms of regional variations, women members as a percentage of all members in their county varied from a high of 29 per cent in Kentish London, through medium levels of 10–12 per cent for the Midlands counties of Derbyshire and Nottinghamshire, to lows of below 2 per cent in Huntingdonshire, Cambridgeshire and Gloucestershire. Women members as a percentage of the adult female population of their areas was highest in Kentish London, Devonshire, Cornwall, Essex, the West Riding, the rest of Kent, Westmoreland and Wiltshire in that order, and lowest in Huntingdonshire followed by Bedfordshire, Gloucestershire, Dorset and Oxfordshire, which suggests little in the way of analysis in terms of either industry or urbanity. However, chart 11.2 illustrates gender membership as a proportion of the adult population in each area, and does suggest that the greatest gender contrast in participation rates was in London, followed by the six industrial counties. The first case is explained by the number of metropolitan national associations which were large and exclusively male bastions. But the second may reflect a more underlying and serious gender exclusion in industrial urban areas. This is suggested by local detail, to which we shall return below.

Table 11.2 Size of Associations and Female Membership

Size Classification	No. Institutions with Female Membership 25% & above in Classification Group	No. Institutions in Classification Group	% Institutions in Classification Group with 25% Female Membership & Above
1–25	8	87	9
26–50	16	187	9
51–100	23	270	9
101–200	20	222	9
201–300	14	88	16
301–400	4	45	9
401–500	0	18	0
501–1000	7	41	17
1001+	1	10	10

Table 11.3 Urbanism and Female Membership

City	% Population Members of Institutions	% Female Membership (where numbers given)
London	1.64	6.42
Manchester & Salford	1.40	12.61
Liverpool	0.91	6.57
Nottingham	1.90	9.06
Birmingham	0.64	5.24
Newcastle-upon-Tyne	1.67	4.23
Sheffield	0.82	0.72
Leicester	1.63	4.05

Tables 11.2 and 11.3 describe female membership in terms of the size of association and in a number of large urban cases. Table 11.2 shows that female participation was more likely either in medium-sized associations or in those larger institutions with membership between 500 and 1000. The even larger institutions of 1,001-plus tended to include a large proportion of those London-based intellectual societies which entirely excluded female membership, principally through some form of 'fellowship' criteria which prohibited women by definition. Table 11.3 shows Manchester and Nottingham as leading cities as measured by the proportion of female to total membership.

Steam Intellect

When Thomas Love Peacock in his *Crotchet Castle* of 1831 so disdainfully labelled as 'steam intellect' all those associations of mechanics and the middling classes which arose in the efforts of Radical intellectuals, politicians and idealogues to educate, inform and influence the industrial nation, he captured a large proportion of our sample. Here we treat steam intellect as embracing all mechanics' institutes, literary and scientific associations and artisan's clubs which included scientific discussion or lectures and charged an annual fee of less than 1 guinea with an allowance for weekly, monthly or quarterly payments. Of 60,357 individuals entered, gender information is given for 58,500, 10.11 per cent of whom were women. Of these, fifty-seven institutions had female membership of 25 per cent or above. Of the total associations in this category, 91 per cent boasted a library, 74 per cent provided lecture courses. Overall, steam intellect incorporated well over half the total of 1851 association membership with a just disproportionate number of women (10.11 per cent as opposed to 7.88 per cent for the other associations).

Chart 11.3 shows the distribution of institutions by cost and region. The industrial counties were the ones most biased towards steam intellect in their associations, the Home Counties and London least so.

Table 11.4 Steam Intellect by Size of Membership and Library Books per Member

Number of Members Classification	No. Institutions in Class	No. Members	No. Female Members	% Female Membership	Library Books per Member (Male & Female)
1–25	67	1,248	88	7.05	18
26–50	151	5,918	385	6.51	10
51–100	221	16,509	1,290	7.81	6
101–200	206	28,834	2,501	8.67	6
201–300	68	16,348	1,651	10.10	6
301–400	35	12,093	863	7.14	6
401–500	10	4,435	231	5.21	11
501–1000	28	18,461	1,946	10.54	6
1001+	6	8,438	1,032	12.23	8

Table 11.4 provides good summary of the character of steam intellect overall. Because the elite London institutions now drop out, the largest institutions of steam intellect show a high proportion of female membership, in contrast to medium-sized provincial associations which show a surprisingly low proportion of female members at an average of 5.2 per cent. In terms of our measure of facilities, books per member show a great consistency amongst the middle-sized associations, with smaller and larger institutions somewhat better situated. Chart 11.4 shows the strikingly inverse relationship between the

degree of female membership and facilities per capita in the steam intellect associations whatever their size category.

A Wider World

There can be little doubt that the census data in fact captures only a large portion of the true population of intellectual associations and their members. Local studies of all sorts testify to the commonality of such institutions, particularly in the industrial urban areas and in London. Very briefly, we here

Table 11.5 Extension of the Census: 'Steam Intellect' in Liverpool

Institution	Foundation Date	Science Provision	Other Provision	Member -ship	Charges
Artisan's Lecture Society	1848–49	S/L	ES/S/PR/N/F	4,000*	1d*
Birkenhead MI	1844	Lib/L/S	Lib/ES/F	-	-
Brougham Institute	1836	L/Lib/P	N/PR/F	500	'lowest'
Catholic Middle School Society	1850	L	L/Lib	-	-
Church of England Institute	1849	L	N	-!	-
Clarence Foundry Mut.Instr. Soc.	1838	L	N/ES	-	2d*
Collegiate Institution	1839	L/S/Lib/ES	S/L/ES/Ex	200'	10/6d
Eclectic Discussion Society	1849	D/Lib	L/Lib	-	Free
Edge-Hill Mechanics' Club	1839	L/D	PR/N	-	'low'
Elms' Reading Room	1847	N/S	N/Lib/PR	-	-
Female Apprentices' Library	1835	Lib	Lib/L	100!	3/-
Franklin Club	1840	D	PR	-	-
Hall of Science	1839	L/ES	L/Lib	-!	3d*
Harrington Discussion Class	1847	P/ES/D	PR/N	50	Free*
Lime St Educ. Institution	1844	L	Lib/F	80!	-
Literary and Scientific Institute	1835	L/S/D/P/M	N/D/ES/Lab/L/ P/Ex/Lib	300!	30/-
Liverpool Commercial Institute	1826	L	N/Lib/L	500!	-
Liverpool Governesses' Inst.	1849	L	L	35!	Free*
Liverpool Mechanics' Library	1823	Lib/L	Lib/L	800!	3/-
Liverpool MI	1825	L/S/Lab/Sch/ Lib/M	S/L/Lib/N/ES	1,300!	5/-
Liverpool Northern MI	1839	P/L/Lib/S	ES/S/L/PR	500!	4/-
Operative House Joiners Assoc	1848	Lib/L/S/D	Lib/S/L	300!	-
Operative News Room	1837	D	PR/N	-	-
Ormskirk Lit & Sci Institution	1850	L	ES/L/N	-	-
Protestant MI	1839	L	S/L	-	-
Roscoe Club	1841	L/D	L/D/P	-	-
School for Natural History	1844	L/S	-	-	-
Scientific Institution	1836	L	-	100!	'cheap'
Seamen's R R. and School Soc	1844	Lib/L	S/Lib	-	-
South End Disc Class & Sci Soc	1847	D/L	D/PR	300!	1d*
St Patrick's School Society	1844	S/Lib	S/Lib	100!	Free
Sunday School Institute	1846	Lib/D	Lib/D/N/S	384!	-
Tradesmen's Institution	1839	Lib/L/D	D/Lib/L/PR/S	-	-
Tuckerman Institution	1848	L/Lib	L/PR/Lib	160!	'cheap'
W. Derby Soc Prom. Intell.Imprt.	1844	L	ES/L	700!	-
Wavertree MI	1849	S/L	L/N/ES/F	150!	-
Woolton MI	1847	M/S/L/D/Lib	S/D/L/Lib	187!	6/-
Y.M.'s Estab. Church Soc.	1839	L/D	-	100!	-
Young Men's Christ Assoc	1847	L	L/S	-	-

Key:
Science Provision
D Discussion
ES Exhibitions or Soirees
L Lectures
Lab   Laboratory Facility
LibLibrary
M Museum
P Papers read by members
S Schools and Class Teaching
Sch   Scholarships

Other Provision
ES Exhibitions or Soirees
Ex Exercises (Gymnastics etc.)
F Separate Female Provision
N   Newsroom, reading room etc
P Papers read by members
PR Political and Religious debate allowed
S Schools and Class Teaching
Sch   Scholarships

Membership
* quoted estimate/enumeration of audiences or lectures
' pupils in classes (regular)
! reportedly rising throughout period to 1850

Charges
quarterly charges have been recalculated per annum
* charge per lecture or per class

begin an extension of the approach by centering on steam intellect in the city of Liverpool. Very detailed independent local research yields table 11.5 which lists no less than thirty-nine formal associations in the city which existed prior to and during 1850–51.[4] As may be seen, these ranged from the census-listed institutions such as the Liverpool Mechanics' Institute with its 1,300 members, library, lectures and classes to the amazing Artisans' Lecturing Society which could boast meetings of up to 4,000 at 1d a time for exhibitions, large classes, separate facilities and meetings for women, and the admittance of subjects of religious and political controversy. Of the entries under Liverpool in the census, four may properly be entered under the heading of steam intellect, suggesting an under-registration of reality in a ratio 1:10 (4:39). A multiplier of ten for steam intellect may be somewhat dramatic, although other areas supply supporting evidence. On the other hand, the non-steam intellect associations are unlikely to be as underestimated, many members would be involved in more than one association, and areas of small population are perhaps less likely to have spawned significant associations which would not be registered in the census. Again, the largest institutions would be more likely included, although the Liverpool list does give some very large unrecorded institutions.

At the same time, estimates of the intellectual public sphere which focus on formal associations at all, will tend to undervalue such platforms as independent lecture courses, reading societies, evening classes, and so on, precisely the informal associations which might have a higher female ratio and a more practical and political purpose. Thus, the highly informal intellectual and technical associations which met in Nottingham's public houses cannot be

[4] For this research base see Ian Inkster, *Studies in the Social History of Science in England during the Industrial Revolution*, PhD thesis, University of Sheffield, 1976, pp. 353–60

found in any census although they seem to have gathered a relatively high proportion of women as members and encouraged controversial political debate.[5] On the other hand, the census data gives a membership of steam intellect in the Huddersfield area of 2 per cent of the population, or 8 per cent of total adult males, which seems low when given the reputation for intellectual self-help this area had. Again in Huddersfield the census data suggests a female to male ratio in steam intellect membership of only 2 per cent, which would be very low compared to the national average. From all of which we might quite moderately suggest that the real extent of steam intellect was around five times that recorded, and posit a somewhat lesser multiplier for the more expensive associations of intellectual and technical discourse and information. This would yield a membership in this public sphere of approximately half a million persons in mid-century.

Conclusions

The associations of intellect and of technique were more widespread in 1851 than often thought, and acted as a solid base to the Great Exhibition of that year and to the subsequent twenty years of Golden Age machinofacture. It seems possible that these representatives of a very large public sphere of a Habermas type were of as much importance as Bagehot's 'education' factor in promoting elements of that civic culture of emulation which was supposedly demanded by the British form of constitutionalism as elaborated during 1867–69.[6] On the whole, women were effectively locked out of this new sphere despite a great growth of female scientific education, the female evening school movement and the known commonality of the private participation of women in the cultures and coteries of intellect and technique – whether this be Mary Sommerville in a private parlour or George Elliot in the strange and wayward Rosehill seminar at Coventry. Outside of London, in those locations where women were present in larger numbers they probably obtained less in the way of serious resources for learning or intellectual debate. Finally, despite enormous efforts made by both middle-class radicals and activists amongst worker groups from the 1820s, the industrial areas in 1851 were not especially well served by the associations of intellect and technique, although somewhat better so in terms of what we have designated as 'steam intellect'. If we accept

---

[5] Ross Quarrell, *Middle Class Science to Pub Culture*, dissertation submitted as partial fulfilment of BA Hons History, Nottingham Trent University, 1998.

[6] J. Habermas, *The Structural Transformation of the Public Sphere: An Inquiry into a Category of Bourgeois Society* (1962), translated by Thomas Burger with the assistance of Frederick Lawrence, Polity Press, Cambridge, 1989; Walter Bagehot, *The English Constitution*, Humphrey Milton, London, 1867.

the possibility of a great underestimation of real facilities in the census of institutions, it might be that more informal and less middle-class infiltrated associations served as both media of civic life and information networks in the urban industrial areas at the outset of the Golden Age.

INTELLECTUAL AND TECHNICAL ASSOCIATIONS 173

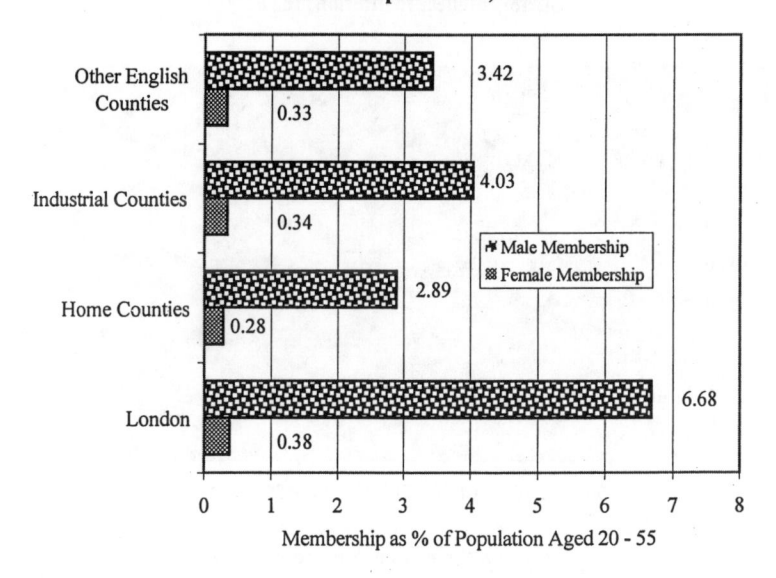

Chart 11.1: Regional Distribution of Institutional Membership, 1851

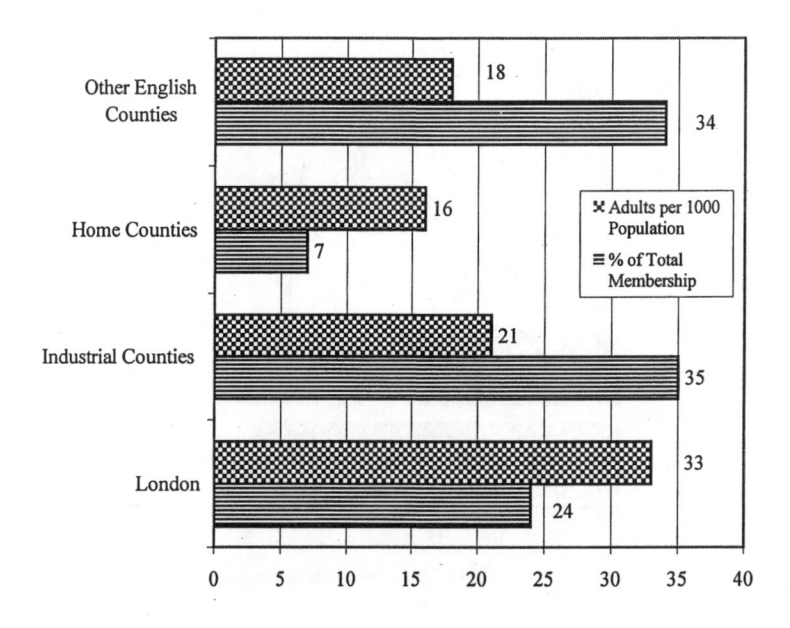

Region	Adults per 1000 Population	% of Total Membership
Other English Counties	18	34
Home Counties	16	7
Industrial Counties	21	35
London	33	24

Chart 11.2: Male and Female Membership of Institutions in Per Capita Terms, 1851

Membership as % of Population Aged 20 - 55

Region	Male Membership	Female Membership
Other English Counties	3.42	0.33
Industrial Counties	4.03	0.34
Home Counties	2.89	0.28
London	6.68	0.38

Chart 11.3: Distribution of Institutions by Cost

Region	Over 1 Guinea	Under 1 Guinea
Wales	5%	
Other English Counties	41%	42%
Home Counties	11%	10%
Industrial Counties	19%	36%
London	30%	7%

% of Total Institutions in Category

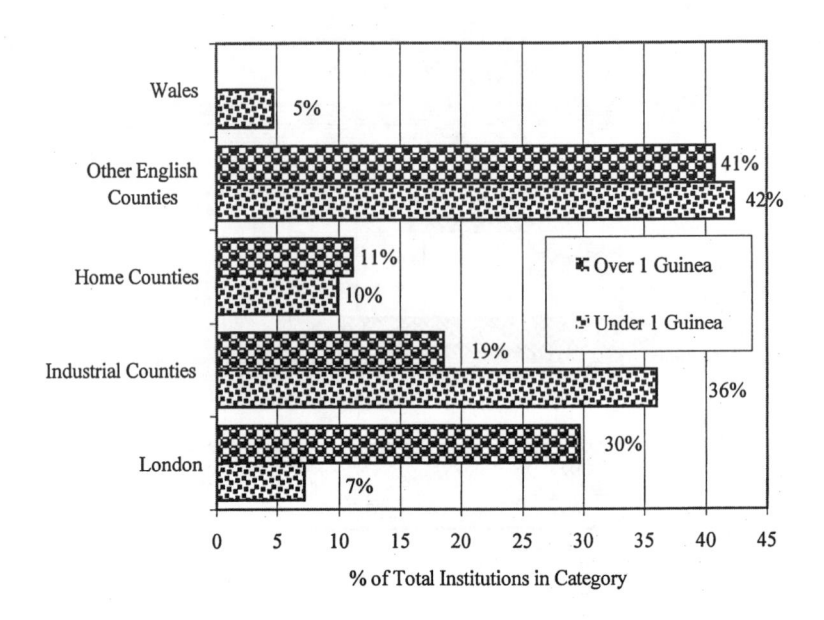

Chart 11.4: Female Membership and Books per Member by Size of Steam Intellect Institution, 1851

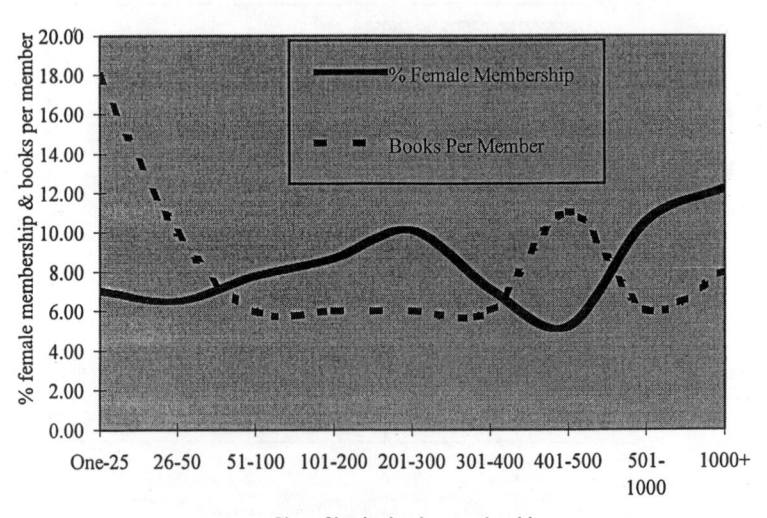

Size of institution by membership: One-25, 26-50, 51-100, 101-200, 201-300, 301-400, 401-500, 501-1000, 1000+

Legend: % Female Membership; Books Per Member

Chapter 12

'Golden Age' and 'Better Days': Narratives of Industrialism in the Cotton Trade of North-East Lancashire, 1860s to 1920s

Jeff Hill

The image of a 'Golden Age' figured in the culture of the Lancashire cotton industry for over a century. It was usually something looked back on. 'The history of the weavers in the nineteenth century', E.P. Thompson once famously remarked, 'is haunted by the legend of better days'.[1] In his dialogue with the 'optimists', who insisted with an array of statistical evidence that industrialization had improved things, Thompson reminded us why nineteenth-century weavers idealized a past before proletarianization. In the transition to steam looms and factories, in the change of status from 'artisan to depressed outworker', something, it was commonly felt, had been lost. The memory persisted of a time when things were better, not only in material security but in the way life was organized. The sense of loss became an important myth sustaining the world-view of weavers in the mid-nineteenth century, helping them to fashion a critique of the industrial system that had now arrived. Nor did this folk memory necessarily diminish with the great leap forward in cotton production. Handloom weavers were still numerous. Geoffrey Timmins has remarked that 'in Lancashire, as in other British textile districts, the handloom weaver remained a familiar sight for a surprisingly lengthy period of time'.[2] When, eventually, mechanized production took over, the ghost of the past continued to weigh on the brain of the living. There continued in the culture of the factory a pervasive sense of present uncertainties, of bad times being either with us or just around the corner. There was little to persuade Lancashire people inthe 1850s and 1860s that a new Golden Age was being experienced.

Quite the contrary: this period in the development of the industry in Lancashire was characterized chiefly by the problems of the 'cotton famine',

---

[1] E.P. Thompson, *The Making of the English Working Class* (Penguin Books, Harmondsworth, 1968), p. 297.
[2] Geoffrey Timmins, *The Last Shift: the Decline of Handloom Weaving in Nineteenth-Century Lancashire* (Manchester University Press, Manchester, 1993), p. 187.

and was remembered as a time of deprivation. It formed the context for what became one of the most celebrated of all Lancashire dialect poems, Samuel Laycock's 'Welcome, Bonny Brid', in which a new-born babe is welcomed into a family whose size is already beyond its means:

> Tha'rt welcome, little bonny brid.
> But shouldn't ha' come just when tha did;
> Toimes are bad.
> We're short o' pobbies for eawr Joe,
> But that, of course, tha didn't know,
> Did ta, lad? [3]

Through such maudlin cultural products the legend of the cotton famine was perpetuated. It haunted the weavers of these times as that of better days haunted the previous generation. At the end of the 1860s Ralph Holden, 'Anybody's Tailor' of Burnley, also went into verse, in his case to advertise himself and his wares. He took as his text what by then had become a popular refrain:

> The cotton trade is bad
> And long it has been so;
> And when it will be good again I'm sure I do not know. [4]

Ralph's only consolation to his readers was that he could offer them better and cheaper clothes than his competitors. (He offered no apology for his verse.)

At the same time and in the same town David Holmes, later President of the Amalgamated Weavers' Association, one of the country's largest trade unions at the start of the twentieth century, was making his way in the world and beginning to fashion a narrative of personal struggle and achievement that would be told and retold on countless occasions by Holmes himself and his admirers. Whether he availed himself of Ralph Holden's bargains we do not know, but he had little cause to look back with affection on this time. His trade as a handloom weaver had finally disappeared, after threatening to for so long, and he had made the change to powerlooms. These were years of self-denial and hardship for Holmes, who was victimized by employers for attempting to form a weavers' trade union. He succeeded – the Burnley Weavers' Association was established in 1872, though initially had only a tenuous existence – and gradually made for himself a considerable reputation in town and region as a trade union leader and Liberal politician. But the legend of bad times stayed with him all his life. It affected his industrial and his political stance and confirmed his belief in mutual association allied to liberalism as the

---

[3] Quoted in Patrick Joyce, *Visions of the People: Industrial England and the Question of Class 1848–1914* (Cambridge University Press, Cambridge, 1994 edn), pp. 386–7.
[4] *Colne and Nelson Guardian*, 9 January 1869.

basis of material and moral life. At the same time, it underscored his violent opposition to socialism. In 1894, after attending an international conference of textile workers in Manchester, he delivered a withering attack on the 'socialistic' methods of Continental workers which, he felt, had 'left those workers today in a pitiable condition compared with those of this country'.[5] There were no easy solutions to the job of providing for a worker's material security. The Golden Age had taught Holmes a bitter lesson.

Holmes had come in the 1840s as a boy from Manchester to Burnley. It was a new frontier. His whole working life was lived in a period of immense growth in the cotton trade in north-east Lancashire, made possible largely because thousands like Holmes forsook the old ways and methods and settled for life in the mill. The cotton industry, which some have seen as at the very epicentre of Britain's industrialization, was well established by the third quarter of the nineteenth century. Like other sectors of the British economy it was beginning to show signs of uncompetitiveness, especially in relation to producers in Europe and the New England states of the USA. But it was still capable of generating new initiatives. Notable among these was the late flourishing of industrialization in north-east Lancashire, which saw the expansion into factory-based weaving towns of the old medieval boroughs of Burnley and Colne, and the creation from mere hamlets of an urban network centred upon the new town of Nelson. By the end of the century these towns formed the hub of a manufacturing economy of global proportions. After Blackburn and Burnley, Nelson was Britain's most productive weaving town, where a far greater proportion of economic activity was given over to cotton weaving than in the other two places.[6]

When, in 1912, six years after Holmes had died in Burnley, the manufacturer James Nelson passed away in nearby Nelson, it seemed to the local newspaper that his life epitomized the changes that had occurred in the district. Born in Colne in the early 1830s, 'Jimmy' Nelson had only set up as an independent producer in the 1880s, with money saved from his previous employment as a manager of a local mill. In other words, his years of greatest activity came after he had 'retired'. Initially he rented space in a mill under the local 'room and power' arrangement, which involved a division of the means of production between the owners of premises and power, and the owners of looms and labour. Nelson joined the ranks of the latter in a small way, as many others were doing from the 1860s onwards. He successfully introduced the

---

[5] See W. Bennett, *The History of Burnley* (4 vols) (Burnley Corporation, Burnley, 1951), ch. 7 also *Cotton Factory Times*, 4 January 1906; *Burnley Gazette*, 1 August 1894. See also Jeffrey Hill, 'Lib-Labism, Socialism and Labour in Burnley, c. 1890-1918', *Northern History*, 35 (1999), pp. 187–204.

[6] Jeffrey Hill, *Nelson: Politics, Economy, Community* (Keele University Press, Edinburgh, 1997), chs 1 and 2.

profitable trade of sateen weaving to the area and never looked back. By the mid-1890s he had his own mill, and by the time of his death his firm employed some 2,000 workers and possessed over 3,000 looms and a doubling plant. By now he had combined the devolved functions of the 'room and power' system into one operation under his own control. His firm was the largest employer of labour in the town. 'His life', said the *Nelson Leader*, 'affords an interesting study to those who claim that a man's own dogged determination has largely to do with his own destiny'.[7] Looking back on the progress of Nelson, town and man, from this vantage point – and fully expecting that the progress would continue –it seemed to many that a unique phase in history had been witnessed: when 'small' men had struggled to take advantage of opportunities, had built a new economy, and in the course of so doing had also constructed a new society by their own efforts. It had been a phase of economic, social and moral improvement which might have recreated that lost Golden Age of the previous century. Looking back on this period from the 1920s Ben Bowker saw it as one of seemingly inexorable progress, a time of success that might even have anticipated the coming Nemesis.[8]

A cluster of villages in the remote north-east corner of Lancashire became the site for a remarkable economic and urban transformation. Great and Little Marsden, Barrowford, and a string of hamlets in the shadow of Pendle Hill, had a tradition of domestic textile production allied to pastoral agriculture going back to the sixteenth-century. It was a marginal economy, well off the beaten track of British industrial advance. These settlements were that type of upland village, referred to by Thompson,[9] where folk were clannish, suspicious of townspeople, and where the lure of the new mills in the valley bottoms of Colne and Burnley was resisted until the mid-nineteenth century. From about this point in time the villages, together with others farther afield across the county border with Yorkshire, began slowly to disgorge their population into the newly-developing centre of Nelson, a town formed around the villages of Great and Little Marsden and taking its name from the inn close by the railway halt installed there by the East Lancashire Railway in 1849. The census of 1851 noted the decrease in population of some of the nearby villages, 'in consequence of the removal of families to other parts to seek employment'.[10]

What prompted this removal was the establishment in Nelson of cotton factories. The process came about in what might be regarded as a classic Victorian *laissez-faire* manner: several individual business initiatives – in finance, property, steam power, transport and production – converged to

---

[7] *Nelson Leader*, 4 October 1912.
[8] Ben Bowker, *Lancashire Under the Hammer* (L. and V. Woolf, London 1928), p. 19.
[9] E.P. Thompson, *Making*, p. 340.
[10] *Census of Great Britain* (1851), division viii, the North-Western Counties (HMSO, London, 1852), p. 47.

produce a new town. A number of advantages were exploited. These may be summarized as: the adjacent expertise and capital available in nearby Colne and Burnley, whose industrialization had occurred earlier; the reserve army of labour to be tapped in the outlying villages, whose marginal agriculture was the only source of alternative employment; the presence of a transport system based on road, canal and, latterly, railway, which made possible the economical movement of goods; and, most important of all, the availability of cheap land for both industrial premises and housing. On such bases the town of Nelson grew rapidly, from a population of some 10,000 in the 1860s (when the name first came into common use) to 17,000 in 1881, and (following boundary changes to incorporate adjoining districts) over 30,000 in 1891.[11]

For both entrepreneurs and workers weaving was, as Farnie has shown, a most accessible trade. Its growth in this area was facilitated by the system that eased James Nelson's entry into the trade with 160 looms at the Brook Street Mill – 'room and power'. It was devised in the 1860s when entrepreneurs in the foundry and building trades speculatively bought land on which they built mill premises, hoping that cotton producers would rent the spaces available for the installation of looms. Ten such mill companies were registered in Nelson in the twenty years after 1875. All were owned by local businessmen, often with the help of mortgages from local building societies. The provision of room and power in this form enabled relatively small producers to start up. Many producers operated on such small-scale bases. A large manufacturing concern would have over 1,000 looms. Though this could scarcely be rated a big undertaking by comparison either with other Lancashire weaving centres like Blackburn and Burnley, or with other sectors of British industry, it was big by local standards where twenty looms could be enough to run a successful business. For workers also weaving provided opportunities. It was, as H.A. Turner has said, an 'open' trade.[12] Although Nelson subsequently developed specialisms in weaving which enabled the town's producers to sell high-quality goods in dear markets, there was little basic training required to enter weaving. Apprenticeships and other methods of controlling the supply of labour, such as were applied through trade union action in cotton spinning, were absent in weaving. There was little evidence of an aristocracy of labour. Moreover, in contrast to other towns where it was a major employer of labour, weaving in Nelson was never subjected to a sexual division of labour which defined it as 'women's work'. To be sure, a large proportion of the workforce was female, but males were equally represented. The absence of any serious alternative

---

[11] See Hill, *Nelson*, ch. 2; D.A. Farnie, *The English Cotton Industry and the World Market 1815–96* (Clarendon Press, Oxford, 1979).

[12] H.A. Turner, *Trade Union Growth, Structure and Policy: A Comparative Study of the Cotton Unions* (George Allen and Unwin, London, 1962), 3, ch. 2.

forms of labour meant that a gender equality existed in the local economy, and this had repercussions in the social life of the town.[13]

Those involved in these developments of the two decades from the 1860s were immigrants of quite local origin. The vast majority had moved only a few miles to their new home. Like migrants everywhere their initial response to their new location was to re-create their old territory, either in distinctive institutions such as chapels, or in neighbourhoods where families hailing from the same villages settled. Like all immigrants, they were ambivalent about their new surroundings, which represented both promise and threat: the chance of a new and better life, perhaps, but also disruption and a more regimented work routine. Gradually, however, some form of common identity through the new town was created. It was based on a realization that what had happened to them was novel, perhaps unique, but it was still hedged around by other tensions – of religion, occupation and social class. Nor did this civic identity arise naturally. It was fostered through a discourse which arose from the initiatives of specific power groups, and was negotiated in relationships of inequality, much like the notion generally of a Golden Age. Educated and better-off groups of people – economic and social leaders – were important in the shaping of a local patriotism based on a sense of town.

By the 1880s a feeling was emerging among such people that in looking back over the previous twenty years, a special time in their lives had been experienced. This sense of the past had much to do with an awareness of local history. Local histories, municipal yearbooks, commemorative guides all contributed to the historicizing of the recent past. One of the most evident of such forms was the publication of accounts and reminiscences in the local newspapers. At the beginning of the 1880s, for example, the *Preston Guardian* printed a lengthy and anonymous article which surveyed the rise of Nelson and its adjoining settlement, Brierfield. It was a factually detailed narrative which contained little explicit interpretation. Beneath the surface of the story, however, was an emphasis which told much about the impact that economic change had wrought on local people. Particular features stood out. Firstly, the pace of change, which was outlined in the account's opening statement: 'Nelson is undoubtedly a modern town. Fifty years ago it was entirely unknown, and no mention of it appears in any book dealing with the ancient history of the county'. Secondly, the worthies whose energies had provided the motive force of change. And thirdly, the great works – especially in the provision of gas and water – which had made possible the basis of civic life. It was a story of pioneering endeavour by people committed to the building of a new society.[14]

---

[13] Hill, *Nelson*, chs 2 and 4.
[14] *Preston Guardian*, 12 February 1881.

To a large extent it was the story of the creation of an identity. What was felt to characterize Nelson was a set of values best expressed in the term frequently used in the 1850s and 1860s to describe the struggle for civic progress: 'improvement'. It applied to both material and moral development. It included the building of waterworks and gasworks to enable industry to flourish and its labour to be sustained, but it also involved the elimination of vice and the getting of education. Mechanics' Institutes, which both Marsden and Burnley possessed in the 1850s, epitomized the ideal:

> The advantages of such Institutions no one can deny, nor are there many who need to be told of the beneficial effects which they confer upon the middle and lower classes of society. A large number of boys and young men [sic] cannot congregate together for purposes of mutual improvement without exercising an influence for good, both upon themselves and the circles by which they are surrounded; and hence any assistance which may be rendered towards increasing the facilities for such intercourse becomes both a public and a private good. It enables the masses to improve themselves both mentally and physically and the healthy reaction so produced ultimately benefits the whole community.[15]

Such a spirit extended to the world outside. There was support for a moral foreign policy by Britain, which was why so much of the local press was concerned with Tsarism and the Crimean War in the early 1850s. Domestically, it expressed itself in the plethora of social activities recorded in the advertisements of local newspapers: for drapery, proprietary medicines, carpets, gardening requisites, tea, insurance, false teeth, commercial training, literary and philosophical societies, band concerts, excursions to the seaside, savings banks, building societies, choral societies (seemingly forever performing the 'Messiah'), and many others. There is no clearer memorial of the daily life of new communities than the proliferation of these social activities and institutions. It was out of the everyday energy vested in all this activity that the idea of a Golden Age was created.

In July 1890, after a short period of lobbying, Nelson received its Charter of Incorporation. On a sunny Saturday at the end of August the achievement was celebrated with a display of civic pageantry that drew some 80,000 people into the town. Nelson was bedecked with the bunting, mottoes and symbols of 'history': in the very centre of the town, for example, at the junction of the main roads, a castellated arch posing as an ancient gateway had been erected. The climax of the day was a 'monstre' (sic) procession through the town by groups representing all the main societies, trades and institutions of the district, including some 8,000 children from the local Sunday schools dressed in what the press reporter described as 'historic garb'. It followed a carefully planned

---

[15] *Burnley Advertiser*, 25 August 1855.

route which encompassed the main parts of the town and thus described the physical extent of Nelson. When the procession concluded, attention shifted to the Town Hall, where crowds gathered to hear the reading of the charter, and to listen to speeches from some of the notables who had been in the forefront of local government and who had orchestrated the campaign for borough status.[16]

Events of this kind were not uncommon in towns which had received Incorporation.[17] In the new town of Nelson, however, they assumed an even greater significance than they would have elsewhere. To be sure, they marked the passage from one form of local government to a new one which signified a higher civic status. But the celebrations served also to register the existence of a community with an identity and a social hierarchy. The procession played an important part in this, acting as a symbolic representation of the people of Nelson, their component parts – the trades, the schools and institutions – fused together into a united and planned whole. This ritual allowed individuals to identify themselves as part of a community, as part of 'their' town. The speechmaking ceremonies, however, which foregrounded the leaders held to have been responsible for the progress the town had enjoyed up to that point, fulfilled a slightly different function. Here, there was a clear demarcation of elite and mass, as a social leadership was accorded its place of eminence. What emerged from the speeches themselves was an emphasis on the dramatic changes that had taken place over the previous twenty-five years, and the sense that this pattern of growth would continue into the future. Thus, John Wilkinson, the chairman of the Incorporation Committee, noted:

> most of those present knew that within a recent period Nelson was comparatively a small place. The town had grown very rapidly indeed. It had no parallel either in Lancashire or Yorkshire in their time ... They would thus see that in the 25 years the number of looms had multiplied seven times and the rateable value had gone up fourteen times, the population had increased seven times ... and he did not see why they should not progress as well in the future as they had done in the past.[18]

The local newspaper, the *Nelson Chronicle*, expressed the same sentiments, but set in a yet more ambitious historical sweep: 'the incorporation of the town is not so much an end in itself as a means to an end ... incorporation is no final goal, not even a half-way house, but a post-station on the way to social, moral and material progress'.[19] In these ways Nelson, through a particular discourse of history and change, was being invented.

---

[16] *Nelson Chronicle*, 5 September 1890.
[17] On town patriotism see Marie Dickie, 'Town Patriotism in Northampton, 1918–1939: An Invented Tradition? *Midland History* (1992), pp. 109–17.
[18] *Nelson Chronicle*, 5 September 1890.
[19] 5 September 1890.

The proclamation of its legal and cultural existence was accompanied by its being invested with a set of spiritual values, as a place of thrusting energy and achievement. In short, Nelson was synonymous with 'progress'. It was a profoundly optimistic celebration of a Victorian ideal which it was envisaged would flourish in the future. There was an element of utopianism about it. It presented an image of the capitalist order that differed fundamentally from that envisioned by Dr Marx, and which some members of the audience who had never heard of the German scholar might also have had difficulty in accepting. But the idea was compelling. This invention of Nelson carried a sense of Golden Age which was immediate, specific in place and time. It related to the particular conditions of this part of the country which had produced a late industrialization; when the rest of Britain looked back to a Golden Age in the mid-century, the inhabitants of north-east Lancashire could imagine that conditions of growth and prosperity were still flourishing in their part of the world at the end of the century. When such optimism was shattered by the economic problems of the post-war world, people were even more inclined to look back to the prosperity of these years as a Golden Age. But reflections of this kind contained more than a touch of pathos. It had been planned to re-create the celebrations of Charter Day in Nelson on its fiftieth anniversary, but this could not happen in 1940 because hostilities had intervened. The event was postponed and eventually staged over an entire week in August 1946, in a manner that outshone the original.[20] It appeared a fine augury for the resumption of progress. But within ten years the local cotton trade, which had withstood the worst depredations of the inter-war years, was in the doldrums. By 1962 Nelsonians were processing again, not this time to celebrate their past, but through the streets of London to petition government to keep their industry alive. Weavers were once more haunted by the legend of better days.

---

[20] Nelson Corporation, *Jubilee of the Incorporation of the Borough, 1890–1940* (Coulton and Co., Nelson, 1946).

Chapter 13

Popular Culture and the 'Golden Age': The Church of England and Hiring Fairs in the East Riding of Yorkshire c. 1850–1875

Gary Moses

During the mid-Victorian period, the East Riding of Yorkshire witnessed a vigorous campaign for the reform and abolition of hiring fairs. Not surprisingly, the most persistent protagonists in this campaign were drawn from organized religion, with both Nonconformist and established churches playing a role. However, whilst the ecumenical nature of the movement deserves to be noted, it is also clear that the most organized and expansive aspects of the campaign was promoted by the Church of England. Churchmen were the most prolific letter-writers and tract publishers, they attacked hiring fairs at church and social science congresses, and it was they who, along with their wives, daughters and other laity, organized and publicized the major initiatives taken against East Riding hiring fairs. The most active clergy were those with hiring fairs in or close to their parishes, but the issue became one of general concern to the Anglican Church in the East Riding during the mid-Victorian years.[1]

The ultimate aim of the campaign, for its most zealous advocates at least, was to secure the end of hiring fairs, annual contracts and eventually, farm service itself. The actual means of achieving this goal varied over time and place but, although abolitionist feeling never completely disappeared, most opponents of hiring fairs recognized that their economic function as labour markets and their popularity as rural festivals meant that immediate, outright suppression was unlikely to succeed. As a consequence, the main focus of the mid-Victorian campaign by the Church of England concentrated upon developing an approach that was designed to secure abolition through stealth.

---

[1] On this campaign see also M.G. Adams, 'Agricultural Change in The East Riding of Yorkshire 1850–1880', unpublished PhD thesis, University of Hull 1970, pp. 405–8; S Caunce, *Amongst Farm Horses: The Horselads of East Yorkshire* (Alan Sutton, Stroud, 1991), esp. pp. 54–65, 175–180; G. Moses, '"Rustic and Rude": Hiring Fairs and Their Critics in the East Riding of Yorkshire c. 1850–75' *Rural History*, 7 (1996), pp. 151–75.

What emerged was a strategy of eroding and undermining the functions of hiring fairs by providing a substitute system.[2]

This substitute system had two main aspects. The first centred upon introducing a 'moral nexus' as the basis of relations between masters and servants. Consequently, farmers were urged to agree to hire only those servants who possessed a written testimony of their moral character and conduct over the past year. It was envisaged that these written characters would, by reintroducing moral criteria at the point of engagement, become the basis of a 'moral compact' between masters and servants in the East Riding. In order to promote this new system, and guarantee its moral rigour, the written characters were to be collated and administered through a network of registration societies formed and managed by permanent organizations in the form of committees of gentlemen.[3] The task of each registration society was to gather lists of vacancies and servants seeking employment and, through the establishment of offices in the towns and villages of the Riding, to encourage both employers and servants to make enquiries and reach agreements away from the hiring fairs. The ultimate scope of these proposals was, therefore, somewhat wider than the initial aim of reasserting moral criteria as the basis of the relationship between master and servant. By shifting the act of hiring away from the hiring fair and encouraging the adoption of a contract that was to be more flexible in terms of its creation and duration, it was hoped that the current practice of hiring servants from Martinmas to Martinmas (23 November) would fall into decay. It was also envisaged that as a consequence of this, the custom of all farm servants being released simultaneously from their contracts on 23 November, for a week's annual holiday, would also fall into decay as the farm servants' holiday moved with the point of engagement to any time of the year. The result of both measures, it was anticipated, would mean that over time the number of servants attending hiring fairs to be hired would decline and the festive and carnivalesque excess that was especially characteristic of those hiring fairs located in Martinmas Week would be purged. Stripped of their legitimate economic functions and denied their annual influx of rural youth, the

---

[2] Examples of the criticisms directed at hiring fairs by Anglican opinion include Rev. Nash Stephenson, *On The Rise and Progress of the Movement for the Abolition of Statutes, Mops, Or Feeing Markets: A Paper Read Before the National Association for the Promotion of Social Science at Glasgow, 1860* (London, 1861); Rev. Greville J. Chester, *Statute Fairs: Their Evils and Their Remedy* (York, 1856); Rev. James Skinner, *Facts and Opinions Concerning Statute Hirings, Respectfully Addressed to the Landowners, Clergy, Farmers and Tradesmen of the East Riding of Yorkshire* (London, 1861); Rev. F.O. Morris, *The Present System of Hiring Farm Servants in the East Riding of Yorkshire with Suggestions for its Improvement* (London, 1854); Rev. J. Eddowes, *Martinmas Musings: Or Thoughts about the Hiring Day* (Driffield, 1854).

[3] *Yorkshire Gazette* letters: 28 January 1854, Rev. J. Eddowes, 'The Abolition of Hiring Fairs'; Rev. F. Simpson, 'Hiring of Farm Servants'; 4 February 1854, Rev. F.O. Morris, 'Agricultural Servants'.

hiring fairs would, it was hoped, effectively wither away and abolish themselves. For their more zealous and earnest advocates then, the establishment of a system of written characters was not only important in its own right but was also a stepping stone towards a more fundamental change which involved the abolition of annual contracts, hiring fairs and even the institution of farm service itself.

This first stage of reform predominated during the 1850s. A second phase developed from the early 1860s. This aspect of the substitute system was related to and overlapped with the first. It revolved around direct measures designed to compete with, and erode, outdoor hiring and the carnivalesque transgressions that were regarded as integral to the atmosphere of a hiring fair. This second stage of reform was in some respects a response to the limited success of the first. Registration societies had been formed but although they enjoyed some support from employers and a minority of female farm servants they had failed to offer a credible alternative to the traditional verbal contract made at the hiring fair. Hiring fairs had continued to be viable as labour marts and popular as rural festivals and probably attracted more people at the end of the 1850s than they had at the beginning. Consequently, a number of local Anglican clergy decided to instigate a more direct means of challenging the customs and practices of the hiring fairs. In an attempt to draw servants away from the traditional location for hiring – the streets, market places and public houses of market towns – they offered alternative indoor venues for the hiring of servants in large rooms nearby. These rooms also offered registration facilities, warm fires and cheap refreshments and local clergy were also on hand to supervise the proceedings and offer religious guidance to those present. There was also an attempt to compete with the commercialized pleasures and entertainments offered by publicans and showmen at hiring fairs by offering alternative 'rational recreations' in the form, for example, of brass-band music and indoor concerts. These indoor hirings and their associated rational recreations performed therefore the multiple roles of enabling a more assertive religious presence to be established at the hiring fair itself; providing a new means of promoting the alternative system of hiring contracts; and offering alternative 'rational' recreational activities which competed with the carnivalesque excesses associated with the entertainment provided in public houses and at the large pleasure fairs that gathered at the hirings.

This second phase of reform was more successful than the first. Although hiring fairs continued to be popular, and although the system of written characters failed to establish itself, reformers did succeed in attracting female servants to these indoor hirings which resulted in segregated hiring for male and female servants. From the early 1860s gender segregation became the norm at most of the East Riding's larger hiring fairs. For some opponents, however, this gain was too limited. Hiring fairs remained popular and, though most females no longer stood in the open market to be hired, once hiring was finished they rejoined male servants in enjoying the usual festivities of the hiring day. This, and the passing of legislation facilitating easier abolition of fairs, prompted a third phase of agitation in the 1870s which saw calls for the adoption of a variety of measures designed to secure either more fundamental reform or the total demise of the hiring fair. Critics now called for separate hiring fairs for men and women and for parliamentary intervention to secure the legal suppression of hiring fairs in the East Riding. This final phase had no success, principally because local farmers, who had become increasingly lukewarm in their support for the earlier initiatives, now cold-shouldered these latest proposals, preferring instead to mount their own (equally unsuccessful) campaign for moderate secular reform of hirings. Farm servants had consistently resisted all reforms other than the indoor hirings for females. Such, briefly, is the general nature and course of the campaign waged against hiring fairs in the East Riding of Yorkshire during the mid-Victorian period. The remainder of this chapter will seek to examine the motives behind this attack upon a form of rural popular culture.[4]

Events of this nature had always had their critics, hostility towards hiring fairs existed in the East Riding prior to the mid-Victorian years, and in some respects this campaign can be regarded as a continuation of that earlier tradition.[5] However, there was a noticeable increase in the level of critical comment and reforming activity from the 1840s and especially the early 1850s in the East Riding. Why was this the case and why was the Church of England at the centre of this reforming zeal? A full explanation would require more space than is available here but two factors are of particular significance: the Church of England as an institution and the 'Golden Age' of English agriculture. It will be argued here that a contradictory relationship between the institutional aims of the Church and the nature of East Riding agriculture in its Golden Age was central in prompting this campaign against hiring fairs.

One might expect the Church of England's clergy to be concerned about an event like the hiring fair. Criticism of, and interventions against, popular culture were an integral aspect of the Church's role as an national institution responsible for the encouragement of morality and social order. This tradition was as old as the Church itself but enjoyed something of a revival as the influence of evangelicalism increased within the Church of England during the

---

[4] For a fuller evaluation of the impact of these attempts to reform and abolish hiring fairs during the mid-Victorian period see Moses, 'Rustic and Rude', pp. 164–9.

[5] On the opposition against hiring fairs in England including the East Riding, see R.W. Malcolmson, *Popular Recreations in English Society 1700–1850* (Cambridge University Press, Cambridge, 1973), pp. 103–4, 149–51. Joseph Dent, Esq. of Ribston Hall, writing in 1850, claimed to have campaigned against Yorkshire hiring fairs for over forty years, *Hull Advertiser*, 8 November 1850.

first half of the nineteenth century. Evangelicalism encouraged a more active and crusading approach towards events perceived as encouraging immorality. The influence of the Oxford Movement, with its emphasis upon the importance of the pastoral role of the parish priest, was also significant. Although they disagreed on many issues, both evangelicals and High Churchmen believed that increased energy, activism and intervention on the part of parish clergy should be an essential element of the ideal Victorian parish. They also emphasized the importance of confronting sinful behaviour and promoting the values of hierarchy, paternalism and deference as the basis of social relations between rich and poor.[6] A number of historians, including Brian Heeney, Hugh Cunningham, Alan Gilbert, James Obelkevich and Alun Howkins have emphasized that as a result of these ideals and their organizational correlation the parish system enjoyed a new salience for rural clergy during the mid-Victorian period which, in many respects, can be regarded as marking the onset of a more energetic and professional phase in the Church of England's relationship with rural communities.[7] The East Riding appears to conform to this pattern. Here also, the mid Victorian period was a time of renewed enthusiasm and clerical professionalism which resulted in a determined effort to revitalize parish life and revive the fortunes of the established Church. It is this attempted rejuvenation of Anglicanism that, in part, explains the growing antagonism towards hiring fairs in the York and East Riding area at this time. In particular, it will be argued here that this hostility towards hiring fairs on the part of the Church of England was rooted in a contradiction between aspects of East Riding agriculture in its Golden Age and the Anglican ideal of a revitalized parish based upon a stable, resident population united by ties of deference and a common faith. Before exploring this further however, it is necessary briefly to delineate the Golden Age of English agriculture and especially how it manifested itself in the East Riding.

The phrase 'the Golden Age of English agriculture' was first coined by the distinguished writer on and observer of English agriculture Roland Prothero (Lord Ernle). In his book *English Farming Past and Present* published in 1912, he defined the years 1852–63 as the 'Golden Age of English agriculture', largely because it was a time of sustained prosperity for landlords and farmers.[8]

---

[6] D. Roberts, *Paternalism in Early Victorian England* (Rutgers University Press, (New Jersey), 1979, pp. 61–3; A.D. Gilbert, 'The Land and the Church', in G.E. Mingay, (ed.), *The Victorian Countryside*, vol. 1 (Routledge and Kegan Paul, London, 1981).

[7] B. Heeney, 'On being a mid-Victorian Clergyman' *Journal of Religious History*, 7:3 (1973); H. Cunningham, *Leisure in the Industrial Revolution, c. 1780–c. 1880* (Croom Helm, London, 1980); Gilbert, 'Land and the Church'; J. Obelkevich, *Religion and Rural Society: South Lindsey 1825–1875* (Oxford University Press, Oxford, 1976); A. Howkins, *Reshaping Rural England 1850–1925: A Social History* (Harper Collins, London, 1991).

[8] R.E. Prothero (Lord Ernle), *English Farming Past and Present* (Heinemann, London, 1961, 6th edn).

Prothero suggested that this prosperity began to ebb away from 1864 and finally ended with the onset of the agricultural depression in 1874. Most historians however, use the expression more broadly than this, referring to a period lasting from the early 1850s to the mid-1870s. However for both Prothero and more recent historians the Golden Age is more than a mid-Victorian boom sandwiched between the depressions of the early and late nineteenth century, it is also the moment when a particular mode of production – 'high farming' – reached its apogee.[9] The characteristics of this system of agrarian capitalism were large farms which utilized capital-intensive forces of production, including high inputs of investment in farm buildings, land drainage, machinery, fertilizer, animal feed and labour. The result was a progressive capital-intensive system of large-scale agrarian capitalism which combined an extensive arable acreage with high supplementary feeding of livestock within an integrated system of capitalist production. The relations of production usually associated with this system were also firmly capitalist, in that it is generally associated with the maturation of a rural social structure characterised by a tripartite model which divides society into landlords (who owned the land), tenant farmers (who 'managed' the land) and landless labourers (who worked the land).[10]

From the late eighteenth and early nineteenth century the East Riding developed a particularly dynamic and prosperous form of high farming which enjoyed its Golden Age during the mid-Victorian period. In the mid eighteenth-century the East Riding had been something of a backwater in terms of agricultural progress. From the late eighteenth-century, however, the Riding experienced an 'agricultural revolution' which transformed both the character of its agriculture and the nature of its landscape. By the mid-nineteenth century the East Riding of Yorkshire was establishing itself as one of the more progressive and modern regions in English agriculture. Although something of a latecomer the East Riding's 'agricultural revolution' displayed similar characteristics to the capitalist modernization of agriculture elsewhere – enclosure, increased farm size, new methods of cultivation and advances in mechaniation. Indeed, partly because of this lateness – in terms of the size and layout of its farms, its utilization of modern technology, and the productivity of its labour force – the East Riding offered a prime example of the intensive mixture of arable and livestock farming that is associated with English

---

[9] J.D. Chambers and G.E. Mingay, *The Agricultural Revolution 1750–1880* (B.T. Batsford, London, 1966), pp. 170–78; F.M.L. Thompson, 'The Second Agricultural Revolution 1815–1880', *Economic History Review* 21 (1968), pp. 62–7; B.A. Holderness, 'The Origins of High Farming' in B.A. Holderness and M.A. Turner (eds), *Land Labour and Agriculture 1700–1929, Essays for Gordon Mingay* (Hambleton Press, Hambleton, 1991).

[10] A. Howkins, 'Peasants, Servants and Labourers: the marginal workforce in British agriculture, c. 1870–1914', *Agricultural History Review*, 42 (1994), p. 49.

agriculture during its Golden Age.[11] The progressive nature of agrarian capitalism in the East Riding is best exemplified by the Yorkshire Wolds. This was an area of late enclosure which, due to the application of progressive forms of crop rotation, land use, animal husbandry, natural and artificial manure, technology and feedstuffs, developed large capitalist farms with integrated systems of livestock and cereal production. Its farmers were commended by writers such as James Caird for their liberal and progressive attitudes, their willingness to embrace the use of modern technology, and their wealth.[12] Although the two other agricultural regions of the Riding – Holderness and the Vale of York – were less progressive in their development of high farming than the Wolds, agriculture in both areas was relatively advanced in comparison with the rest of England and high farming practices advanced in both areas after 1850.[13] The East Riding of Yorkshire conforms, therefore, to the established model of Golden Age farming. It developed an advanced form of capitalist agriculture which laid the foundation for two decades of relatively sustained prosperity after 1850 which lasted until the onset of the agricultural depression in the later 1870s.

There is, however, one aspect of high farming in the East Riding during the Golden Age that might be regarded as relatively however: its reliance upon the labour of male and female farm servants. The difference between these workers and the 'typical' agricultural labourer usually associated with high farming was that the farm servants had yearly contracts and 'lived in' on the farm, whereas the 'typical' agricultural labourer was hired by the week, day, or piece and lived off the farm. Farm service was a long-established means of recruiting and retaining farm and farm-domestic labour in England. In the late eighteenth century approximately half of the agricultural labour force in England and Wales was made up of farm servants.[14] The East Riding's uniqueness lay in the fact that in most, if not all, of the Midlands and south of England, the transition

---

[11] O.Wilkinson, *The Agricultural Revolution In The East Riding of Yorkshire* East Yorkshire Local History Series, 5, 1956; A Harris, *The Rural Landscape of the East Riding of Yorkshire 1700–1850* (Oxford University Press, Oxford, 1961); Adams 'Agricultural Change'; Caunce, *Farm Horses*.

[12] In 1851 Caird thought them to be amongst the wealthiest farmers in the country, J. Caird, *English Agriculture in 1850–51* (London, 1852), p. 310; See also G. Legard, 'Farming in the East Riding of Yorkshire' *Journal of the Royal Agricultural Society*, 9 (1848); W. Wright, 'On the improvements of the farming of Yorkshire Since the date of the last Reports in the Journal' *Journal of the Royal Agricultural Society*, 22 (1861). A useful comparative discussion which emphasizes the early development of high farming on the Yorkshire Wolds is provided by G.G.S. Bowie, 'Northern Wolds and Wessex Downlands: Contrasts in Sheep Husbandry and Farming Practice, 1770–1850' *Agricultural History Review*, 38 (1990) pp. 117–26: The advanced nature of East Riding agriculture is also emphasized by Caunce, 'Farm Horses' and Moses, 'Rustic and Rude'.

[13] Adams, 'Agricultural Change', p. 45.

[14] A. Kussmaul, *Servants in Husbandry in Early Modern England* (Cambridge University Press, Cambridge, 1981), pp. 4; 18.

to large-scale agriculture of the type increasingly evident in the East Riding had coincided with the abandonment of farm service.[15] There, a labour surplus had made the use of day-labourers hired off the farm on short-term contracts both economically and socially more attractive to farmers. Economically, it was less costly to hire day-labourers when required than to employ and keep farm servants for the whole year. Socially, the wealthy and increasingly status-conscious capitalist farmers found the practice of sharing their households with servants increasingly distasteful. In contrast, a number of factors ensured that farm service remained integral to East Riding agriculture throughout its Golden Age. Two were particularly significant. Firstly, there was the settlement pattern of much of the East Riding after enclosure. Enclosure had promoted, in many areas, a settlement pattern in which large isolated farmsteads away from village settlements were common.[16] These farms, with their increased arable acreage, required a larger permanent workforce than the smaller pastoral farms that had prevailed prior to enclosure. This dispersed and isolated settlement pattern and East Riding landlords' reluctance to build cottages meant that these labour requirements could not be met entirely from the local resident population. Secondly, the close proximity of alternative industrial employment in the West Riding and Cleveland, and the emigration it encouraged, meant that farmers could experience labour supply problems, particularly at peak times of demand.[17] It also meant that day-labour was relatively expensive in the East Riding. Taken together, these factors reinforced the value of having a large proportion of the farm workforce resident on the farm, paid partly in kind, and tied for the whole year by the annual contracts created at hiring fairs. As a consequence, many East Riding farmers continued to regard farm service as the most reliable and practical means of procuring and retaining the labour they required for the whole year. Farm service was most prevalent in those areas of the East Riding characterized by large isolated farmsteads located away from established centres of population. For example, on the Yorkshire Wolds, farm servants were the largest single category of the male agricultural work force in 1851.[18] Although this pattern is most apparent on the Wolds where high farming and isolated farmsteads were particularly common, a similar correlation between large isolated farmsteads and large numbers of resident farm servants can also be found in both the Vale of York and Holderness.[19]

---

[15] Ibid., p. 130; K.D.M. Snell, *Annals of the Labouring Poor. Social change and agrarian England, 1660–1900* (Cambridge University Press, Cambridge, 1987), ch. 2, p. 2.

[16] M.B. Gleave, 'Dispersed and Nucleated Settlement on the Yorkshire Wolds, 1770–1850', *Institute of British Geographers, Transactions and Papers*, 30 (1962), pp. 105–18.

[17] Adams, 'Agricultural Change', pp. 344–7.

[18] J.A. Sheppard, 'East Yorkshire's Agricultural Labour Force in the Mid nineteenth-century' *Agricultural History Review*, 9 (1961), pp. 43–54.

[19] Ibid.

That it was the most modern form of agricultural production, high farming, that proved to be most deeply attached to farm service was significant because it meant that in the East Riding the Golden Age of agriculture was also a Golden Age of farm service. However, East Riding farmers did not merely retain what has been termed 'classic farm service' whereby young males and females lived in the house of their employers. Here, farm service was adapted and reshaped in accordance with the rising social status of the Riding's wealthy tenant farmers. With enclosure and the expansion of the arable acreage many new farmhouses were built whilst others were modified and enlarged. The prevailing trend in both new and enlarged farms was for the farm servant accommodation to be located in a separate and distinct part of the farmhouse away from that occupied by farmers and their families. This often involved almost total segregation, with all farm servants housed in separate dormitory units located in adjoining outbuildings with only limited, discrete and controlled access to the main body of the farmhouse. In some cases both innovations were combined, with the female servants occupying a wing of the farmhouse with its own access to the wash-house and kitchen, and the men occupying a separate dormitory outbuilding or 'men's end' attached to the farmhouse but with its own external doorway and yard. A further significant change was the development of the 'hind house' system, whereby the role and functions of a single foreman who lived with and oversaw the younger servants were transferred to a hind – a married man who, with his wife, female servants and/or daughters housed and fed the servants in a separate 'hind house' away from the main farmhouse.[20] Each of these changes had precedents prior to the mid-Victorian period. But whereas they had once been exceptional, by the 1880s they combined to form a near universal pattern of social segregation between capital and labour on the larger tenant farms of the East Riding. What this suggests is that the symbiotic relationship between farm service and the development of high farming did not represent the continuity of a pre-capitalist labour system alongside a capitalist mode of production. Instead, the expansion of large capitalist farms promoted changes in farm service, changes which meant that the relationship between masters and servants lost most, if not all, of

---

[20] For contemporary details of farm servants' accommodation in the East Riding see: BPP, 1868 *First Report From the Commissioners on The Employment of Children, Young Persons and Women in Agriculture* Parl. Papers (1868), p. 368; W.H.M. Jenkins, 'Show Farm "Eastburn" – Driffield', *Journal of the Royal Agricultural Society*, 5:5 (1869); M.C.F. Morris, *Yorkshire Folk-Talk, with Characteristics of Those who Speak it in the North and East Ridings* 2nd edn (A. Brown, London, 1911); describes the hind system as largely prevailing in the East Riding by the late Victorian period, especially on the Wolds. For a recent analysis of these changes which emphasizes the social distancing between farmers and servants during the nineteenth century see C. Hayfield, 'Farm Servants Accommodation on the Yorkshire Wolds', *Folk Life*, 33 (1994), pp. 7–28.

the close familial associations once regarded as characteristic of farm service in its classic form.[21]

This continuation of farm service in the East Riding had important ramifications for rural popular culture. As custom determined that the majority of farm servants' contracts were created at hiring fairs, the health of the hiring fair as a popular event was intertwined with the well-being of the institution of farm service as a labour system. As long as East Riding farmers sought farm servants, hiring fairs were likely to remain important both as labour markets and as popular festivals. In short, a possible explanation for the increased hostility towards the hiring fair during the mid-nineteenth century is its continuation and even its revitalization at this time. Contemporary attacks upon the fairs were certainly informed by a mixture of outrage and incomprehension that an event of this nature not only survived, but appeared to be enjoying something of renaissance at a time when nationally, society was thought to be experiencing unprecedented material and moral progress. This was the first level of the contradiction between the ideals of the Church of England and the pattern of agricultural development in the East Riding. By perpetuating and even expanding farm service and thereby underpinning the hiring fair, agricultural improvement raised a barrier against the Church's agenda of moral and spiritual improvement. However, if one examines the Church of England's campaign in more detail, one can see that this contradiction between the ideals of the Church and the nature of East Riding agriculture during its Golden Age is in fact deeper than a mere hostility towards a boisterous event. Equally important was the manner in which the pattern of agricultural modernization had encouraged a reshaping of farm service. This, when combined with the dispersed and isolated settlement pattern of the East Riding (itself also largely a by-product of agricultural change), posed particular problems for the Church of England and its aim of reasserting itself as a social institution. In order to illustrate this, the Church opposition to hiring fairs will be examined in a little more detail.

Church opposition to hiring fairs can be analysed in terms of two analytical categories: 'internal' and 'external'.[22] By the 'internal' critique is meant those criticisms directed at the fairs' actual content. Here the evangelicals' concern with the dangers of 'the world' were most salient. From this perspective the hiring fairs offered a concentrated example of the temptations that so often

---

[21] A fuller discussion of these and other changes and their implications for the analysis of nineteenth-century farm service can be found in G. Moses, 'Proletarian Labourers? East Riding farm servants c.1850–75' *Agricultural History Revue*, 47 (1999) pp. 78–94. On this and related themes see also Howkins 'Peasants' S. Caunce, 'Farm Servants and the Development of English Capitalism', *Agricultural History Review*, 45 (1997).

[22] Moses, 'Rustic and Rude', p. 159. The following analysis of the roots of the Anglican critique of hiring fairs is based upon the arguments offered in this article, esp. pp. 158–64.

precipitated a life of sin and eternal damnation. In this respect the Church critics focused upon the carnivalesque turbulence of the hiring fair, its reputation for crime, violence, disorder, drunkenness and sexual promiscuity. In the words of a memorial issued by the York Diocesan Board of Education in 1856, it was 'an assembly from which moral control is simply excluded'.[23]

This 'internal' critique also focused upon the hiring practices of farmers, particularly the criteria used by farmers in selecting their servants. Farmers were consistently attacked for neglecting moral character when choosing and negotiating with servants. Reverend Greville Chester summed up the views of many of his colleagues when he attacked the employers' propensity to 'hire their servants, of both sexes, solely on the recommendations of brute strength', adding 'that to ask a character is so exceptional an event, as to be almost unknown'.[24] Like many other Anglican critics of hiring fairs, however, Chester also regarded the hiring fair as a structural problem due to its strategic location in the life cycle of the young farm servants. In a telling piece of moral sociology, Chester linked the internal dynamics of the hiring fair to a more general parochial difficulty – the mobile, independent and allegedly immoral and unstable lifestyle of the farm servant. He argued that the hiring fair corrupted farm servants and that the farm servant system then ensured that those who had been corrupted propagated yet more immorality throughout the farmhouses and villages of the East Riding: 'the system under consideration is injurious, because, by its means, bad and profligate persons are disseminated over the country far and wide, to work as much evil as a bad example can work in a year's time in one parish then move onto pursue the same work of corruption in another'.[25] In part, therefore, Chester's critique of the hiring fair was 'external' in that he regarded it as an event that should be viewed in systemic terms, that is within the context of its relationship with the two other pillars of farm service – annual contracts and living in.

This 'external' critique of the hiring fair regarded the problematic behaviour at hiring fairs as less an exceptional transgression of everyday norms and more a reflection of the farm servants' normal behaviour; behaviour that reflected a moral crisis, precipitated in part by changes in the nature of farm service. Two aspects of the farm servants' life away from the hiring fair created particular anxiety. First, there were concerns regarding changes in farm servants' living-in arrangements. Second and relatedly, there was the manner in which annual hiring encouraged most farm servants to change employers each year. In general, the clerical view of farm service was that it had lapsed from a prior

---

[23] Memorial of the York Diocesan Board of Education, 'Hiring of Farm Servants', appendix to Chester, *Statute Fairs*; this memorial was published in the local press, see for example, *Yorkshire Gazette*, October 1856.
[24] Chester, *Statute Fairs*, p. 7.
[25] Ibid., p. 15.

Golden Age when it had acted as a force for good in rural society. In this pre-lapsarian Golden Age farmers had acted *in loco parentis*; they had exerted discipline over the farm servants' everyday lives and therefore cared for their moral and spiritual welfare. Perhaps most importantly, they had encouraged and in some cases insisted upon regular attendance at the parish church. In acting in this manner, farmers had maintained the chain of connection between rich and poor that was so central to the Anglican ideal of a stable parish community unified by a common faith. Now, however, it was suggested, farmers neglected this moral dimension of farm service and instead were concerned only with their servants performance as factors of production. In the words of one Anglican clergyman, they now treated their farm servants 'as mere machines who must get through a certain amount of work'.[26] Increasingly, the Church of England's view was that a farm service system that had once been characterized by a close and caring paternalism had in recent years degenerated and was now based upon nothing more than the 'cash nexus' reinforced by the legal controls provided by the laws of master and servant. An early articulation of this discourse, which delineated a decline in the degree of moral regulation exerted by farmers over farm servants and regarded this as a major cause of increasing immorality and declining social cohesion, was offered by R.I. Wilberforce, Archdeacon of the East Riding in the 1840s.[27] Wilberforce proffered an idealized vision of society as family –hierarchical, paternal and bound by mutual affection and a common faith. He contrasted this ideal with what he regarded as the reality of parish life in the East Riding. He focused in particular on the degenerate condition of the farm servant population. He voiced his disquiet that farmers neglected the moral character of their servants and were seemingly unconcerned at the possibility of immoral sexual conduct between male and female servants boarded on the same farm. He noted that farm servants in rural parishes were spoken of as 'bad to do with' and that farmers themselves complained that their servants 'grow every year more headstrong, wayward and selfish'. He suggested, however, that farmers should look to their own lack of control as the main cause of this moral and social malaise, complaining that they permitted their farm servants to 'roam abroad' and 'neglect public worship' and that as a consequence villages on a Sunday were full of 'unemployed loiterers'.[28] Wilberforce contrasted the improvements that farmers had achieved in the economic sphere with the moral decline of their farm servants:

---

[26] Rev. F.D. Legard, 'The Education of Farm Servants', in F.D. Legard, ed., *More About Farm Lads*, London, 1865.
[27] Rev. R.I. Wilberforce, *A Letter to the Gentry, Yeomen, and Farmers of the Archdeaconry of the East Riding* (Bridlington, 1842).
[28] Ibid., 2–6.

> My complaint is that your crops are better educated than formerly but not your fellow Christians. Your Soil is kept clean, that nothing vile may grow in it, but how many of your fellow creatures have to be plucked out as weeds fit only for the burning.[29]

He then called upon farmers to restore the paternal authority, kindness, concern for moral character and regular attendance at church which he believed had once been characteristic of farm service.[30] Similar criticisms of the unruly, independent and immoral character of farm servants, and farmers' failure to exercise care and authority over them, continued into and throughout the mid-Victorian period – the common refrain being that farmers' neglect had precipitated a breakdown in social relations between themselves and their servants which had in turn promoted a decline in morality and social cohesion.[31]

A related criticism of farm service was that annual hiring facilitated and encouraged farm servant mobility, and thereby removed farm servants from the influence of the national Church. A number of points were made in this regard. Firstly, farm service encouraged the young to leave their parishes at an early age. When asked in the periodic visitation returns conducted by the Archbishop of York to comment on the factors that most impeded their ministry, many clergy cited farm service as a major difficulty because it drew the young of both sexes away from their parish when they entered the system.[32] This growing antipathy to the institution of farm service was amplified by the difficulties experienced when clergymen attempted to reach the farm servants that the system brought into their parishes.[33] Their mobility and isolated existence and their employers' lack of control had, it seemed, created a rootless, delinquent and amoral sub-group outsiders whose very lifestyle and culture frustrated and challenged the Anglican vision of the ideal parish community. The salience of rural clergymen's antipathy towards farm service and farm servants was heightened by the fact that that if farm servants did attend public worship it was

---

[29] Ibid., p. 3.
[30] Ibid., pp. 10–13.
[31] Moses, 'Rustic and Rude', pp. 161–2; Eddowes, 'Agricultural Labourer'; Chester, *Statute Fairs*; Legard, 'Farm Servants'; M.E. Simpson, *Ploughing and Sowing; Or, Annals of an Evening School in a Yorkshire Village, and the Work that Grew out of it*, ed. Rev. F.D. Legard, 1861; 'M.E. Simpson', The Life and Training of a Farm Boy, in Legard (ed.), *Farm Lads*.
[32] Borthwick Institute of Historical Research, 1865 Visitation Returns (V/Ret. 1865), examples include: Boynton, Rev. F.O. Simpson 'Children leave the parish at an early age for farm service'; Bridlington, Rev. H.F. Barnes, 'Boys and girls go early into service'; Foston on the Wolds, Rev. W. Bayles, 'All go to service at 12 or 13 years of age'.
[33] Eddowes, 'Agricultural Labourer'; Chester, *Statute Fairs*; Simpson, *Ploughing and Sowing*; Simpson, 'Life and Training'; Morris, *Yorkshire Folk-Talk*, pp. 206–11; Rev. M.C.F. Morris, *The British Workman Past and Present* (Oxford University Press, Oxford, 1928), pp. 121–2.

not usually in the parish church but at the meetings of the Primitive Methodists who, with their circuit system and religious populism, were able to reach out and into the lives of those farm servants located on isolated farms away from the village settlements and their under-used churches.[34]

These contradictions between the parochial ideal of the Church of England and the mobile and independent lifestyle of the farm servants informs much of the Church campaign against hiring fairs in the East Riding during the mid-nineteenth century. Despite their increased energy and enthusiasm, East Riding clergymen found that the combined forces of religious indifference and Nonconformity posed obstacles of considerable magnitude before their project of reasserting the profile of the Church of England in rural communities. The pervasiveness of farm service in the East Riding was regarded as being important in enabling and encouraging both obstacles. From the Church of England's point of view, farm service drew the young of both sexes out of their parishes into a mobile way of life which was outside of its control and influence. Once, in an imagined Golden Age this system had complemented the aims of the Established Church. It had operated as a form of paternalism maintaining social hierarchy and regular attendance at Anglican worship. Although excessively gilded, this prelapsarian vision may have contained a glint of truth. Prior to enclosure, farms were more likely to have been smaller, located in village settlements, and there was little Wesleyan and no Primitive Methodism in rural communities. Perhaps, as clergy believed, farmers had once adopted the role of surrogate parents and had enforced attendance at the parish church. There were still isolated examples of such farmers in the mid-Victorian period. By this time, however, farms were larger and more likely to be located away from village settlements. Farmers were often Wesleyan and presided over a reshaped proletarianized form of farm service which, partly through acquiescence before the independent culture of the farm servants, lacked the degree of control that clergy expected. This neglect on the part of farmers had, it was believed, prompted a breakdown in master– servant relations which had resulted in a situation in which Anglican clergymen found themselves confronted by a delinquent rural working class outside of their influence and, worse still, susceptible to the excitement and enthusiasm offered by Primitive Methodism.

---

[34] Rev. H. Woodcock, *Piety Among the Peasantry: being Sketches of Primitive Methodism on the Yorkshire Wolds* (London, 1889); R.W. Ambler, 'Attendance at Religious Worship, 1851' in S. Neave and S. Ellis (eds), *An Historical Atlas of East Yorkshire* (University of Hull Press, Hull, 1996); D. and S. Neave, *East Riding Chapels and Meeting Houses* (East Yorkshire Local History Society, C. Ward and Co., Bridlington, 1990), pp. 4–6; D. Neave, *Mutual Aid in the Victorian countryside; Friendly Societies in the rural East Riding, 1830–1912* (Hull University Press, Hull, 1991), pp. 10–11.

There was, therefore, a contradiction between the institutional moral and social concerns of the Church of England and the development of East Riding agriculture in its Golden Age. At the centre of these moral and social concerns was a desire to re-create paternal social relations as the basis for a revitalized Anglican parish. This 'new' or 'revived' paternalism had been pioneered in the south of England.[35] There, an absence of farm service and an agricultural workforce comprised mainly of married day-labourers resident in nucleated settlements, provided conditions that were more conducive to this project of attempted renewal. In contrast the expansion of large-scale capitalist agriculture in the East Riding tended to create adverse circumstances for such a project. Firstly, it encouraged an expansion of dispersed and isolated farmsteads away from village centres. Secondly, it sustained a labour system and with it a form of popular culture that nurtured a rural sub-culture that was regarded as a barrier against Anglican renewal. The Church of England therefore sought to erode, and ideally end, hiring fairs, living-in and annual contracts because they were integral aspects of a farm service system that frustrated the Church's attempts to reassert itself. The southern model of married day-labourers resident in their parish was increasingly seen by East Riding clergymen as both morally and institutionally more compatible with Church aims. It is not surprising to find, therefore, that those clergymen who called for the abolition of hiring fairs and attacked farm service also advocated the building of cottages in village settlements so that, in the words of the Archbishop of York:

> The young labourer, instead of living with a knot of loose companions in the farmer's house, may look forward to a separate dwelling, to marriage, to a long connection with the same employer; and may be softened by a wife's sympathy, and the humanising touch of childrens' fingers.[36]

---

[35] See Gilbert, 'Land and Church'; and especially Howkins, 'Reshaping', pp. 60–92.
[36] Archbishop W. Thomson, *Work and Prospects: A Charge* (London, 1865), p. 15.

Chapter 14

In Defence of Respectability: Financial Crime, the 'High Art' Criminal and the Language of the Courtroom 1850–1880

Sarah Wilson

By 1850, contemporaries had new levels of societal protection offered by a revolution in law and order, and believed that they could look forward to a period of continuing economic prosperity. Accordingly, this was a time both for positive reflection, and to herald the arrival of a 'Golden Age'. The progress of capitalism and the establishment of more dynamic sources of wealth creation alongside traditional ones actually had much in common with the promises made for the protection of communities during the modernization of policing. Not only were many of their intended beneficiaries people who were respectable, but they also represented important triumphs for respectability itself. In contrast, this period also generated unease within respectable society.[1] During these sensitive times, Victorian society believed also that it discovered, through the revelation of financial crime, a type of criminal activity which was not confined to the lower 'dangerous' classes.[2] Historically, this marked the birth of modern white-collar crime. For contemporaries, however, it meant consideration of respectability in a new, less positive light, as deep discomfort about a rapidly changing society ran through the ranks of the esteemed, at a time when the need to preserve the integrity of respectability, and indeed the 'goldenness' of the era, was correspondingly acute.

In more modern times, white-collar crime is viewed as a 'cancer' within society.[3] While this sentiment reflects the social damage it is believed to cause,

---

[1] F.M.L. Thompson, *The Rise of Respectable Society: A Social History of Victorian Britain, 1830–1900* (Fontana, London 1988).

[2] The revolution in policing itself was driven by concern about crime and depravity principally associated with the lower classes, see notes 40 and 41.

[3] Many authors allude to this, but August Becquai made direct reference in *White Collar Crime: A 20th Century Crisis* (D.C. Heath and Co., Lexington Massachusetts and Toronto, 1985).

financial crime, in its true modern sense, was discovered over 150 years ago.[4] Moreover, its revelation was shocking, and had implications for contemporary society which, in the late-twentieth century, are almost too disturbing to contemplate. While it gave rise to a panic of the kind which was not untypical of mid-Victorian Britain, responses to it were actually quite distinct.[5] The atmosphere of helplessness induced did not become the driving force for the enactment of panic-driven legislation which characteristically followed such occurrences.[6] Instead, the moral outrage generated was initially channelled into new aggression in prosecutions using existing legislation and common-law concepts already established in the legal cultures of England and Wales, and Scotland. Significantly, this was coupled with recognition that they were being tested in a new economic climate, and perhaps most remarkably, in an entirely new legal context.[7] It is by exploring some of these factors that it will be suggested that financial crime actually became inextricably linked with Britain's 'goldenness' through coincidences and ironies which accompanied its appearance.

This study draws on a cluster of cases, recognized as the first large-scale trials of fraud and related crimes of commercial dishonesty, which occurred throughout the 1850s and into and beyond the 1870s. While the body of material is small, its significance cannot be overstated. Along with the surrounding commentary it generated, it will be used to suggest a number of propositions. Firstly, that the impact of financial crime, and the ambiguous position of respectability exposed by its discovery, provide a most suitable front from which historians can question how 'golden' these times really were. Although it is well established that contemporaries were not uncritical of their own times, this exploration of respectability and criminality suggests that early encounters with business crime will actually reveal that the sections of contemporary society who were most conscious of the age's glittering triumphs

---

[4] In the 1930s Edwin Sutherland identified two broad types of white-collar crime, of which financial crime was one. The other was manipulation of political power; i.e. crimes of corruption. See E. Sutherland, *White Collar Crime* (reprinted New Haven, Yale University Press, 1983).

[5] This idea was first explored in the context of financial crime in S.J. Wilson, 'Reconstructing the Middle-Class Criminal: Aspects of Respectability and Criminality 1800–1914' MA dissertation, University of Wales, Swansea, 1996 (hereafter 'Reconstructing'), and continued in 'Invisible Criminals?: Middle-class Criminality in Britain 1800–1914' (hereafter 'Invisible Criminals') University of Wales, Swansea PhD thesis approaching completion. As a quite separate development, Behaving Badly – Socially Visible 'Crime' and the Law, based at the Nottingham Trent University, is a large inter-disciplinary project in progress.

[6] It is the concept of a Victorian 'social panic' which has set the agenda for the Bahaving Badly project.

[7] S.J. Wilson, 'Invisible Criminals' has looked at how and why the appearance of this new legal culture was such an important point of departure. This line of discussion is beyond the scope of this paper, wherein it is suffice to say that the direction taken in early responses to financial crime has actually required some detailed consideration.

were also most acutely aware of its shortcomings. Indeed, it will be suggested that it was the social climate of achievement itself which was responsible for drawing out a darker side of wealth creation, optimism and progress, and for precipitating belief that some parts were actually so ugly they required intervention from the criminal law. In essence, financial crime provided an important link between what was 'golden' about mid-Victorian Britain, and what was not.

Although modern criminology had to wait for American sociologist Edwin Sutherland to provide the famous coupling of dishonesty and occupation with respectability, which became immortalized in the term 'white-collar' crime, these ideas were being explored in a criminal context almost one hundred years earlier.[8] Writings from the mid-nineteenth century recorded the novel occurrence of activity which was financial in nature, and thus committed in commercial circles, and characteristically perpetrated by persons of esteem and repute. Most significantly, contemporaries clearly realized that it was beginning to attract the attention of the criminal law. This activity was christened 'high art' crime by *Times* financial journalist David Morier Evans. Evans's ideas on its economic and financial context, as well as its occupational and class influences, suggest this was early terminology for financial or 'City' crime. His thoughts on dishonesty itself, and the relationship between public opinion and legal responses, were particularly interesting. He observed that crimes of dishonesty were not new, as 'From time immemorial clerks have been discovered embezzling the property of their employers', while at the same time insisting that the late 1840s marked the arrival of something which was clearly distinct from the pilfering from employers which had been characteristic of life since 'time immemorial'.[9] Further, he claimed that a general adulation that forgery no longer carried the capital sanction had been 'qualified by the increase in dishonesty' which had 'followed mitigation of the law'.[10] Evans' commentary recorded the movement of the criminal law into business and commercial circles during the 1850s, and this was itself indicative that a fundamental shift in attitudes had started to take hold. This was also well documented in commentary in both the professional and general press.[11] Indeed, at the very time that crime statistics were encouraging society to feel its

---

[8] Criminologists and social scientists regard Sutherland, *White Collar Crime* as a classic work.
[9] D.M. Evans, *Facts Failures and Frauds. Revelations Financial Mercantile Criminal* (Groombridge, London 1859). See especially, pp. 1–4.
[10] Evans, *Facts*, p. 1.
[11] George Robb's use of *The Times* is well known. See G. Robb, *White Collar Crime in Modern England: Financial Fraud and Business Morality, 1845–1929* (Cambridge University Press, Cambridge 1992).

most comfortable, the financial criminal *cause célèbre* trials were starting to emerge.[12]

The parallels between initial responses, and those which are characteristic of more modern times, can provide insight into the horror of the discovery of financial crime. In the late twentieth-century, the serious fraud trial is symbolic of a societal determination that 'City' crime should not pay.[13] The symmetry is very real, as the *cause célèbre* was the most important single development in the construction of social and legal responses to financial crime when it first came to light.[14] In turn, true appreciation of the magnitude of financial crime in the mid-nineteenth century can be related to the endurance of a trial which is resource-intensive, and fraught with procedural difficulty and uncertainty.[15] It thus follows that the arrival of the *cause célèbre* represented an important point of departure in terms of consolidating attitudes to commercial dishonesty, and in framing institutional responses towards business crime.

In his writings, Evans insisted that the growth in crimes of dishonesty during this decade was not simply a product of a more liberal penal regime, and he viewed economic influences as pivotal. He observed that capitalism itself was central, by identifying a very close relationship between its essential features and opportunities not only for wealth, but also for the dishonesty which had come to light.[16] By the 1840s, as Britain's 'drive to maturity' was approaching completion, dishonesty within business circles also reached a pinnacle.[17] Notwithstanding this, Evans went considerably further than suggesting a coincidence of capitalism and commercial crime. Instead, he focused his commentary on an entirely new alliance between dishonesty and financial sophistication. This was itself attributable to opportunities provided by a newly and greatly expanded investment market. Evans claimed that at the

---

[12] Professor David Jones put this point well when he suggested that even in the late 1830s, Edwin Chadwick found himself having to sell greater efficiency and proactivity in policing against a backdrop of declining crime statistics. D. Jones, 'The New Police, Crime and People in England and Wales 1829–1888', *Transactions of the Royal Historical Society*, 33 (1983), pp. 151–68. For an excellent discussion on patterns of crime see V.A.C Gatrell, 'The Decline of Theft and Violence in Victorian and Edwardian Britain' in Gatrell et al., *Crime and the Law: The Social History of Crime in Europe since 1500* (Europa, London 1980). For discussion of patterns of crime and the emergence of institutional responses, see Wilson, 'Invisible Criminals'.

[13] For consideration of the ancestral relationship between the early *cause célèbre* and the modern-day serious fraud trial, as part of a larger consideration of modern day institutional responses, see Wilson 'Invisible Criminals'.

[14] The discussion in Wilson 'Invisible Criminals' accordingly focuses on the use of criminal proceedings, but also considers other aspects, such as policing, prosecution and sentencing.

[15] For consideration of Victorian and modern parallels see Wilson 'Invisible Criminals'.

[16] Evans, *Facts*, p. 2.

[17] W.W. Rostow, *The Stages of Economic Growth* (Cambridge University Press, Cambridge 1971).

core of this inflammable new combination was the mania for investment caused by speculation in railway companies during the 1840s.[18]

At the time, contemporaries realized the essence of investment activity for progress and the generation of wealth, and in many ways the panic generated mirrored achievement of greater understanding of the factors which influenced economic growth.[19] Britain's expanding programme of rail construction became the epicentre for meeting the needs of a growing capitalist economy, and also for appreciation that within the capitalist doctrine, opportunities for attaining personal wealth were substantial.[20] There are many contemporary accounts of how this had seduced the reckless and respectable alike, as the belief that everyone who became involved in the railways made money gathered momentum.[21] The ability to attract widespread private investment was crucial to the success of a project with national significance, but one which was entirely a creation of private enterprise.[22] In addition, many others realized that this appetite for speculation could be directed towards purposes which were not always legitimate. Although this railway boom was neither the first nor the last of the nineteenth century, it was the one which transformed the share market in joint-stock companies, capturing an entirely new and much expanded investment interest.[23] It was also the period of dynamism which exposed what Evans described as an 'evil and plethoric speculative mania'.[24]

Within this new atmosphere of excitement and intense stock-market activity and, to a large extent, precisely because of it, a number of less positive things were occurring. While for the operators of legitimate railway schemes this new investment interest provided the basis for an entirely new and sophisticated

---

[18] Evans, *Facts*, p. 2.
[19] Realization that investment was an essential prerequisite for progress and economic success comes across very clearly in the trials which followed in the 1850s, but during the 1840s, much was made of this not only in more obvious professional publications, but in the parliamentary addresses of ministers. The appearance of financial works of observation were also indicative of greater understanding of the factors which influenced economic growth, and perhaps especially, contemporaries had greater awareness of the trade cycle, and understanding of the significance of periods of economic instability. This can be seen during this period through the works of D.M. Evans, *The Commercial Crisis 1847–1848* (Groombridge London 1848) and *Commercial Crisis 1857–1859* (Groombridge, London, 1859).
[20] Evans, *Facts*, p. 2.
[21] Evans refers to this explicitly in *Commercial Crisis 1847–1848*, p. 11.
[22] Indeed, Sir Robert Peel remarked with pride that the railways had been built 'without any pecuniary aid or assistance from the Government', *Hansard*, Third Series, LXXVI (11 July 1844), col. 668.
[23] Most succinct account of this, and the reasons behind it can be found in M. Reed (ed.), *Railways in the Victorian Economy: Studies in Finance and Economic Growth* (David and Charles, Newton Abbot, 1969). See especially M. Reed 'Railways and the Growth of the Capital Market', in Reed, *Railways*, pp. 162–84.
[24] Evans, *The Commercial Crisis*, pp. 12.

system of available capital and credit,[25] this period also drew attention to deliberate and wholesale setting up of fraudulent bubble companies.[26] Often less overtly deliberate and calculating, but frequently no less worrying, and particularly prevalent in prospectuses and publicity literature, were misrepresentations of the assets and the potential profitability of prospective companies.[27] In addition, collected deposits were often put to 'alternative use', and occurrences of price-fixing, and delay and strategic releasing of allotments of shares in order to capture maximum interest, were also found to be commonplace.[28] As the full extent of these occurrences became apparent, the negative effects of new investment opportunities could not be ignored. Victorian society was forced to respond, and did so by alleging that some actually amounted to crimes.[29] In so doing, contemporary society acknowledged the presence of 'high art' crime, and bestowed on the earliest criminal trials the task of defining its remit, and its scope, by working through important deliberations. Although these concerns were exposed by the railway boom, the presence of the railways in these discussions became subservient to a far stronger overriding objective: to distinguish between practices which were not to be encouraged in a modern commercial society, and others which were actually so hostile to contemporary business culture and ethos, that they should actually attract criminal liability. Within this, the single most important consideration which came from the railway boom was a direct consequence of the opportunities in investment practices that had been exposed. At one level, it is clear that there was perceived the need to establish parameters for the operation of legitimate investment. However, from the trials, the need to address a number of very worrying secondary problems, which were seen to

---

[25] How this was achieved is explained with great conciseness and precision in Reed 'Railways and the Growth of the Capital Market'.

[26] Rande Kostal coined the term 'assetless shells' to describe bubble companies. See R. Kostal, *Law and English Railway Capitalism 1825–1875* (Clarenden Press, Oxford, 1994), p. 12.

[27] Although in reality, many railway companies, and not just ones which were fraudulent or even 'marginal', had far less financial potential than their publicity literature would suggest, Evans documented the worst 'cases' in *The Commercial Crisis*, p. 12.

[28] Evans, *The Commercial Crisis*. Evans suggested that this amounted to, in many cases, conduct which was least reprehensible. Harold Pollins, 'The Marketing of Railway Shares in the First Half of the Nineteenth Century', *Economic History Review*, 7 (1954–55), pp. 230–39 gives an extremely interesting insight into the marketing and selling of shares during the railway boom.

[29] This was clearly a reflection of the desire to take more questionable financial dealings out of the realm of private litigation, and to make them matters of public concern, but understanding how this occurred is not without difficulty. The precise workings are beyond the scope of this paper, and have been given quite considerable attention in Wilson 'Invisible Criminals'.

accompany speculative practices which were more marginal, was also apparent.[30]

It was in *Facts, Failures and Frauds* that Evans first suggested a connection between the progress of 'high art' crime and 'irregularities' in business which had occurred during the boom of 1844–45. During the early 1850s several high-profile people found themselves charged as criminals as a result of their conduct in business dealings, and were required to answer charges of fraud, embezzlement and other misappropriations, misapplications, and misrepresentations of assets.[31] Others found themselves the subject of official investigation into their business dealings.[32]

As these early years progressed, it can be seen that even after the immediate proximate context created by the railway boom had subsided, intrusiveness into business dealings not only continued, but it actually gathered considerable momentum. Indeed, it is suggested that towards the last quarter of the nineteenth century, responses to business crime actually became caught up in the larger economic debate which dominated this period. This involved consideration of a number of issues relating to the operation of corporations, and the emergence of modern commercial transactions. The issues concerned principally discussions of financial responsibility for capitalism and progress, and the conduct of individuals engaged in the promotion and management of corporations.[33]

In this immediate context, however, the criminal trials born as a result of these critical events reveal much about the impact of the discovery of 'high art' crime, and the panic it generated, while illuminating how this influenced contemporary appreciation of a somewhat tarnished Golden Age. While many of those charged were respectable, the proceedings indicate that much of the shock precipitated actually attached to the adverse implications for

---

[30] It is here that we can see graphically the legacy of the railways, even in trials which occurred over a decade after the crisis, which itself passed fairly quickly. This has been given extensive consideration in Wilson's 'Invisible Criminals'. In essence, what came out of the railway boom and crisis was acknowledgement that speculation and investment practices required long-term consideration.

[31] Most of the London trials were heard before the Central Criminal Court, although some were located elsewhere. The trial of the directors of the Royal British Bank in 1858, for example was heard before the Court of the Queen's Bench Division. The trial of the City of Glasgow Bank directors (note 38), tried under Scots Law, was heard before the High Court of Justiciary in Edinburgh. Certainly, insofar as the Central Criminal Court trials were concerned, it is likely that *Facts* provides the most comprehensive record of the proceedings, as the Court Records themselves lack documentation of the proceedings themselves, and comprise instead depositions and statements of witnesses, and verdict and sentence.

[32] This has been explored more fully in Wilson, 'Invisible Criminals'.

[33] It was here that the modern law relating to companies was born. Elsewhere it has been suggested, thus that responses to financial crime not only became caught up in, but are best explained, as being part of a larger and far more ambitious articulation of capitalism and the law. See Wilson, 'Invisible Criminals'.

respectability itself.[34] Capitalism was itself a creation of the respectable, yet the trials suggest that by the close of the 1840s there was a growing appreciation of a very different legacy of the progress of new wealth and the rise of the middle classes. Accordingly, during these years, the rhetoric of affluence and progress, and especially that of respectability itself, became transformed into the language of criminal proceedings, and even criminal charges.

From the early 1850s, the courtroom became an important location for discussions of respectability, which occurred on a number of different levels, and did so at least partly in response to the motives of a society trying to establish a new type of crime, which clearly had to be based on dishonesty. It is clear from indictments that many of the charges themselves were not new, but instead that the novelty attached to a determination to respond. In the trial of the former directors of the Royal British Bank in 1858, Lord Campbell remarked that in the current climate of laxity in commercial dealings, existing law was not defective, it was simply not being used effectively, thereby indicating the need for enforcement.[35] However, it is also the case that many of the earliest criminal charges were actually combined with civil liability for fraud (based on the doctrine of deceit) and other non-criminal misdeeds, most notably breach of trust, which suggests some recognition that 'high art' crime might actually require a new approach.[36] In the trial of the former directors of the City of Glasgow Bank in 1879, all parties in this case without precedent agreed that if the charge of falsifying a balance sheet with intention to deceive were proven, this would represent an important landmark in the criminal law of Scotland.[37] But even at their most inventive, responses to these new directions in financial dealings reveal a determination to establish roots in dishonesty, desirous, of course, to establish a clear role for the criminal law in relation to what might otherwise have been viewed as essentially 'private' matters in business.

While an intention to recognize some business conduct as criminal activity was evident from the willingness to use criminal proceedings against

---

[34] There were also a number of frauds which while involving spectacular sums of money, were perpetrated by clerks in fairly menial positions. Evans's observations on cases like that of W.J. Robson show with remarkable perception, I feel awareness, of the concept of a 'barrow boy' long before anyone had heard about Nick Leeson.

[35] Evans, *Facts*, p. 385.

[36] The observations of R.R. Formoy in *The Historical Foundations of Modern Company Law* (Sweet and Maxwell, 1923), can be illustrated with reference to a number of these cases, and were appreciated certainly at the time by writers for the *Law Times*. This has been considered extensively in Wilson, 'Invisible Criminals', as part of the discussion intimated in note 7.

[37] Couper's Report on *The Trial of the Directors and Manager of the City of Glasgow Bank, before the High Court of Justiciary* (Edinburgh, 1879). See especially submissions made by Mr McIntosh on behalf of Taylor, during petitions for bail p. xx.

businessmen, the precise details had still to be worked through. By testing old concepts in a new environment, and in a new aggressive public forum, policy-makers clearly anticipated that unacceptable business practices could be identified, and any attendant criminal liability declared. Although the scope and limits of criminal liability were not defined, the message of a lowering of tolerance was clear. Throughout the trials of many respectable people, the defendant's social status, along with the accusation that his conduct actually amounted to abandonment of the very virtues held by reason of this position, predominated. However, the use of respectability was probably more significant, and even interwoven with a wider social determination to restore, if not actually recapture, Britain's self-image of goldenness.

In a more narrow context, use of the language of respectability almost certainly reflected insecurity, and prosecutors might well have anticipated at least some difficulty in establishing the idea of commercial crime in the legal culture of a society predisposed to the benefits of capitalism. In addition, the business community could not easily be ignored. Indeed, it is clear that as the limits of 'high art' crime remained undefined, those concerned with the administration of criminal justice recognized the necessity of cooperation with, and even contribution from, those who were representative of the interests of commerce. Lord Campbell acknowledged, in summing up the case against the former directors of the Royal British Bank, that the members of the jury knew far more about 'the subject' than he, and that as they had clearly understood the evidence presented, the judge's only duty was to state the questions of law which may arise.[38] Several addresses made to jurors by prosecuting counsel, and those who defended, indicate awareness that the City clearly had interests it wished to protect, but was unlikely to welcome perceived 'outside' interference, while having to accept that close cooperation with those in commercial circles was crucial.[39] It could be for this reason that it was believed that articulating values of respectability against those most likely to rely on position and status would be effective in helping to identify immoral, if not illegal conduct, and thereby tapping into the panic which had been generated. Nevertheless, the proceedings indicate that no one was quite sure where things would lead.

The aura of uncertainty was perhaps most acute in the treatment of those on trial, rather than in the conceptual difficulties raised by the idea of a

---

[38] The trial of the former directors of the Royal British Bank, 13 February 1858, for charges relating to conspiracy and intention to misrepresent and deceive. Recorded in Evans *Facts*, p. 375.

[39] This has been discussed further elsewhere. Wilson, 'Invisible Criminals', has suggested that many addresses appeared skilfully to suggest to juries that business community involvement was actually crucial for safeguarding the interests of the commercial community, that such involvement was not only necessary to achieve protection from financial misdeeds, but also from the possibility of more intrusive forms of regulation.

'commercial crime', as this was probably the fundamental difficulty raised by its discovery. Here, attention was drawn to activity which was not confined to the so-called criminal classes, and appeared to be something different from the crimes of violence of which even the virtuous found themselves accused from time to time, and which were often 'explained' on grounds of passion, human emotion and provocation.[40] It was, instead, a type of crime characteristically perpetrated by respectable people, it was dishonest in character, and by the 1840s, it was occurring on an increasingly large scale. In turn, the proceedings indicate that it was plainly understood that such people bore little if any resemblance to Edwin Chadwick's archetypal 'habitual depredators'.[41]

The new model criminal and his attendant peculiarities dominated throughout trial proceedings, and even intruded into pre-trial events, thus encouraging focus on respectability and position.[42] When opening the case against London bankers Strahan, Paul and Bates, initially before Bow Street Magistrates, Sir Archibald Bodkin remarked that those who now stood at the bar were 'not prisoners such as are usually seen in that position, but gentlemen.[43] In the Central Criminal Court, the prosecuting Attorney-General reminded those assembled that standing accused were gentlemen 'known to most of us', who had 'hitherto maintained a high position in society'.[44] The addresses themselves, and their progression throughout the hearings, suggest that the close association of position with alleged criminality was quite deliberate, and accompanying emphasis on social expectation an important part of this. At one level, this might well have reflected wariness, and inability to predict societal reactions to the respectable criminal and his activities. While it might be very obvious to suggest that because so much attention was given to the defendants' social position their criminal conduct was being correspondingly downplayed, this is probably far too simplistic. Esteem and probity were always very closely aligned with confidence and trust enjoyed, as a result of social status, either as bankers in the case of Strahan *et al.*, or as respected businessmen. The treatise invites that this was actually regarded as adding to the gravity of the alleged criminality, and it is possible that this was

---

[40] Chapter 16 in this volume. For an overview of changes in law and order see D. Eastwood, *Tradition and Transformation in Local Government 1780–1840* (Clarendon Press, Oxford, 1994).
[41] *Report of the Rural Constabulary Commissioners*, Parl. Papers, XIX (1839) p. 169.
[42] This has been given greater consideration in Wilson 'Invisible Criminals'.
[43] The trial of Strahan, Bates and Paul, 26 October 1855, for offences relating to their illegal disposal of securities entrusted to them as bankers. Again, the full trial was documented by Evans, *Facts*, p. 117.
[44] Evans, Facts, p. 126.

itself an important part of trying question the social and cultural assumption that only the 'criminal classes' committed crimes.[45]

The references to respectability were probably indicative of a number of very different concerns and objectives and most strikingly, at times, respectability itself appeared to be 'on trial'. At this point, allusion to the more general fragility of respectability characteristic of this period becomes hard to dismiss. On many occasions, prosecuting and defence counsel alike, and even presiding judges, entreated that following a criminal conviction, an additional price would be paid. In the trial of Strahan, *et al.*, the prosecution alleged that the charge against the men was one which not only involved penal consequences of great magnitude, but that it also affected their honour and character, thereby identifying some important wider perspectives of a 'guilty' verdict. Although both the idea of additional gravity and the suggestion of far-reaching social implications plausibly explain uses of respectability, the actuality was probably far more complex. The Attorney-General continued to amplify the social consequences of conviction by stressing that the defendants' positions had actually placed them in circumstances of extraordinary irony: that their characters of 'unquestioned honour and integrity' had 'prevented them from being supposed capable of the offence with which they are now charged'.[46] While this amounted to a warning that acceptance of dishonesty among the respectable was unavoidable, the tenor was also reserved, and probably indicative of difficulty in predicting reactions to the message being conveyed.

It is not surprising, perhaps, that the language of status featured strongly in defences, but perhaps more surprisingly, as in prosecutions, this was essentially constructed around the presumption that respectable people would not commit crimes. In these circumstances, counsel sought to draw attention away from their clients' deviance, while emphasizing their considerable esteem, and even suggesting that the use of proceedings to allege criminal conduct was a most 'unfortunate' outcome of surrounding events.[47] Moreover, many learned figures insisted that these events were often ones which were controlled by external forces of market and economy.[48] While this was designed to negate the inference of calculation and deliberate behaviour, the idea of misfortune was almost always directed towards questioning the scope of criminal liability, and the limits of criminality; looking at the boundaries between bad practice and

---

[45] Such ideas became crystallised during the 'crime debate' of the first half of the nineteenth century. See David Jones, 'The New Police, Crime and People in England and Wales 1829–1888', *Transactions of the Royal Historical Society* (1983), pp. 151–68.
[46] Evans, *Facts*, p. 126.
[47] One orchestration by Sir Frederick Thesiger on behalf of Strahan can be found in Evans, *Facts*, p. 132.
[48] See statements made on behalf of Lewis Potter and Robert Salmond in the City of Glasgow Bank trial; pp. 427 and 439 respectively.

management on the one hand, and conduct which was criminal on the other. Acting on behalf of Robert Stronach, former Manager of the City of Glasgow Bank, the Dean of the Faculty of Advocates presented an extremely elaborate articulation of the ambiguities inherent in the rhetoric of business and criminality, in which he submitted that while the legal distinction between the two might exist, reality would render it impossible to operate.[49] Clearly, all defences sought to distance particular clients from any taint of criminal conduct, but implicit in many, especially that the dean, was an understanding that there was a point beyond which bad practice would descend into criminality.

Accordingly, on some notable occasions, even defence counsel were prepared to acknowledge the existence of 'high art' crime, and that in some circumstances respectable people would be criminals. This unhappy acceptance was also evident in judicial comment. Following Joseph Windle Cole's conviction for the Dock Warrant frauds in 1854, Chief Baron Pollock insisted that there was no justification for respectable people to act dishonestly, when comparing Cole's actions with those who had poverty, want, or bad example as some extenuation of their offences.[50] In passing sentence for crimes committed out of indulgence, the Chief Baron acknowledged that regrettably, respectable people would sometimes act dysfunctionally.

These trials were not simply about establishing new crimes, and recognizing new types of criminal. The use and articulation of respectable virtues in criminal proceedings illuminate many Victorian fears and concerns. They also appear to endorse the view that contemporaries themselves realized their age of achievement was not so golden, and that as a result, there was the need to restore its integrity. Moreover, early encounters with financial crime unveil not only important parallels with more modern times, but also some deeply-rooted ironies about contemporaries' perceptions of their own. One which is particularly striking follows from the earlier suggestion that Britain's early encounters with financial crime revealed a society aware of its limitations as well as triumphant of its achievements. Within the Victorian experience of crimes of fraud and financial dishonesty, it is significant that the key actors were bankers, either on trial, or concerned to protect their assets and repute; businessmen, as alleged criminals, and as promoters of buoyancy and integrity in commerce; and lawyers, either alleging criminal behaviour, or trying to protect clients from criminal liability. Thus, those who were actually best placed, socially and culturally, to appreciate the achievements of the new

---

[49] The Dean made these remarks while dismissing fraud itself as an arbitrary and highly subjective concept, and one which was particularly susceptible to manipulation by 'expert' witnesses. The Glasgow Bank Trial pp. 519–20.

[50] The trial of Joseph Windle Cole, 24 October 1854, documented in Evans, *Facts*, p. 209.

capitalist society also became responsible for questioning the manner of its progress, very publicly. It fell to bankers, businessmen and lawyers simultaneously to condemn subversive practices, and to endorse the capitalist ethic, not only in the City itself, but actually in the courtroom, as part of a criminal trial. In addition, these ambassadors for capitalism had to allege that practices which were at best irregular, and at worst criminal, were also alarmingly common.[51] In this respect, concerns about undesirable practices in the railway boom and subsequent crisis, became transformed not only into the need to respond, but also to deter.

Notwithstanding these considerable difficulties, it is clear from the records that the panic generated surpassed the realization that respectable people engaged in dishonest activity, and even the confirmation that dishonesty was rife in the market-place as it was known to be in the streets. Other concerns were becoming visible, and once again those who had reason to be most positive – the commercially literate, and the legally competent – had the task of drawing out the darker side. Policy-makers could not ignore the fact that some of these pillars of polite society were prepared not only to risk their good names and social standing by acting without integrity, and sometimes without honesty, but would actually go to considerable lengths to attain personal wealth as well as commercial success. Far too many would not stop short of using positions of repute and honour in a more ruthless manner, and with little regard for the ruinous consequences of their actions. The railway boom had identified this new virile business culture; one not always characterized by probity and integrity, but instead motivated by greed, and sustained by people who should have known better, and the trials embodied a perceived societal need to respond.

While the trials suggest, unsurprisingly, that calculation and deliberate action were regarded as factors which should operate to distinguish improvidence from criminality, abuse of position was also publicly condemned. When passing sentence on Strahan, Bates and Paul, Baron Alderson insisted that such conduct from respected London bankers, and by analogy all perpetrators of financial crime, would not be tolerated. Focusing on the prisoners' astonishing social metamorphosis, he addressed the serious ramifications of a criminal conviction for well-placed persons. Alderson expressed his regret at having to pass sentence on the prisoners, recollecting that he was actually well acquainted with at least one of them under very different circumstances – by his side 'in high office', instead of 'being ... in the prisoners' dock'.[52] In making such a statement, Alderson did not seek to diminish the significance of conduct which was firstly alleged, and

---

[51] Lord Campbell in the Royal British Bank trial. See Evans, *Facts*, p. 385.
[52] Ibid., p. 145.

subsequently proven as criminal. In explaining that the prisoners had 'moved in a position of society', he also gave equal, if not greater, emphasis to the adverse implications for respectability. Certainly, there were appeals to the responsibility which accompanied high status and rank, during which it was noted that society was entitled to expect more from such people. There was also a more powerful suggestion that being in breach of this unwritten code of social conduct would have ramifications outlasting penal consequences. Alderson remarked that the punishment administered 'would be much more keenly felt than it would probably by persons in a lower condition of life'. Further, those connected with the gentlemen would feel their 'present situation with great severity', as the social costs would extend to family and even acquaintances.[53] This was clearly a warning to others to preserve their honour, and the position of respectability itself, and to do so at a time when the respectable felt least sure of their own positions in social order.

The significance of the trials in public attempts to restore Britain's goldenness becomes even sharper through the emphasis placed on the economic implications of commercial dishonesty. In the Royal British Bank case, leading prosecuting barrister Sir Frederick Thesiger identified 'Widespread ruin, and loss of the 'hard earnings of industry',[54] and in sentencing Strahan et al., Baron Alderson alleged that 'A greater or more serious offence could hardly be imagined in a great commercial community like this.[55] Many addresses indicate a perception of the need to protect the nation's commercial interests, as it became apparent that its 'architects of progress' might also be responsible for its destruction.

The trials show that 'high art' crime had drawn attention to a new side of respectability, while the 'high art' criminal himself caused difficulty for a society not accustomed to habitual dishonesty among those occupying high social rank. The need also to appraise and uphold the integrity of respectability appeared in another very important issue related to the discovery of commercial crime. The focus on regulation of business was perhaps not surprising given the background to the trials, but its discussion of respectability concerned more than a need for tighter control. Many addresses alluded to a traditional perception of commerce as a world of dealings among gentlemen, and to recognition that respectable virtues, and wider fiduciary obligations, had a quasi-regulatory role in determining behaviour among businessmen. The regulatory question hovered throughout the trials, and implicit in many references to entrustment and position was a warning that if dishonourable behaviour were found to be too commonplace, then a more official regulatory

---

[53] Ibid., p. 145.
[54] Ibid., p. 289.
[55] Ibid., p. 145.

framework might have to replace existing conventions.[56] However, the addresses also suggest that within events which had exposed behaviour which clearly was not that of gentlemen, attention was being paid to the 'outside' discussions on the more precarious position of respectability, and its ability to endure.[57]

The trials in which respectable people from business and financial backgrounds were charged in criminal courts as a result of their professional behaviour, represented awareness of a critical point of convergence between old ideas dating from 'time immemorial', and transformations in social and economic conditions. In establishing financial crime as a new application of dishonesty, and by testing its operation and its limits, these trials focused not only on individual depravity, but they also sought to defend both the integrity of the economy, and its health. The single most important symptom of the moral panic generated by the discovery of financial crime lay in willingness to bring criminal charges against businessmen. The 1850s marked the beginning of an important social journey in the calibration of responses towards commercial crime.[58]

This essay has looked at the events surrounding the discovery of financial crime, and the responses it precipitated, while weaving through social panic, the significance of the 1840s, and their links with respectability. For the Victorians, the railway became the symbol of progress; the symbol of what was great about Britain, and thus it represented the core of its goldenness.[59] For this reason, the railways themselves, and not just the boom of the 1840s, were crucial. In turn, the railways represented the true source of the panic generated by the discovery of financial crime. By encouraging dishonesty, as well as facilitating wealth creation, the railway was able to fuse Britain's 'goldenness' with the darker side of the Golden Age.[60]

In short, the popularity of railways during the 1840s gave rise to an uncomfortable symmetry between their national significance, and the dangers which were latent in their construction. Although it was the most expansive and also most expensive operation the early Victorians had experienced, the project of constructing the railways was not in receipt of State finance, and thus

---

[56] This was first explored in Wilson, 'Reconstructing', and has been developed in Wilson, 'Invisible Criminals'.

[57] Such a discussion is beyond the scope of this essay, and has been considered in Wilson, 'Invisible Criminals'.

[58] Wilson, 'Invisible Criminals' details the chronology and significance of institutional responses.

[59] J. Simmons, *The Railway in England and Wales: 1 The System and its Workings 1830–1914* (University of Leicester Press, Leicester, 1978), p. 82.

[60] This theme has been explored extensively in Wilson, 'Invisible Criminals'.

entirely dependent on private investment and enterprise.[61] Moreover, the opening of the railway share market in the 1840s was recognized as shaping future investment patterns, and providing corporations with new forms of credit facilities.[62] Regretfully, this period was also the one which exposed the practices which formed the basis for social and legal recognition of financial crime. While it was an unfortunate coincidence of two diametrically opposed occurrences, the imperative to address the continuing needs of early capitalism, and to remedy what the 1840s had exposed, was forced on society. The trials sought to send clear warnings that this could not happen again. They show decisively that Victorian society believed that it had to be seen to be making investment safe, and had to reinforce with criminal sanctions its rejection of business practices which were at odds with the progress of capitalism. In so doing, it also perceived the need to clarify that those who were in positions of trust and esteem would be judged especially harshly, making plain that society would not look kindly on respectable people who acted in a manner which was wholly inconsistent with their status.

Within this, there were of course a number of different considerations interwoven, and by publicly chastizing bad business practice, attacking improvidence, and establishing criminal liability, Victorian society concerned itself with restoring confidence in the market. Thus, the focus became continued investment and progress, and the encouragement of more formal regulation and greater professionalism in business, rather than seeking to address the problem of dishonesty among the respectable. At several points, the concept of the financial crime, and the protection of economic interests, became almost impossible to distinguish. Undoubtedly, this was compounded by the moral gloss of a focus on respectability, although ironically, early emphasis on trust and esteem almost certainly sought to amplify the repugnance of financial crime, rather than to diminish it. The rhetoric of 'less criminal' which is often closely associated with financial crime, and white-collar crime more broadly, suggests that over time, these difficulties have become more acute. While 150 years later, focus on the social influences as distinct from criminal ones in modern discourse on white-collar crime, has come to connote something very different, it would, however, be foolish to forget that its origins lie in the shocked responses of nineteenth-century society. It lies in the reactions of a society having to come to terms with a discovery which cut across many social and cultural perceptions of crime and deviance. Not only did this raise a number of issues, but many actually mirrored rather too closely questions relating to social and class structure

---

[61] For further consideration of 'railway history' and the relationship between the railway and the Victorian economy see Wilson 'Invisible Criminals'.
[62] Reed, *Railways*.

which dominated the more general social discourse, outside the confines of polite commercial society and the world of the 'gentleman's agreement'.

Part IV

Gender

Introduction to Part IV: Gender

Judith Rowbotham

The questions that interest historians when investigating past periods are generally dominated by those that have a current resonance for them. This has, over the last thirty years in particular, seen a shift in the way in which the past has been interrogated to produce answers relevant to the late twentieth century. Thus, questions of social class have been joined by questions of gender. Initially, the focus was firmly on reclaiming female experience from historical obscurity, starting with Sheila Rowbotham's critique of social history writing in *Hidden From History*, but increasingly the interest has broadened to include issues of both femininity and masculinity.[1] Historical investigations using gender considerations as a methodological tool have challenged a number of conventions, including those relating to the operation of social systems and the role of class as the a priori tool for comprehending social relations. It has given prominence to questions about the processes involved in socialization and thereby revived aspects of the debate over social control.[2] Also, just as gender considerations have forced practitioners to question the links between biology, social constructs and language, so they have forced questions about the competing claims of the importance of types of stratification in social hierarchies and about the agencies that produce such hierarchies. Employing the lens of gender, as well as of class, it very swiftly becomes apparent that the identifications of the period as a 'Golden Age', highlighted elsewhere in this volume emanate essentially from respectable, predominantly middle-class, masculine interpretations of the era, and that while the most easily accessible of contemporary definitions they constitute far from the only viewpoints that historians should utilize. This leads to a position where it must be asked how widespread, in terms of geographical, cultural and gender locations, was the golden gloss identified by both certain contemporaries and later scholars.

---

[1] See, for example, Sheila Rowbotham, *Hidden from History: Rediscovering Women in History from the Seventeenth Century to the Present* (Pluto Press, London, 1973). Catherine Hall, *White, Male and Middle-class: Explorations in Feminism and History* (Polity, Cambridge, 1992). Michael Roper and John Tosh (eds), *Manful Assertions: Masculinities in Britain since 1800* (Routledge, London, 1991).

[2] R.W. Connell, *Masculinities* (Polity, Cambridge, 1995); Hall, *White, Male and Middle Class*; Roper and Tosh, *Manful Assertions*, especially pp. 8–11; Sonya Rose, *Limited Livelihoods: Gender and Class in Nineteenth Century England* (Routledge, London, 1992).

As previous contributions in this volume have demonstrated, there was undoubtedly an important strand in contemporary thinking about the Golden Age that, in summary, identified the period as one of achievement, witnessing the genesis of modernist, mechanistic economic, commercial and social thinking, despite the undoubted continuation of, or even increase in, certain social problems such as those highlighted by Harold Perkin in his chapter. This perspective relates, in various ways, to those individuals or groups or occupations that were, by and large, successful according to this thinking. In other words, it is an interpretation that considered it had good reason for promoting a positive profile of the period. In view of these links, class considerations must form an important part of any exercise in comprehending these decades. However, any more profound insights into the nuances of social relations, including the pressures for and against conformity to this perspective, must go beyond the rigidities of class, and take gender issues into their considerations. As Joan Scott has commented, gender is 'a way of denoting "cultural constructions"' or the 'entirely social creation of ideas about appropriate roles for men and women'.[3] Since, in contemporary discussions about the nature of the period 1850–70, significant assumptions can be shown to emanate predominantly from a masculine perspective, this raises the question of how that perspective relates to other contemporaneous comprehensions of the period.

Essentially, the Golden Age concept was one expression of a hegemonic masculinity reflecting the interpretations of the world around them of those men who were socially and economically successful, and so saw themselves as justified in seeking to assume a cultural dominance in society over both women and what John Tosh has described as 'subordinated masculinities'.[4] So far as this particular element in the project is concerned, the Golden Age was a time when ideas about Britain's global and imperial role were coalescing, against a background of the establishment of the supremacy of 'respectable' society and its mores. Yet contemporaries did also feel that the equipoise of the period was no more than, at best, a precarious balance – one that would survive only if it was merited as a result of human effort at all levels of society. If there was confidence, then continuing upsets and agitations, at home and abroad, during the 1850s and 1860s ensured a persistent consciousness of a nearby precipice. Rather than forgetting Chartism in a belief that its threat had passed, there was a continuing consciousness that Britain had only narrowly averted disaster. Arguably, fear of a resurgence of some similar force was one of the distinguishing features of these two decades. After all, the gloomy events immediately succeeding the victory over Napoleon showed how vulnerable the

---

[3] Joan Scott, *Gender and Politics of History* (Colombia University Press, New York, 1998).

[4] Roper and Tosh (eds), *Manful Assertions*, p. 76.

nation could be to problems and disasters unless the population as a whole laboured together to continue to deserve God's blessings.

This attitude of quintessentially masculine complacency underscored by doubt can, perhaps, best be summed up in the immortal lines so much quoted during the period: 'You should feel some danger nigh, / When you feel most content'. A consciousness of recent events on the Continent, culminating in the 1848 revolutions, had convinced large numbers of those in positions of power and influence that an essential aspect of Britain's ability to continue to withstand the pressures represented by mass discontent expressing themselves in riot or revolution, was the relatively 'orderly' nature of the British system. Many of the British establishment's perspectives in 1850 on recent European history were, in other words, comparative in nature. Having so recently started a process of constitutional reform, and conscious of its continuation at local (if not, until the 1860s, metropolitan) levels, most British authority figures were, by 1850, at heart reasonably complacent about the stability of the political system. But they did have certain concerns about the social system, and the ways in which this was perceived by groups within the British population, and the importance assigned to these in contemporary thinking is highlighted by gendered interpretations of the age. From the perspective of respectable masculinity there were still many problems in the British social system. These were usually problems associated with the unwillingness of too many individuals and groups to give up unprofitable pastimes and preoccupations, from easily identifiable offences such as drunkenness, prostitution and a variety of crimes, to less readily-defined resistance to 'respectability' from women seeking greater individual rights, say, or trade union members. So ways were sought of passing on information to convince these groups of the advantage to them of adopting, individually, crucial concepts like 'respectability' which would militate against such behaviour. This section encourages questioning of the extent to which this proselytizing exercise was successful, and assessments of the reasons for either success or failure with certain target groups.

In a chapter building on work already done by historians such as Amanda Vickery and Jose Harris, challenging both the extent of widespread popular acceptance and practical operation of the concept of gendered 'separate spheres', Tina Parratt considers the contributions towards domestic ideologies and practices made by working-class males in Lancashire.[5] The group she identifies has certain characteristics in common with both respectable masculinity and resistant masculinity as expressed through trade unions and Chartism, for example in terms of the emphasis laid on literacy, education and self-help of a form. But, more in line with Chartist attitudes, these men

---

[5] Amanda Vickery, 'Golden Age to Separate Spheres? A Review of the Categories and Chronology of English Women's History', *Historical Journal*, 36 (1993) pp. 383–414; Jose Harris, *Private Lives, Public Spirit: Britain 1870–1914* (Penguin, London 1993).

demonstrated a comprehension of domesticity that was far more collaborative than that associated with the respectable masculinity of both the middle and the working classes. She highlights an acceptance of the contribution of women to the family economy, and a related masculine willingness to share in the work of domesticity. Parratt's group were, on the whole, more financially secure and had better economic prospects than that examined by Nicola Verdon, in her chapter on women farmworkers in Norfolk. In an interesting commentary on earlier chapters on agricultural labour in this period, Verdon underlines the practical problems faced by women workers in an area suffering from both de-industralization, in terms of traditional craft occupations, and a decline in available regular agricultural employment. But she also points to how individual difficulties were compounded by assumptions relating to respectable masculinity and its economic responsibilities, and to consequent expectations about appropriate work for women and female skill levels. If, as shown elsewhere, the gilt associated with the age can be held to have clung to agricultural workers in the north, there was little that was positive about the period for this particular group of female workers in this locale.

In a chapter highlighting another way in which participation in the advantages associated with the Golden Age was profoundly gendered, Kim Stevenson focuses on the experience of men and women encountering the legal system as a result of expressions of male sexual violence. There was, in this period, a will to make the legal system ever more apparently responsive to contemporary needs and problems as a way of guarding against internally-generated discontents within society. The ways in which Parliament and the newly developed legal profession sought to interpret the existing laws of the land were to lead to the evolution of new formulae aimed at the socio-legal adjustment of disconcerting industrial and urban formations, with their impact on the workings of society. Their concerns promoted a determination to examine the need for either a regularization of existing legal provisions, or new legal solutions for the control of social problems which seemed to conflict with the maintenance of respectability. This perspective involved sacrifice to the needs of society as a whole of the actual needs of individual women who fell foul of stereotypical expectations, to safeguard the underpinnings of the age. My own chapter further explores the issues of respectable masculine responses to perceived threats to the status quo. The gilt-edged male triumphalism found amongst the middle classes in particular was accompanied by fear that this could be short-lived unless action was taken to export their perspectives on individual behaviour and so eradicate unacceptable expressions of masculine behaviour in particular, something that, in an earlier chapter, has been underlined by Sarah Wilson's explorations of white-collar crime – offending that was quintessentially masculine and middle class. It was this sense of insecurity about the enduring nature of respectable masculinity that helped to fuel the rapid adoption of new printing technology by patriotically-minded commercial publishers such as John Cassell, James Nisbet and John Snow, producing what could certainly be termed a Golden Age of publishing, given the financial profits thereof. These men, and others like them, sought to develop a publishing industry that produced high-quality wares that could peddle in an attractive form the various messages associated with respectability and its profits. They were successful to the extent that they came to dominate Victorian publishing, but it is open to question how far the messages they peddled were accepted at anything beyond the superficial levels of socialization by their target masculine audiences – and this chapter leaves unexplored the pressures placed on women through similar texts to conform to certain standards of gendered behaviour, and their reaction.

These chapters also contain within them a comparative element, identifying current gender behaviour with past precedents, for good or ill. The procedure of characterizing an age primarily through reference to the past, and so identifying it as in some way an age of optimism or pessimism, with all the implications that had for the regulation of social relationships, was neither new nor exclusive to the mid-nineteenth century. To this day, historians' assessments, both contemporary and subsequent, of the character of a period depend heavily on linked estimations of the nature of a previous period. The whole historiography of 'declinism' in Britain from the 1870s is dependent upon the comparison with this previous Golden Age, for instance. The point has already been made that commentators of these mid-century decades sought to define, explain and perpetuate the prosperity of their time by reference to a set of incidents and beliefs against which contemporary cultural complacency could be justified, and threats to its maintenance identified. A sense of the relation of history to the present was thus central to mid-century comprehensions of their good fortune, as was particularly explored through the emphasis on biographical stereotyping of respectable masculinity.

However, more than anything else, these chapters underline the fragmentary nature of the Golden Age. As shown earlier in this volume, there were undoubted locales of success, and this was associated strongly with masculinity, modernity and industrialization. But for those individuals or groups in society who did not, or could not, participate in maintaining the edifices of respectability according to the parameters of gender expectations, the period could be harsh in the extreme. This volume has sought to pursue a reconsideration of the Golden Age as the genesis of modernist, mechanistic economic and social thinking through a number of major themes – technological, economic, cultural. The contributors to this section have sought to utilize aspects of the more recently-established discourse surrounding gendered perspectives on the past, while linking their assessments to earlier areas (agricultural labour) and themes (respectability) in order to provide a fuller picture of the complexities of the period.

Chapter 15

'Physically a splendid race' or 'hardened and brutalised by unsuitable toil'?: Unravelling the Position of Women Workers in Rural England during the Golden Age of Agriculture

Nicola Verdon

By the middle decades of the nineteenth century the rural woman worker had emerged as a distinct social problem in the eyes of many contemporary commentators. In particular, those who participated in outdoor field labour – most notoriously gang-labour – were vilified as unnatural and brutalized women. Richard Heath, a prominent writer on the nineteenth-century countryside, commentating on the Sixth Report of the Children's Employment Commission which had exposed the workings of the gang system in eastern England, wrote:

> Let anyone look at the Sixth Report ... and he will find a tale of horror as to some facets of social life amongst the labouring people of the fen districts...By such authoritative testimony the cause is shown to be mainly due to the destruction of the material instinct in women whose lives are hardened and brutalised by unsuitable toil and continual contact with moral corruption, and by the neglect that must ensue when they are obliged to leave their babies to the care of others.[1]

Not all women fieldworkers were condemned in this way, but there is a good deal of evidence to suggest that the prevalent Victorian attitude to women fieldworkers was one of forthright disapproval.[2]

---

[1] Richard Heath, *The Victorian Peasant*, 1st edn 1893 (Alan Sutton, Gloucester, 1989), p. 156.

[2] For example, J.E. Henley, who reported on Northumberland and Durham in the late 1860s, was more approving of female agricultural labourers. He argued that, 'The Northumbrian women who do these kinds of labour are physically a splendid race; their strength is such that they can vie with the man in carrying sacks of corn, and there seems to be

Rural working women had become visible in the public arena at the height of the agricultural Golden Age. As Stephen Caunce has already argued in this volume, the period between the repeal of the Corn Laws in 1846 and the onset of depression in the mid-1870s represents a clearly defined era of high farming in English agriculture. Some nineteenth-century observers were clearly uneasy that the demands of the increasingly specialized agrarian system partly relied, in some areas at least, on the labour of women. They drew upon middle-class constructions of femininity, domesticity and sexuality to demonize working women who stood outside their ideal of the rural paternalist social system.[3] Economic dependence on men was just one of a number of attributes assigned to the model woman of the Victorian age. But how far this ideology impinged on the lives of women who lived and worked in the mid-nineteenth-century countryside is unclear. Indeed, it is yet to be established how far rural women benefited at all from the agricultural Golden Age. Did a rise in agricultural output and improved profits for large-scale farmers translate into wider opportunities for women to work in mid-nineteenth-century agriculture? Or was the attention that rural labouring women received at this time totally out of proportion to the actual numbers involved in agricultural labour? This chapter will attempt to address these issues by focusing on the employment opportunities available to labouring women of the English countryside in the period between 1846 and 1875. It will attempt to establish the types and amount of work that women participated in, and how far, women were able to contribute to the rural family economy in the mid-nineteenth century. Evidence from two counties will be utilized: the East Riding of Yorkshire and Norfolk. Both were located east of James Caird's division between the corn and grazing districts of nineteenth-century English agriculture.[4] The two counties had undergone sweeping agricultural change so that by the middle decades of the nineteenth century both were characterized by enclosed fields, arable production and innovative methods of farming. Such districts of eastern England were seen to epitomize the best features of farming in the Golden Age. Opportunities for women to work in the formal economy of both counties were essentially restricted to agriculture. There was little industry in either county. However, distinct differences between the areas remained. In particular, the persistence of divergent methods of hiring rural farm labour meant that the types and amount of work available to women varied. An exploration of the

---

no work in the fields which affects them injuriously, however hard it may appear'. *First Report from the Commissioner on the Employment of Children, Young Persons and Women in Agriculture*, Joseph J. Henley, Report on Northumberland and Durham, Parl. Papers XVII (1867–68), p. 55.

[3] See Karen Sayer, *Women of the Fields: Representations of Rural Women in the Nineteenth century* (Manchester University Press, Manchester, 1995), for a full discussion of the ways rural women were depicted in the nineteenth-century media.

[4] James Caird, *English Agriculture, 1850–51*, 2nd edn (Longman, London, 1852).

regional distinctions in rural women's work at this time is therefore made possible by this approach.

As we have already seen in this volume, the East Riding of Yorkshire was unique in the nineteenth century, being the only arable county to persist with the system of hiring yearly farm servants – both male and female – as an essential element of its labour force. The area was also renowned as a high-wage one.[5] The hiring of servants remained the most reliable means of procuring and retaining labour for the whole year in East Yorkshire, a county distinguished by large farmsteads and sparse population settlements. June Sheppard's analysis of the 1851 census indicates that farm servants were the largest single category of the agricultural workforce on the Wolds in that year, and their share of the total agricultural workforce actually increased during the mid-Victorian period.[6] The year-round labour of hired servants was supplemented by the day-labour of married men, women and children from local villages. Norfolk stood in stark contrast to this. It was a low-wage region, with an increasingly casualized workforce susceptible to spells of underemployment. The system of hiring living-in servants had broken down in this region in the early nineteenth century. Labour surpluses after the cessation of war in 1815 meant farmers could dismiss servants in large numbers, hiring labourers by the week, day and even hour in some circumstances. The casualization of the rural Norfolk workforce by the mid-nineteenth century is shown clearly by Alun Howkins. He has calculated that in the 1840s a ratio of two regular to three casual workers at nominal times of the year could alter to three casual to one regular worker at harvest.[7] It was only in the 1870s, as the agricultural depression set in and real wages rose, that the labour market began to shift in favour of the rural agricultural labourer in Norfolk.

Farm service in the East Riding represented a distinct phase in the life cycle of rural women. Girls left home to go into farm service between the ages of twelve and fourteen. This system of hiring farm labour continued to offer young single people an important opportunity to leave the parental home and establish themselves as independent wage-earners, relieving 'their parents houses from overcrowding'.[8] As a consequence, few unmarried women were employed as day-labourers in agriculture as most had entered service by the age

---

[5] George Legard argued, 'There is no part of the Kingdom where the wages of the agricultural labourer rule higher than in the Riding'. G. Legard, 'Farming of the East Riding of Yorkshire', *Journal of the Royal Agricultural Society*, 9 (1848), p. 103.

[6] J.A. Sheppard, 'East Yorkshire's agricultural labour force in the mid-nineteenth century', *Agricultural History Review*, 9 (1961), pp. 48–50. This analysis is based only on the male workforce, however.

[7] Alun Howkins, *Poor Labouring Men: Rural Radicalism in Norfolk, 1870–1923* (Routledge and Kegan Paul, London, 1985), p. 9.

[8] Report by Hon. E.B. Portman, on the Counties of Yorkshire and Cambridgeshire. Evidence to Portman's Report, Parliamentary Papers, XVII (1867–68), p. 366.

of fourteen. Young women hired into farm service in East Yorkshire were responsible for a number of tasks. These included preparing food for the household, cleaning the farmhouse, washing clothing, running the dairy, taking care of poultry and other small animals, and assisting in the fields at peak seasons. The tasks individual women performed depended on the size of the farm, the number of male servants employed and whether other female servants were engaged on the same farm. However, contemporary descriptions of female servants' work in the mid-nineteenth century highlight the juxtaposition of household and outdoor tasks for women of all ages. One contemporary wrote that a dairygirl was expected to 'milk five or six cows every day and assist in the harvest-field and other out-door work'.[9] Specialization in Norfolk agriculture meant there were few opportunities for single young women to find permanent work in the region. This contributed to a drift into domestic service in the county in the second half of the nineteenth century. The onset of agricultural depression in the 1870s exacerbated this process.[10] Thus, women employed in Norfolk agriculture were engaged only if and when needed, as a complement to male and child labour. There was not the absolute distinction between married and single workers that existed in East Yorkshire. Indeed, it was this aspect of agricultural work in Norfolk – the seemingly indiscriminate mixing of ages and sexes – that most outraged mid-nineteenth-century commentators.

Farm servants in East Yorkshire seem to have benefited from the prosperity of mid-Victorian farming. The *Hull Advertiser* reported in 1854:

> The condition of the farm servants generally, from their outward appearance in dress etc., the conclusion would be that they are in comfortable and thriving circumstances especially when compared with their costume and general demeanour a few years since.[11]

Servants' wages began to rise briskly in the 1850s and continued to be buoyant across the 1860s and 1870s. At Driffield in 1856 'Servants stood out for higher wages' and at Bridlington in the same year wages were 'rather high for all descriptions of servants'.[12] In this period of high wage demands farmers often responded by substituting less experienced and cheaper labour for the more experienced, more expensive male servants. According to newspaper accounts

---

[9] Rev. M.C.F. Morris, *Yorkshire Reminiscences* (Humphrey Milford, London, 1922), p. 311.

[10] In Norfolk the number of servants increased by some 75 per cent between 1871 and 1891. Young women were also migrating further afield: by 1881 there were over 57,000 Norfolk born women working as servants in London and the South East. See Alun Howkins, *Reshaping Rural England: A Social History, 1850–1925* (Harper Collins, London, 1991), p. 13.

[11] *Hull Advertiser*, 18 November 1854.

[12] *Hull Advertiser*, 15 November 1856.

the demand for female servants was sustained throughout the mid-nineteenth century across the county. In Howden in 1858, for example, it was reported that 'female servants hired well' with the 'demand for foremen and older class of farm labour less brisk than usual'.[13] By 1870 the older experienced female servants at Driffield could expect to be paid up to £20 a year, general servants £14 a year and girls first entering service between £7 and £8.[14] It was only in the late 1870s, with the onset of depression, that farmers began refusing wage demands and servants submitted to reduced wages. Throughout the mid-Victorian period, experienced and general female servants earned around 60 per cent to 70 per cent of their male equivalent's wages, whilst younger girls were paid yearly wages nearer to their male counterparts than any other class of female servant: the gap between men's and women's wages thus began to widen with age and experience, although it was influenced by other external factors such as the state of the agricultural and labour markets in any particular year.

Local newspaper accounts suggest that female farm servants in East Yorkshire during the peak of high farming were very much in demand and well paid for their services. Interestingly, census figures for the county convey a different picture. According to census returns the number of female farms servants peaked in 1851 and declined significantly in the period 1851 to 1871. It is probable that general female servants on farms are one likely source of under-renumeration in the mid-nineteenth-century census returns, and we cannot take these figures at face value. However, a number of changes in the nature of female farm service were taking place in the 1860s and 1870s and this certainly was not an unchanging institution.[15] There is some evidence to suggest that farmers may have been pushing women out of the dairy and away from other outdoor work, encouraging a more segregated workforce on the farm, with women increasingly confined to indoor labour. In 1869 at Driffield, the practice of turning the work of the milkmaid over to male servants was noted, which had led the dairymaids to 'make a stand against carrying the milking pail in the future'.[16] This process, as Gary Moses argues, helped to facilitate the acceptance by women servants of segregated hirings at the Martinmas fairs, indoor hirings mirroring the 'already established trend

---

[13] *Eastern Counties Herald*, 18 November 1868.
[14] *Eastern Counties Herald*, 17 November 1870.
[15] Edward Higgs has questioned the distinction between farm and domestic service and the former's decline in favour of the latter. General servants on farms, he argues, were probably officially recorded as occupied in the final tables of the occupational censuses, but placed in other economic categories. E. Higgs, 'Occupational censuses and the agricultural workforce in Victorian England and Wales', *Economic History Review*, 68 (1995), p. 707.
[16] *Eastern Counties Herald*, 18 November 1869.

towards gender segregation within the farm service labour force'.[17] Women were not passive bystanders in this process, however. By the late 1870s and throughout the last quarter of the century, there seems to have been an increasing unwillingness on the part of women to be hired into rural service at all. Virtually all newspaper accounts by the 1880s note the scarcity of female servants; instead, they were attracted into town service. Evidence suggests that as farmers increasingly moved female servants into the domestic sphere of the farmhouse, women themselves were moving away from farm service altogether. Although this process accelerated after the agricultural depression of the 1870s set in, the beginning of the process can be seen at the end of our period and may have been one case where prevailing middle-class notions of 'respectability' and 'fit' work for women were impinging on rural women's consciousness and affected how they viewed their work.

In the East Riding, the prosperity of mid-Victorian farming induced farmers to increase their workforce of servants to ensure maximum cultivation of the land. The same prosperity in Norfolk underpinned the evolution of the most casualized – and most notorious – aspect of female agricultural labour, the gang system. This system originated in the Norfolk parish of Castle Acre in the 1820s. Historians have argued that as more and more land was brought under arable cultivation during the mid-Victorian boom, the need for huge amounts of extra labour rose, fuelling the persistence of ganging to the 1870s. The best surviving evidence we have on ganging comes from the parliamentary reports of the 1860s. A close reading of these reveals that the participation of adult women in agricultural gangs in mid-nineteenth-century Norfolk was probably much less than both contemporaries and historians have suggested and was avoided by labouring women whenever possible. If we look at James Fraser's report for the 1867–70 Royal Commission this becomes clear. He collected data from four Poor Law unions in the county: St Faith's in central Norfolk; Depwade, bordering Suffolk to the east; Docking in the north; and Swaffham in the west. In the evidence attached to his report, covering in total 127 parishes, only nine mention the existence of public gangs within their borders. Only in Swaffham, the most purely agricultural region, was the gang system found to prevail extensively. In the Docking Union, a district of large farms, sparse population settlements and light soils – factors which perpetuated ganging around Swaffham – Fraser writes, 'The gang system exists, but to a much smaller extent than might have been expected under the circumstances'.[18] In the Depwade Union the system was reported to be dying out and a number of respondents argued that residents of this areas would not understand the

---

[17] Gary Moses, '"Rude and rustic": hiring fairs and their critics in East Yorkshire, c. 1850–75', *Rural History*, 7 (1996), p. 168.

[18] Report by Rev. James Fraser on Norfolk, Essex, Sussex, Gloucester and parts of Suffolk, Parliamentary Papers, XVII (1867–68), p. 7.

meaning of the term. Thus, by the late 1860s the existence of ganging was very regionally based in the western portions of the county. Whilst it is clear that gangs did exists in mid-nineteenth-century Norfolk and were important to the cultivation of large farms in some instances, it seems the system became a *cause célèbre* at this time and its existence – both in terms of the areas in which it was adopted and the number of women (and children) employed under it – has been exaggerated ever since. Ganging in fact represented an employment opportunity to only a small proportion of women in rural Norfolk and the controversy it aroused appears to have been significantly disproportionate to its extent.[19]

Female farm service and gang labour can be seen to represent two extremes on the spectrum of women's agricultural work. These systems of hiring labour show how diverse rural women's experiences of work could be during the Golden Age. In contrast, one of the most striking features of women's day-labour across the two regions is its similarity. In both Norfolk and East Yorkshire a rigid sexual division of agricultural labour is evident. Women were utilized for a number of specific agricultural operations. These were mostly associated with cleaning the land by weeding, stone-picking and hoeing, and in planting and harvesting root crops such as potatoes, swedes and turnips. In both counties women were employed in haymaking. There were other specialist crops such as flax which women tended. Women were seen as being more suited to cleaning and planting jobs. George Legard noted in 1848, for example, with regard to turnip-hoeing, 'women are employed for the purpose, and it is thought that they are more adroit at this work than men'.[20] The sexual division of labour was not identical or unchanging, however. The employment of women in the corn harvest was more widespread in East Yorkshire, whilst women's main role in Norfolk revolved around the customary right of gleaning the harvest fields once the crop had been gathered. Other agricultural operations such as ploughing, hedging, ditching and taking care of livestock were specifically male jobs in both counties. Another persistent feature of female agricultural labour was the male–female wage gap. Across both counties, the wages paid to women for agricultural labour were considerably less than men's, and usually varied at between a third and a half of the male wage. For instance at Sewerby Home Farm, near Bridlington, in 1861, women were paid 10d a day, men 2s 4d. At the Old Hall Farm, Hoveton St Peter in Norfolk in 1871, women received 8d a day; men between 1s 8d and 2s 0d.[21]

---

[19] For a fuller discussion of this see Nicola Verdon, 'Changing patterns of female employment in rural England, c.1790–1890', unpublished PhD thesis, University of Leicester, 1999, ch. 4.
[20] Legard, 'Farming of the East Riding', p. 111.
[21] Brynmor Jones Library, University of Hull, DDLG 43/5–15, Farm and private accounts, Lloyd-Greame family of Sewerby, 1821–93; University of Reading, NORF 9.1/1–75, Farm account books, Neatishead, 1859–1938. Although these accounts are classified as

However, there were regional differences in day-labour rates between the two counties: Norfolk women earned on average a third less than their northern counterparts.

With some exceptions, the tasks female day-labourers in agriculture were engaged to perform were fairly uniform across both counties in the mid-nineteenth century. However, the amount of work available to women varied both between counties, and between different parishes of the same county. Regularity of employment was influenced by local farming patterns and crops grown on individual farms, by male work patterns, by local routine and custom, as well as by women's individual life-cycle circumstances. East Yorkshire women were more extensively employed as day-labourers in agriculture in the Wolds and the Vale of York. In the latter region, large-scale potato cultivation meant full-time employment year round for women who wanted it. However, in Holderness, women's agricultural labour was utilized to a smaller extent and seems to have been declining over the mid-nineteenth-century period: at Sewerby Home Farm in 1861 women performed 19 per cent of total days worked on the farm, men 74 per cent and children per cent. By 1881, women performed just 7 per cent of total days worked.[22] Evidence from farm accounts also suggests that women's day labour was also becoming increasingly marginalized on Norfolk farms after the 1850s. At Flitcham Hall Farm over the period of the mid-Victorian Golden Age the utilization of female labour becomes more circumscribed and a shift in favour of the male worker is evident: 71 per cent of the overall farm labour expenditure in 1851 went on male workers, 10 per cent on women. Child labour and task-work payments accounted for the remainder of the expenditure. By 1872 only 6 per cent of labour expenditure for the year is spent on women, 84 per cent on men.[23]

In terms of agricultural day-employment at least, it could be argued that the mid-nineteenth century was no Golden Age for women workers in rural England. Women who lived in the East Riding had more security of employment, especially unmarried women hired as farm servants. On the whole, however, the expansion of arable production in the mid-nineteenth century was not accompanied by an increase in work opportunities for rural women. Indeed archival records from both counties suggests that on some farms, the female-day labour force was being increasingly disposed of at the time. However, despite the restrictions on women's agricultural labour in the period, it is clear that the money women earned from paid work could still

---

belonging to a farm in the parish of Neatishead, by cross-referencing details with relevant census returns it became clear the farm was actually the Old Hall Farm in the neighbouring parish of Hoveton St Peter.

[22] Brynmor Jones Library, DD43/5–15.

[23] University of Reading, NORF P429/3, Farm records of Flitcham Hall Farm, Flitcham, including account books, August 1847–October 1852.

make a considerable difference to individual family incomes. This impact can be highlighted through a reconstruction of agricultural family earnings from mid-nineteenth-century Norfolk farm records and census material.

Table 15.1 Female Labourers on Flitcham Hall Farm, 1851

Name of Woman	No. Days Worked in 1851	Age	Marital Status	Occupation of Husband	No/age of Children at Home	Amount Earned in 1851	Occupational Description in 1851
Bet Bridges	n/a	46	Married	Blacksmith	8 (21,18,16, 12,10,8, 5,2)	£10 9s 6d	Blacksmith's wife
Ann Bridges	n/a	30	Widow	-	3 (11,10,7)	£1 2s 1d	On parish allowance
Susan Fickle	n/a	32	Married	Farm lab.	None	£3 11s 10d	Labourer's wife
Mary Fulcher	n/a	37	Married	Farm lab.	3 (12,10,8)	£3 17s 10d	Labourer's wife
Peggy Howard	n/a	31	Married	Farm lab.	None	£4 12s 6d	Labourer's wife (farm)
Fanny Thistle	n/a	33	Married	Farm lab.	None	£5 6s 5d	Labourer's wife

Source: University of Reading, NORF P429/3; Norfolk Local Studies Library, Census enumerators book, Flitcham parish, 1851.

Table 1 shows women regularly employed on Flitcham Hall Farm in west Norfolk in 1851 and information recorded about them by the census enumerator. With the exception of Bet Bridges and Ann Bridges, all women were married to workers on the same farm: Peggy Howard to John, Susan Fickle to William, Fanny Thistle to Frances and Mary Fulcher to John. The women who worked alongside their husbands earned between £3 11s 10d and £5 6s 5d over the year. This amounted to between a fifth and a quarter of their husbands' total earnings on this farm. Ann Bridges, a widow aged thirty, was recorded as being 'on parish allowance' although she earned £1 2s 1d at harvest time. The wages of both men and women are likely to be an underestimate of their actual earnings, however: men worked by the piece at hay and corn harvests and women earned additional sums stonepicking by the acre, although these figures were not noted under individual labourers in the accounts. It is clear that none of these women workers at Flitcham were returned as being employed, the enumerator adding the word 'wife' to the occupation of the spouse. This evidence points to the omission of the part-time, casual agricultural work of women in the second half of the nineteenth century and adds more weight to the notion that the real size of the female agricultural workforce in the nineteenth century was significantly under-recorded in official sources.[24]

At Flitcham in the 1850s, the evidence suggests that where women were employed in agricultural work, their contributions to annual family income could be quite substantial. Three of these women had no children living at home and demands on both time and money would have been less pressing. However, the woman who worked most consistently is the one with the greatest number of children at home. Bet Bridges was nearly fully employed year-round at Flitcham and earned £10 9s 6d on that farm. The fact that Bet worked year-round suggests that financial necessity was the most important consideration here. It is difficult to make any conclusions about the impact of individual women's life-cycle patterns from this one example. But this research suggests that although women's agricultural work accounted for only a small portion of overall annual labour expenditure on this farm, the amount they earned – no matter how small – was still significant to individual family budgets. The importance of these sums tends to be obscured by general observations and women's earnings often made the difference between deprivation and subsistence.

So although women's economic participation in the formal rural labour market of the mid-nineteenth century was small, in terms of individual family economies it was often vital. Moreover, there are grounds for speculating that women's involvement in the rural economy during the Golden Age of agriculture was actually wider than it seems. Women's labour as part of the family group on task-work often went unrecorded as payments were made to the male head of the family. Women's work on smaller, family-run farms also tends to be overlooked as few records from such businesses have survived. It is also impossible to know how far women workers moved between farms in the same locality looking for employment. The fact that a woman was only very casually employed on a farm for a few weeks during the summer does not mean she was not working at all for the remainder of the year. Women also contributed to the rural family household by an undoubtedly large (but incalculable) involvement in the informal, makeshift economy of the mid-nineteenth-century countryside. Women's access to economic resources in the nineteenth-century countryside did not always readily translate into wages or formal employment. Women gleaned, took in washing, ran errands, looked after the sick and elderly, helped with the upkeep of gardens and allotments,

---

[24] See also Judy Gielgud, 'Nineteenth-century farmwomen in Northumberland and Cumbria: the neglected workforce', unpublished PhD thesis, University of Sussex, 1992; Celia Miller, 'The hidden workforce: female fieldworkers in Gloucestershire, 1870–1901', *Southern History*, 6 (1984), pp. 139–55; Helen Speechley, 'The employment of women and children in agriculture in Somerset, c. 1675–1870', unpublished PhD thesis, University of Exeter, 1999; Verdon, 'Changing patterns of female employment'.

tended animals, and gathered fruit, fuel and other valuable resources. In the household they cooked, cleaned, made and mended linen and clothing, took in lodgers, minded children and took charge of domestic budgeting. Few of these tasks are readily measurable in terms of monetary value. Nor is it easy to assess how far they helped to contribute to the rural labouring family's income in the mid-nineteenth century. It is clear, however, that women were central to these survival strategies, and it is only when women's participation in the whole range of informal tasks is considered alongside formal work-patterns that the full extent of women's employment opportunities in the mid-nineteenth-century countryside can be established.

Male agricultural wages only started to increase in the period after 1875 and rural labouring budgets remained extremely inflexible during the period of the mid-Victorian Golden Age. This was even the case in the East Riding of Yorkshire where wages were generally higher than in parts of south-eastern England. Thus, whilst it is important to remember that women often displayed ambivalence to their work and were often active in withdrawing themselves from the formal labour market when they were given the opportunity, any intermittent earnings women procured could still make a significant difference to an individual family's survival. Rural women were becoming conscious of domestic service as a more appropriate alternative to farm service or field-labour, especially for young women. The expansion from the 1850s of domestic service and the employment of other women as laundresses, washerwomen and charwomen may have cultivated a more pronounced outlook of domesticity amongst rural women. Flora Thompson maintains that 'Victorian ideas ... had penetrated to some extent, and any work outside the home was considered unwomanly', in her Oxfordshire village in the 1880s.[25] Yet the ideal of a decorative, non-working woman remained just that: an ideal. Any aspirations rural labouring men and women held to emulate an urban, middle-class lifestyle of gentility and domesticity, continued to be severely undermined by the realities of continuing poverty and struggle in the mid-nineteenth-century countryside.

This chapter has revealed the complexity and contradictions implicit in rural women's working lives. Women who worked in the mid-nineteenth-century countryside were faced with a number of barriers. Some of these were explicitly gendered. The sexual division of labour, for example, ensured that women were marginalized in the rural labour market, and this trend seems, in some areas, to have been escalating at this time. For those women who did work, they could hope to be paid only between a half and a third of the male wage. They could also expect to be increasingly denounced for their labours:

---

[25] Flora Thompson, *Lark Rise to Candleford*, 1st edn 1939 (Penguin, Harmondsworth, 1984), p. 114.

the ideological condemnation of women who transgressed the dominant Victorian stereotype of womanhood gains momentum in the period after 1850. Women were also disadvantaged by their domestic and reproductive roles, although it is clear that any money they did earn could substantially alter the well-being of the family. It is difficult to generalize about the position of rural labouring women during the Golden Age because of the regional variations not only between different counties, but also between different areas within counties. However, this chapter has, on the whole, further added to the pattern of exclusion already established by this volume. That is, whilst the Golden Age of English agriculture in terms of productivity is not in dispute, rural labouring women were another group in society who benefited little in terms of employment opportunities or wage increases.

Chapter 16

The Respectability Imperative: A Golden Rule in Cases of Sexual Assault?

Kim Stevenson

Stimulated by economic prosperity and international superiority, societal confidence in the period 1850–70 collectively portrayed the impression of a 'Golden Age'. However, at a more personal level, individuals felt less secure and assured of their place in the social and moral order. Underneath the surface glitter a considerable unease permeated through society: a concern that the gilt should not rub off. In order to sustain the 'greatness' of the nation, society's moral integrity needed to remain untarnished. As 'respectability' had permeated through all levels of society, it had promulgated the concept that there were underlying moral codes and standards of behaviour for all to follow if they were to maintain their place.[1] For such social manipulation to work, masculine social conformity was practically interdependent on equal feminine adherence, and it was as part of this equation that women came to represent society's aspirational 'icons' within these codes. An ideal woman was presented as virtuous, sexually submissive, and the major conduit for respectability within her family unit. As executor of the domestic idyll she should prevent not only her husband, but also the male populace in general, from being tempted to fall into moral disrepute, thus ensuring against public disorder and instability.

The potency of such definitive codes and stereotypes pervaded society and the establishment quickly assimilated them. Law, in particular, willingly embraced this respectability imperative. The inferior legal position of women and their practical powerlessness meant that law, as a self-referential system, could maintain its integrity and remain true to its inherent patriarchal ideals by absorbing such commonly accepted feminine stereotypes as practical measures of good behaviour. Nowhere is this better displayed than in the operational manner in which the law dealt with sexual assaults, essentially in line with the prevailing concepts of respectability and gender stereotypes. In the courtroom a

---

[1] See F.M.L. Thompson, *The Rise of Respectable Society: a Social History of Victorian Britain 1830–1900* (Fontana, London, 1988).

two-dimensional approach was evident. At first sight, the traditional legal doctrines in relation to the interpretation and application of the law according to its formal rules and procedures were observed, but underneath, more informal rules and presumptions operated, reinforcing the respectability doctrine. As a 'golden rule' the law required, as a matter of practical importance, that both male and female participants conform outwardly to the respectability code. In a context such as the domestic sphere, private negotiation could operate, as Vickery and Davidoff have shown.[2] Consequently, women could establish themselves as competent to negotiate and command respect. Thus, the operation of such stereotypes can be viewed as part of complex and nuanced gender relationships. But in this chapter it is argued that where legal procedures were invoked to deal with cases of sexual violence, the operation of such stereotypes were all too often two-dimensional and so frequently served to victimize women. While it is acknowledged that much work has been undertaken confirming that Victorian women were not passive victims, there were certain areas, notably the law, in which the historian must still accept their victimhood.

The study draws upon a selection of reported cases of rape and indecent assault from *The Times* newspaper between 1850 and 1875. These usefully present not only an impression of the underlying influence of this 'golden rule of respectability' and how it affected trial decisions, but also the way in which certain cases were reported for public consumption. Over 150 cases were examined, mainly from the assize circuits and the London police courts.[3] The use of this type of secondary source is clearly limiting, but as Anna Clark found, from 1796 onwards the Old Bailey suppressed its publication of court transcripts of sexual matters on the grounds of immorality and so press reports are a particularly important source for cases of sexual violence.[4] The *Times* provides a significant historical record. Not only was it highly influential in communicating establishment views and providing an indication of the representation of crime in terms of contemporary ideology, but it was also a significant gauge of the prevailing standards of respectable society.[5] Furthermore, it covered a very wide range of cases, from the police courts

---

[2] Leonore Davidoff and Catherine Hall, Family Fortunes. Men and Women of the English Middle class c. 1780–1850. (Hutchison, London, 1987); Amanda Vickery, 'Golden Age to Separate Spheres? A Review of the Categones and Chronology of English Women's History', *Historical Journal*, 36 (1993), pp. 383–414.

[3] All such references are to *The Times* unless otherwise stated, the date cited refers to the date of publication.

[4] Anna Clark, *Women's Silence Men's Violence: Sexual Assault in England 1770–1845* (Pandora, London, 1987), pp. 17–18.

[5] Rob Sindall, *Street Violence in the Nineteenth Century. Media Panic or Real Danger?* (Leicester University Press, Leicester, 1990), p. 29.

upwards and throughout the kingdom, thus providing a useful summary of the types of cases heard and judgements delivered.

The leading jurist Fitzjames Stephen suggested in his work *Liberty, Equality and Fraternity* that, in relation to the legal status of women generally, 'submission and protection are correlative'. Stephen, reflecting the prevailing attitudes of the time, asserted that if there was a disjunction between these two variables then womankind was more likely to be imperilled by the forces of the social, or 'sexual', contract than by any enacted or general law.[6] The respectability imperative proposed that to enjoy the advantages of (lifelong) masculine protection a woman should invoke that protection through submission. Her 'consideration' in this mutual social contract was to remain pure, chaste and dependent in order to gain the benefit of a promise of marriage. Women who imperilled or relinquished their chastity, especially by wilfully claiming independence, could expect to be stigmatized as personally responsible for any consequences that might ensue, irrespective of cause. Given the emphasis on the woman as the moral custodian of society, this meant that feminine submission could be regarded as the key to solving the problem of public moral order.[7] Stephen had earlier stressed the intrinsic mutuality between the importance of respectability and the reciprocal protection of public opinion in an 1863 critique of Victorian literature written for the *Cornhill Magazine* entitled 'Anti-Respectability', described by Michael Mason as 'one of the period's strongest defences of the double standard'.[8] Any man or woman who attained that 'very indefinite standard of goodness' required of anyone who wished to associate on equal terms with the rest of the human race, could expect to fall under the desired protection of public opinion. Anti-respectability, at least in Stephen's view, comprised the denunciation and trivialization of such societal standards, as often portrayed in Victorian novels of the time.[9] A woman's position in society was intrinsically dependent upon the maintenance of a respectable reputation. The consequences of non-compliance were severe. Any woman who broke this rule risked excommunication from the 'social synagogue', as according to Stephen, all 'the pleasure which her society gives is the only reason why you do associate with her; she stands in no other relation to the world than the social one'. Men, of course, were 'too strong to be held by such bonds', but women, being wholly dependent on others, and 'so much more delicately framed', were

---

[6] Fitzjames Stephen, *Liberty, Equality, Fraternity* (ed.) R.J. White (Cambridge University Press, Cambridge, 1873), p. 209.
[7] Clark, *Women's Silence Men's Violence*, p. 12; Lucy Bland, *Banishing the Beast: English Feminism and Sexual Morality 1885–1914* (Penguin Books, London, 1995).
[8] Michael Mason, *The Making of Victorian Sexual Attitudes* (Oxford University Press, Oxford, 1994), p. 59.
[9] Fitzjames Stephen, 'Anti-Respectability', *Cornhill Magazine*, 8 (1863), p. 282

obliged to obey such social rules ensuring a correspondingly high level of female virtue.[10]

While female conformity ensured, at least to the public gaze, a morally healthy society, in practice such interrelations were less mutual than Stephen supposed because of the impracticability of aspects of this masculine expectation. It was necessary for many, especially working, women to assume certain attributes of independence including physical mobility, without masculine protection, outside the domestic circle. Yet this was not acknowledged and women subjected to sexual assaults could therefore not always rely on the reciprocal protection of public opinion or the law. The plight of women who had broken their side of the respectability contract, often through no fault of their own, attracted less public concern, and in some cases more public opprobrium, than any corresponding societal protection. The underlying social code, as it expressed itself in the operation of the legal system, required that responsible women avoid any compromising situation and be absolutely circumspect about their behaviour. For men encountering the law in such situations, simply the appearance of fidelity to the code was sufficient to satisfy outward conformity: for women, the opposite was generally the case. In the courtroom, the competing strengths of male and female respectability on the part of the defendant and the complainant were often more influential in determining the outcome of the case than any consideration of the actual facts or circumstances. Press representations of the conduct of male perpetrators and female victims reported in *The Times* provide an illuminating insight of the respective male and female stereotypes demanded by the respectability imperative. As these cases suggest, where one party effectively discharged their respectability obligation, through factual evidence and the supporting testimonies of witnesses, the result was more likely to be decided in their favour. However, where competing claims of respectability arose, the outcome was less certain but predictably often weighted in the defendant's favour.

In the public sphere few men of status were convicted of rape, and deference to male respectability was often highly influential in many courtrooms.[11] As Conley found in her study of Kent trials between 1859 and 80, references to respectability were 'ubiquitous in the comments of attorneys, defendants, judges, policemen and journalists'. A charge could be dismissed on the grounds that it was impossible for a 'respectable person' to have committed such a crime, especially if it was a sexual assault.[12] Respectable men might be accused of sexual assault but a rapist only lost his respectability if convicted.

---

[10] Stephen, 'Anti-Respectability', pp. 289–90.
[11] Carolyn Conley, 'Rape and Justice', *Victorian Studies*, 29 (1986), p. 530.
[12] Carolyn Conley, *The Unwritten Law: Criminal Justice in Victorian Kent* (Oxford University Press, Oxford, 1991), p. 173.

Even then his respectability and character were not automatically seriously impugned unless a sentence of hard labour served with common criminals was imposed. Judges often viewed sexual assaults as little more than 'regrettable lapses of self control'.[13] Male respectability appeared to be compromised neither by the commission of sexual misdemeanours nor by the appearance in court on a charge of rape. In contrast, a woman generally forfeited her respectability merely by the fact of having been raped, even though she had neither encouraged nor incited it.[14] A man's high moral character and reputation, his status in the community, profession and position of power in the social hierarchy all served to substantiate his respectability.

Reputation was all. The more witnesses that could testify to this effect, the higher the reputation in the eyes of the court, as in the case of William Davidson, charged with indecently assaulting Annie Dickson aged ten, a 'pretty, intelligent little girl'. Her evidence, that Davidson spoke to her in Victoria Park and took her upon his lap, was confirmed by two park-keepers. (One had detained the prisoner, despite being offered the defendant's watch and 100 guineas to let him go.) But the defence called 'a number of highly respectable witnesses, both male and female' who gave the prisoner a 'most excellent character for morality and general good conduct'. After fifteen minutes' deliberation the jury found him not guilty (*The Times*, 25 October 1860). Conversely, detrimental testimonials could have an adverse effect on the court. Robert Johnson, forty-eight years, approached Sarah Ann Cocking, who was lame, and her three friends in fields near Lincoln one evening, where he 'did violence to her person' three or four times, leaving her senseless. Johnson, a married labourer of 'notorious character', previously acquitted of rape, was swiftly found guilty (10 March 1855). On occasion a bad reputation could also prejudice a defendant even where the victim's reputation was questionable. William Jones appeared at Worcester Assizes for the rape of Sarah Dalton in 1860. The court was informed that five years previously she was a 'common prostitute' but had since lived in 'concubinage' as a mistress. By contrast, Jones's disreputable identity was sworn to by two men of 'good character', ensuring his conviction.

Many a male defendant convinced the court of his respectability simply through judgements invoked by appearance and demeanour. In 1851, Samuel Joseph was alleged to have indecently assaulted a servant in his mother's employ. His description as a young man of 'respectable appearance' accounted for the doubt cast on her testimony (20 June). In 1852, the Associate Institution for Improving and Enforcing the Law for the Protection of Women sponsored one of its first cases of indecent assault on behalf of Mary Burch, aged twelve,

---

[13] Conley, *The Unwritten Law*, p. 91.
[14] Shani D'Cruze, 'Approaching the History of Rape and Sexual Violence: Notes Towards Research', *Women's History Review*, 1:3 (1993), p. 89.

who complained to her mother that Charles Tillotson had 'pulled her about in the hairdressers'. His claim that Burch was mistaken about his intentions was accepted. It seems probable that criticism of the association by the judge for prosecuting the case was linked to the description of the defendant as 'a respectable married man' and painter by trade (11 March).

The nature of the defendant's profession and his employment record were also important factors. Private James Bedford, who had won three medals in the Crimea and had an 'excellent character' and 'good references' was acquitted at Middlesex Assizes of indecently assaulting two girls under ten years (14 July 1857). Defendants of a religious disposition or calling were also likely to have an advantage. Though convicted of assault with intent to commit rape, Job Lawrence, forty-four years, well-to-do and married with a child, escaped transportation because he was a 'God-fearing' Wesleyan Methodist (6 April 1850). As for an allegation of rape upon a twelve-year-old girl in Newport, the defence stressed that it was a 'monstrous impossibility' that the Reverend John Joseph Corley, a Roman Catholic priest, would commit such an act in a room over his chapel (28 March 1852). He was found not guilty. However, even a spiritual reputation could be susceptible to the unwavering attestation of innocents. In 1859, the Reverend Henry John Hatch was accused of indecently assaulting two sisters, aged eleven and eight, at the Old Court. 'Lots of witnesses' testified as to the minister's good character but the girls' stories, which he claimed were 'abominable lies', could not be shaken. The court therefore felt it had no other choice but to convict despite his pleas that two years' hard labour would 'crush and ruin him' (2 December 1859).

Newspaper reporting underlines that individuals in positions of power and influence could more readily establish their respectability. Francisque Michel was charged at Clerkenwell Magistrates Court with raping Ellen Lyons, fifteen years, a domestic servant employed at his lodgings in Bloomsbury. The *Times* reported that the court was told that the 'offence was committed despite her cries and struggle'. Even though Ellen failed to inform her mistress until the next day, Michel, initially described as a foreigner claiming to be a 'professor', was arrested, protesting that he did not fully understand the nature of the proceedings and offering to pay to avoid his name being published and his reputation tarnished. Despite Ellen's tardiness in reporting the assault the magistrates believed her account, and Michel was remanded in custody to the Central Criminal Court (6 September 1856). However, at the higher court, Michel's persona was suddenly transformed from a 'shady foreigner' into 'a man of very gentlemanly appearance', on a mission from the French Minister of Instruction to collate certain manuscripts from the British Museum. The subsequent Times account shows that this had a profound impact on the jury's reaction to Ellen's essentially unchanged evidence and demeanour, both of which thereby became discredited. Seemingly, Michel was acquitted, largely because of who he was (19 September 1956). Thus, the courts were often less concerned with meeting the obligation of discharging the burden of proving or disproving a case on its facts, than with reputational influences.

The respectability imperative could also be used in mitigation to minimize the sentence passed on men. At Lambeth Police Court, Alfred Adams, middle-aged and of 'gentlemanly appearance', was accused of 'disgusting and indecent conduct' towards three little girls, the daughters of 'a respectable tradesman' who witnessed the incident together with a policeman. Adams claimed that he had been 'indiscreet only' and though convicted was saved from the additional disgrace of a public trial because of his 'high character' and sentenced to six months' hard labour (20 September 1856). Similarly, George Tucker was only sentenced to three months' hard labour for indecently assaulting Eliza Bonfield aged fourteen, as he was a 'respectable looking young man' (5 August 1859). Any indication that the defendant had acted in a 'chivalrous' manner might also help his cause. John Robinson was tried at York Assizes for raping a bridesmaid. On admitting that in 'a moment of excitement' he had indecently assaulted her, the judge became more lenient apparently imposing a lesser sentence of two years' imprisonment because her reputation and virginity remained intact, thus, it might be implied indirectly safeguarding Robinson's respectability (21 December 1850). The court could also use the option of finding defendants guilty of the lesser offence of common assault instead of convicting for the more 'disreputable' indecent assault; especially where conflicting claims of respectability arose. In 1851, at the Central Criminal Court, Samuel Candler was charged with indecently assaulting the wife of Thomas Bailey. Both parties 'were respectable', she a 'highly respectable wife' and he a collector of the Queen's taxes who called at her house. Candler was acquitted of sexual assault but found guilty of the 'milder' common assault and fined 1s (28 November).

The contrast with attitudes towards women who alleged sexual violence is acute. Since a female complainant's respectability was dependent upon conformity to social stereotyping, her conduct had to be unimpeachable for her to be regarded as a credible witness. Not only had she to convince a highly suspicious court that she had neither invited, nor consented to, the violation, but also to demonstrate that she had resisted the defendant's advances to the utmost of her physical ability. Judges and juries defined a woman's consent by her character, not by her own desire or discourse.[15] A woman's reputation required constant review and renewal, and often the social practices which secured her neighbourhood reputation might fail to impress a judge or be misconstrued by a potential perpetrator.[16] Practical illustrations of the

---

[15] Clark, *Women's Silence Men's Violence*, p.10
[16] Shani D'Cruze, *Crimes of Outrage, Sex and Violence and Victorian Working Women* (UCL Press, London, 1998), p. 62

respectability imperative can be found in the conduct and imputations of defence lawyers when challenging the complainant's status and actions. A woman who had been 'genuinely attacked' was expected to report the incident immediately – the doctrine of recent complaint. For her story to be believed, corroborating evidence was required in support. A victim was expected to scream and fight back, to make every resistance she could. The expectation was that bruises and other physical signs could corroborate this. Non-compliance with any of these prerequisites threatened her respectability. Whether or not a woman had 'actually' consented depended on her status. If she was found to be unchaste, consent was generally automatically implied unless outweighed by some equivalent aspect of the defendant's bad character. Women who had been 'seduced' often bore the brunt of public censure for the loss of their purity.[17] The respectability pendulum was therefore heavily weighted against the complainant before she even entered the courtroom and the courts were often more preoccupied with scrutinizing her actions and appearance than in examining the rapist's intentions.

The defence, for example, could easily cast doubt on the victim's respectability simply by highlighting the geographical location of an attack. Respectable women did not venture out into public 'unprotected' by their male 'guardians'. Women who did so made themselves vulnerable to charges of immorality. In publicly declaring their loss of sexual innocence as a result of an attack, they were also automatically viewed with suspicion. Judges and jurors frequently concluded that no man should lose his respectability, or freedom, for the mere seduction of such unworthy creatures. As Conley also underlines, while most men probably believed that assaults on respectable women were wrong, the idea that unsupervized women were available was widespread.[18] In 1850 in London, Henry Clifford's lawyer claimed that the eighteen-year-old girl his client had attempted to ravish was 'guilty of great imprudence' in allowing him to walk her home. That she had reluctantly agreed, partly through force and partly through persuasion, and that she was found in an 'insensible state' by her mother with clothes torn and disordered did not prevent an acquittal (12 August). At Clerkenwell Magistrates, a 'respectable young man' called Driscoll received just one month's imprisonment 'for unlawfully attempting to feloniously and carnally know and abuse Emma Gooding as 'although women are to be protected it was partly her fault by going into a dubious area at night' (26 October 1853). Ann Moorhouse's story that she had been seized by her shawl one night in Manchester and dragged up an entry by two young men who tried to ravish and then rape her, was claimed by their

---

[17] Peter Cominos, 'Innocent Femina Sensualis in Unconscious Conflict' in Martha Vicinus (ed.), *Suffer and Be Still – Women in the Victorian Age* (Methuen, London, 1972), p. 165.
[18] Conley, *The Unwritten Law*, pp. 93–5.

defence to be 'implausible' because she had spent over an hour in their company – albeit involuntarily (31 March 1851). Another locale that attracted suspicion was the railway carriage. Women indecently assaulted as the train passed through tunnels found it difficult to convince the courts that they were truthful, even when not travelling alone. Jane Harrison claimed that Edwin Courtenay, 'a gentlemanly-looking man', indecently assaulted her in such circumstances in 1860. She was sitting next to her husband at the time; Courtenay, who denied any involvement, sat opposite. The judge advised the jury that taking into account his 'high moral character' they should consider whether or not they believed him – they did (16 July). Another case that month concerned a complaint against Stephen Holman accused of the 'grossest indecency' upon a young lady travelling with her brother-in-law to Chatham to see her husband. The defence strongly criticized her conduct, claiming it was a fabricated charge because she engaged in indecent familiarities with him. Further condemnation came from another lady who spoke of the 'vulgar levity of the complainant'. There being no other corroborating evidence the defendant was acquitted (14 July).

This provides an interesting contrast with comment on another case where the complainant did fully conform to all stereotypical expectations sufficient to outweigh the impressive status of the defendant. In 1875 a leader in *The Times* commented upon a 'brutal assault inspired by animal passion', the case of Colonel Valentine Baker of the 10th Hussars accused of assaulting Miss Kate Dickenson in a railway carriage (3 August). Baker, alleged to have committed a 'cowardly and unmanly assault', was charged with attempt to ravish, indecent assault and common assault, but having been awarded the Victoria Cross, not only was the colonel clearly a heroic man of valour, but of moral capacity too. For Kate's testimony and conduct to overcome such lofty respectability it would have to be absolutely and stereotypically unimpeachable. Indeed, she gave her evidence 'in a calm, firm and modest way' befitting a 'respectable young lady', testifying that she felt his hand underneath her dress 'on my stocking, above my boot' and that Baker kissed her many times, 'his body was on me and I was quite powerless'. Witnesses confirmed that 'his dress was unfastened' and 'trousers unbuttoned'. Kate reacted in stereotypically correct fashion by trying to throw herself out of the moving carriage and was discovered at the next station clinging perilously to the open carriage door, endangering her life to escape violation. She herself made no formal charge or complaint but her three brothers 'felt they had no alternative' but to try and protect her name and reputation, which was inextricably linked to theirs. Kate's family had put her on the train and had met her at the destination, ensuring she was not travelling 'independently'. For Baker to be convicted of the graver offence of indecent assault, evidence of intent to violate was required. The defence were unsuccessful in trying to shake Kate's respectability as the facts were overwhelming, the complainant of a pretty disposition and her conduct exemplary. In reacting so stereotypically correctly, and backed by her family, her respectability narrowly outweighed Baker's, even given his reputation and position. As it could not be impugned, the jury, all of whom were of a 'highly respectable class', had no option but to convict him – though not of attempted rape, merely of assault. If any of these elements had not been satisfied Baker would have been acquitted and even so, despite the evidence, he was only convicted for a common (not indecent) assault and sentenced to twelve months (without hard labour) and 500 guineas. Mr Justice Brett advised that 'ladies travelling alone must not forget they do incur a certain risk', cautioning that there had been cases of a 'quite different complexion' and many 'men of by no means weak nerves who dread being shut up in a railway carriage with a young woman'. Therefore, however fastidious a person might be about their respectability there was always the unpalatable risk that it might become a hostage to fortune, even in the most commonplace circumstance.

As the Baker case emphasizes, the appearance of the victim was often a crucial factor. Victims were expected to look pretty, modest and virginal. Lack of grace or physical attraction was stereotypically associated with disrepute and sexual experience as shown by a case at York Crown Court. Giles Walmsley and Thomas Ellis were convicted of raping Ellen Johnson, forty years, married with children, who had been forced to leave her abusive husband, a drunken miner. She was 'ravished' until 'nearly insensible' and confined to her bed for three weeks. Described as an 'extremely repulsive looking person' the defence claimed she was a common prostitute and the judge promised that if this were proven, a pardon would be granted (14 March 1853). Prostitutes were, of course, considered as 'public' women and so technically could not be assaulted.[19] On the other hand, an attractive disposition could enhance a complainant's credibility. In the Monmouth case of Philip Lloyd, which was 'too horrible for description', his fourteen-year-old servant alleged that Lloyd had forced her with a carving knife to 'yield to him'. His 'pretty wife' sat on the bedside holding a candle and was a consenting party on one occasion. Lloyd claimed that it was a trumped-up charge and the surgeon confirmed that it was 'doubtful' that rape had occurred. He was found not guilty, perhaps because any such claim of spousely collusion involving his 'pretty' wife could neither be accepted nor believed (2 April 1857). Where the defendant was responsible for the physical protection of the victim the court might be sympathetic. William Wheeler, a landlord, was charged at Middlesex Sessions with attempting to rape Fanny Cracknell, a pretty seventeen-year-old though

---

[19] Ginger S. Frost, *Promises Broken: Courtship, Class and Gender in Victorian England* (Virginia University Press, Virginia, Charlottesville, 1995), p. 110; and see Judith Walkowitz, *City of Dreadful Delight: Narratives of Sexual Danger in late Victorian London* (University of Chicago Press, Chicago, 1992).

'modest-looking', in her bedroom above the public house. Despite the defence's protestations that the court should not be sentimental about the fact Fanny was 'too exhausted' to lock the door after he left, a sentence of four months' hard labour was imposed because, the court stressed, 'he should have been her protector' (1 May 1856). Modest looks could represent respectability but only if supported by unimpeachable conduct. At York Assizes, the 'good looks and extremely modest demeanour' of Martha Powell, aged forty and a respectable married woman, created a favourable impression but did not prevent Daniel Sutherland, a hawker, from being acquitted. Her apparent respectability was called into question when it transpired she had made a previously unsuccessful allegation of rape while in service (17 March 1851).

Complainants were expected to report attacks immediately. Any delay suggested fabrication and was highly prejudicial to any subsequent conviction. Eliza Armitt failed to report immediately that William Page had raped her. She fainted at the time, and on returning to the workhouse attempted to take her own life. However, he was acquitted, as she had made no early complaint (12 June 1853). The requirement of corroboration from witnesses or other means was also highly desirable. Evidence of physical injury through resistance could corroborate a victim's testimony and juries were directed to consider whether or not she 'had made every resistance she could'.[20] Elizabeth Dale, 'a young and respectable' eighteen-year-old, reacted in perfect fashion to her attacker, George Goodwin, a labourer. As she was returning home through fields near Hereford she became alarmed and quickened her pace. Goodwin, on an errand for his employer, ran after her and asked for a 'single kiss', when Elizabeth refused he threw her down and attempted to ravish her. 'She resisted to the uttermost', calling out 'mother', 'murder', scratching his face and in the extremity of terror 'Lord Jesus let me die'. Startled, Goodwin let her go and proposed marriage, but she screamed again and ran away. Finding him guilty his Lordship was concerned that the two-year sentence allowed by the legislature was inadequate. After all, Elizabeth had lived up to the accepted stereotype of an innocent, respectable girl, whereas Goodwin, having three previous convictions, was far from respectable (29 June 1876). Yet in a similar case at the Central Criminal Court, the jury acquitted Joseph Tidy, also a young labourer, who grabbed Susan Mann aged fifteen, and threw her to the ground twice, putting his hand over her mouth and 'committing the outrage'. Tidy, who had been drinking, denied the attack. Witnesses confirmed that she had fainted and walked with evident pain, the surgeon testified that her underclothing was bloodstained and disarranged, 'what had been done must have been done with great muscular force'. The defence dwelt on the 'natural resistance which a woman was capable of making to preserve her chastity' and

---

[20] Per Coleridge J. *R v Hallett* 9 Car & P 747.

submitted that 'her conduct left much doubt as to whether or not she had been a consenting party'. The judge, complimenting the defence, advised that unless the jury was clearly of the opinion that the accused had violated her they ought not to find him guilty. Susan, though satisfying certain elements did not, it would seem, fight back hard enough.

The *Times* cases do not fall neatly and conveniently into particular categories, but they do indicate the existence of a rebuttable presumption, or 'golden rule', that operated against any prosecutrix whose behaviour or conduct stepped outside the ideal stereotype. If a woman compromised her respectability by failing to meet all the demands of the respectability code, her testimony was likely to be regarded as less credible and susceptible to a defence based on masculine repute. Where women, like Miss Kate Dickenson, were presented as completely adhering to the respectability imperative, the court usually had no option but to accept their testimony, even in the face of an esteemed masculine reputation. Hence a subtle shift in the burden of proof is evident; not that the prosecution need prove the facts of the case beyond all reasonable doubt, but that the defence raise any circumstantial element likely to cast doubt on a woman's respectability. Popular opinion, as represented through the press, might dictate a public stereotype of the ideal 'respectable woman', but in practice the law could often be seen to stipulate exactly who or what the ideal typification was. Judges and lawyers were in a position to enforce such legal imperatives, either directly or indirectly. The law can therefore be seen to be reacting to public opinion in a way that condoned those absolutes, thus giving a greater dimension to public expectations and, in the case of sexually-violated women searching for justice, demanding absolute conformity to the 'golden rule' of respectability.

Chapter 17

Keep the 'Whoam' Fires Burning: Domestic Yearnings in Lancashire Dialect Poetry

Catriona Parratt

If, as David Lowenthal suggests, 'historical relativism today makes any golden age an evident fiction', then it is a fiction that has an enduring appeal. Even historians in fields whose hallmark has been exposing the partiality of the kind of interpretation implied when an epoch is accorded the epithet 'Golden', continue to find the construct useful. Feminist historical relativists, for instance, dull the gloss on such treasured periods as the Renaissance by arguing that during these periods women's status and power diminished; but feminist interpretations also rely upon Golden Age narratives. Thus, the mid-1800s are often seen as the nadir of a decline for women from a more felicitous era before industrial capitalism, a Golden Age of the 'family economy' in which they had enjoyed greater freedom and been involved in wider spheres of activity.[1] This interpretation relies heavily upon a particular conceptualization of domesticity and the way in which it figured in the Golden Age narratives of Victorians themselves. As recent work indicates, by the middle decades of the nineteenth century this complex bundle of ideas and practices had a broad currency that extended to working-class people and it was a key element in both reformist and Radical conceptions of a 'better' world. Owenite Socialists, Chartists, short hours advocates, trade unionists and rational recreationists: all conjured up visions of an 'ideal' domesticity that working people had either enjoyed in a previous golden era or would attain in the future once existing wrongs had been righted.[2]

---

[1] David Lowenthal, *The Past is a Foreign Country* (Cambridge University Press, Cambridge, 1985), pp. 23–5, 372 (quote); Joan Kelly, *Women, History, and Theory: The Essays of Joan Kelly* (University of Chicago Press, Chicago, 1984), pp. 2–3, 19–50; Amanda Vickery, 'Golden Age to Separate Spheres? A Review of the Categories and Chronology of English Women's History' *Historical Journal*, 36 (1993), pp. 383–414.

[2] See, for example, Catherine Hall, 'The Tale of Samuel and Jemima: Gender and Working-Class Culture in Nineteenth-Century England' in Harvey J. Kaye and Keith McClelland (eds), *E.P. Thompson: Critical Perspectives* (Temple University Press, Philadelphia, 1990), pp. 78–102; Sonya O. Rose, 'Gender Antagonism and Class Conflict:

Yet for some scholars, working-class domesticity's greatest significance was that it was an ideology that helped immure women in the home and keep them subservient to men.³ The present essay questions this notion by exploring evocations of, and ideas about, domesticity in selected Lancashire dialect poems from the 1850s to the 1870s. This literature has been seen as a particularly important site for the dissemination of an ideology of domesticity that served only men's interests while effacing, ignoring, or denying those of women. I offer a somewhat more complex interpretation here in which dialect poetry is taken to be neither uniformly nor seamlessly ideological. Rather, as Patrick Joyce argues for dialect literature more broadly, there is also to be read in it a yearning for a 'better' way of life, an ideal of domesticity that included gender egalitarianism and mutuality as central elements, that considered women's needs and desires to be as legitimate as men's. My purpose here is to acknowledge this, to balance the recognition of the ideological disposition of mid-Victorian working-class domesticity with an admission of its more idealistic, Golden Age aspirations.⁴

During the Georgian and early Victorian years, self-taught writers from England's artisan and working classes established a distinctive literature of both prose and poetry, but especially the latter. This included dialect poems and songs that were circulated orally, via penny broadsheets and pamphlets, and in a growing periodical press and more up-market collections and anthologies. The urban, industrial culture of Lancashire was a particularly rich fount of this kind of literary endeavour, renowned for 'homely rhymers' such as Edwin Waugh, Samuel Laycock and Joseph Ramsbottom, among others. By the last quarter of the nineteenth century, when the spread of formal education and Standard English increasingly put vernacular culture and language on the defensive, this literature 'of the people', as Brian Hollingworth terms it, lost much of its vibrancy. It became quaint and nostalgic, rather than remaining vital and connected to its working-class constituency. But scholars generally concur that during the mid-Victorian period dialect poetry was one of the more authentic and less mediated expressions of the culture of the northern,

---

Exclusionary Strategies of Male Trade Unionists in Nineteenth-Century Britain', *Social History*, 13 (1988), pp. 191–208; Harold Benenson, 'The "Family Wage" and Working Women's Consciousness in Britain, 1880–1914', *Politics and Society*, 19 (March 1991), pp. 71–108; Anna Clark, 'The Rhetoric of Chartist Domesticity: Gender, Language, and Class in the 1830s and 1840s', *Journal of British Studies*, 31 (1992), pp. 62–98.

³ Susan Zlotnick, '"A Thousand Times I'd Be a Factory Girl": Dialect, Domesticity, and Working-Class Women's Poetry in Victorian Britain', *Victorian Studies*, 35 (1991), pp. 7–27.

⁴ Patrick Joyce, *Visions of the People: Industrial England and the Question of Class 1848–1914* (Cambridge University Press, Cambridge, 1991), pp. 279–304.

industrial working classes, born of the organic unity of its 'author[s], subject matter, and audience'.[5]

Unashamedly 'lowly' in style and subject matter, dialect poetry emerged from a long oral tradition and spoke of and from the experiences of the working masses. Its main concern was the ordinary but vital stuff of their lives and culture – the people, their hard times and good times; life, work, love and death. It celebrated community and asserted the dignity of working women and men, even as it acknowledged and poked fun at their oddities and failings. It admonished and cajoled, celebrated, protested and consoled. Commentators have pointed out, in fact, that dialect poetry did considerably more in the way of consolation than protest, that there was in it a pronounced conservatism and inertia, a tendency to resolve social tensions by resorting to sentimentality rather than direct political action.[6] This conservatism has been seen to extend to gender relations and it is in this regard that feminist scholars have found problematic the genre's frequent descriptions of the pleasures of the home and representations of domesticity. Susan Zlotnick, for one, judges dialect poetry to be a 'pure distillation of working-class domesticity', an ideological mechanism that men employed to endorse a return to 'the natural order of things believed to have existed before the disruptive birth of the factory system'. By this she means that it effaced women's waged labour, refused to grant the services women performed in the home the status of work, saw domestic duties such as childcare as women's sole responsibility, and ignored the oppressiveness of women's situation in marriage.[7]

There is no question that dialect literature, like many other aspects of working-class culture in mid-Victorian England, expressed and reinforced male hegemony and that domesticity was one of the central struts of that hegemony. Zlotnick is able to cite a wealth of evidence of this in the work of well-known writers such as Waugh and Laycock. In addition, she demonstrates that the masculinist and romantic literary traditions in which dialect poets generally positioned themselves muted any potentially oppositional voices. She finds that the few female working-class poets of the period – for example Ellen Johnston, Fanny Forester and the factory operative who wrote under the nom de plume 'Marie' – were equally in thrall to the cult of domesticity as their male counterparts.[8] However, dominant as the ideological tenor of dialect poetry's

---

[5] Hollingworth, *Songs of the People*, pp. 5–7.

[6] Hollingworth, *Songs of the People*, pp. 6 (quote), 7; Vicinus, *The Industrial Muse*, p. 2; Maidment, *The Poorhouse Fugitives*, pp. 13–14, 209, 211, 213, 227, 231, 243; Maidment, 'Prose and Artisan Discourse in Early Victorian Britain', *Prose Studies*, 10 (1987), p. 31.

[7] Zlotnick, "'A Thousand Times I'd Be a Factory Girl'" pp. 10, 25 (quotes), 12–17, 20.

[8] Zlotnick discusses the work of Ellen Johnston, Fanny Forester, and 'Marie', a factory operative who was a regular contributor to *The People's Journal* (1846–51). For biographical detail on these writers and examples of their work, see Hollingworth, *Songs of the People* and Maidment, *The Poorhouse Fugitives*; Ellen Johnston, *Autobiography, Poems and Songs of*

treatment of domesticity is, it is neither as coherent nor as uncontested as Zlotnick implies, particularly if one looks beyond Waugh and Laycock or at a broader sample of their work.[9] Even a moderately wider reading of the literary tradition of which they were a part reveals a number of writers and poems that cut across the grain of the domestic ideology that Zlotnick has identified, and thus disrupts it in important ways.

Edwin Waugh's 'Come Whoam to Thi Childer an' Me' is one of the classics of dialect poetry. First published in 1856, the poem is structured as an exchange between a wife and husband and in it the comforts of home and family serve as a counter to the attractions a man could find with his mates in the alehouse. The husband, having admitted that he likes to indulge in certain pleasures outside the home, nonetheless confirms that:

> ... wherever aw roam,
> Aw'm fain to get back to th' owd ground;
> Aw can do wi' a crack o'er a glass;
> Aw can do wi' a bit of a spree;
> But aw've no gradely comfort, my lass,
> Except wi' yon childer and thee. (ll. 42–8)[10]

As Hollingworth notes, like much dialect poetry 'Come Whoam' tends to the sentimental. It is also typical in its polar configuration of the domestic and public spheres and in casting women as home-makers and men as wage-earners and thus inclines strongly to the ideological. Yet there are other dialect poems in which women do not inhabit only the domestic realm, that show women playing roles other than, or in addition to, those of wife and mother; in which they are producers of commodities for the market and wage-earners.[11]

These works challenge what is often taken to be an uncontested element of the ideology of domesticity: the belief that woman's only fitting social role was as a non-wage-earning, dependent housewife. William Baron's 'Yon Weyver As Warks t'Beam to Me' and 'Hawf Past Five at Neet', Joseph Burgess's 'Neaw Aw'm a Married Man', Samuel Laycock's 'Sewin' Class Song' and 'Bowton's Yard', Joseph Ramsbottom's 'Coaxin'' and the traditional songs 'Th' Owdham Weyver' and 'Rambles in Owdham': all these figure women as

---

Ellen Johnston, the Factory Girl (William Love, Glasgow, 1867); George Milner (ed.), *The Collected Writings of Samuel Laycock* (John Heywood, Manchester, 1908); Edwin Waugh, *Poems and Songs* (John Heywood, Manchester, 1893); Fanny Forester, 'My Poor Black Sheep', *Ben Brierley's Journal* (January 1875), pp. 38–9; 'Homeless in the City' (March 1870), p. 42; 'The Lowly Bard' (November 1873), p. 265; 'Marie', 'Idealise the Real and Realise the Ideal', and 'The Indomitable Will', *People's Journal*, 4 (1851), pp. 175, 63.

[9] For this same point in a less specific context, see Maidment, *The Poorhouse Fugitives*, p. 16.
[10] Waugh, *Poems and Songs*, p. 6.
[11] Hollingworth, *Songs of the People*, p. 139.

men's co-workers in cottage industry, or as waged factory workers. 'Rambles in Owdham', for example, describes the industrial town of Oldham in the 1850s, by which period, Anna Clark suggests, the notion that wives should operate simply within the domestic sphere was pretty much triumphant in working-class culture.[12] The song suggests otherwise, with its glimpse of female operatives hard and skilfully at work in the kind of environment that domestic ideologues considered too hostile for women's 'tender' sensibilities: work that the rambling narrator not only accepts with equanimity, but also seems to celebrate:

> I went into a weavin' shade,
> Un' such o clatter there!
> Wi' looms un' wheels all going so fast,
> I hardly durst go near;
> Then the lasses were so busy
> Shiftin' temples – shuttling cops;
> One shuttle had liked o given me
> O devilish slap o t' chops. (ll. 72–80)[13]

'Coaxin',' a love poem first published in the Lancashire journal Country Words (1866–67), opens with a similar image of a young female weaver absorbed in her work. In this instance, however, it is cottage industry that is depicted, not factory production. It might be argued that this 'domestic' setting is what gives the poem its ideological thrust; but, equally, the woman's action as a skilled, productive and committed worker serves to defuse the power of that ideology:

> Hi thi, Jenny, lyev thi loom,
> There's a bonny sky above;
> Eawt o' th' days we wortch to live,
> We may tak a day to love.
> Wilto stop thi bangin' lathe;
> Come away fro th' neighsy jar;
> Let thi shuttle quiet lie,
> For thi bobbins winno mar.
>
> Fling thi clogs an brat aside;   *apron*
> Let thi treddles rest to-day;
> Tee thi napkin o'er thi yead;
> Don thi shoon an' come away ... (ll. 1–12)[14]

---

[12] Anna Clark, *The Struggle for the Breeches: Gender and the Making of the British Working Class* (University of California Press, Berkeley, 1995).
[13] 'Rambles in Owdham, and Peep Into the Workshops', Hollingworth, *Songs of the People*, p. 84.
[14] Ramsbottom, 'Coaxin', *Songs of the People*, pp. 51–2.

One way in which dialect poetry's treatment of the work identity of women is more consistently in line with an ideology of domesticity is in associating wage labour with younger, unmarried women rather than granting much possibility that wives and mothers might also work for a wage outside the home. Far and away the majority of the female wage-earners depicted in dialect verse are young and single (as they were in real life).[15] It is probably significant, then, that two poems which do feature women who combine marriage and maternity with factory work are by Joseph Burgess, whom scholars judge to have been considerably more radical than most Lancashire dialect poets. Hollingworth points out that Burgess's 'Neaw Aw'm a Married Mon' is atypical for several reasons, one being its assertion that 'woman's work might be a man's work as well'.[16] In this poem the poet accepts the notion of a wage-earning wife and mother and embraces the idea that husbands should assume childcare and other responsibilities in the home:

> Un as hoo's a factory lass   *she*
> Un me a factory lad,
> We'en noather on us brass – *money*
> Aw nobbu' weesh we had;
> Soa we'st booath ha' to work,
> Un it wudno' be so fair
> If aw began to shirk,
> Un didno' do mi share.
>
> Soo aw'st help to mop up stone,
> Help to scrub un skeawr,
> Un do everythin' aw'm shown,
> If it lies within mi peawer;
> Fur, neaw aw'm a married mon,
> Aw'm beawn to be soa good,
> Un do the best aw con,
> To be o' a husbant should.
>
> Aw reckon aw'st ha' t' rock,   *i.e., the cradle*
> Un larn t' mak cinder tay,
> At three or four o'clock,
> When it's happens breakin' day;
> Un other odds un ends,
> Sich as hurryin' eawt foot whot,   *hot-foot*
> When a loife or two depends
> Upo' foindin' Dr. Scott. (ll. 25–48)[17]

---

[15] Louise A. Tilly and Joan W. Scott, *Women, Work, and Family* (Holt, Rinehart and Winston, New York, 1978), p. 126.
[16] Hollingworth, *Songs of the People*, p. 137.
[17] Burgess, 'Neaw Aw'm a Married Mon', *Songs of the People*, pp. 54–5.

Burgess's 'Ten Heawrs a Day' is of a different cast in that it seems to ascribe tragic consequences to maternal wage labour and thus warns mothers away from working outside the home. The poem tells of the death of the child of a female weaver so exhausted by her statutory ten hours of work in the mill that she is unable to care for the infant. But Burgess, far from using this story to inveigh against women's wage labour – as many opponents of female factory employment were wont to do – instead uses the female operative and her child as symbols of the depredations that working-class people as a whole faced under laissez-faire industrial capitalism. He proceeds from recounting her tragic story to making a call on working men to unite in political action that would secure a shorter working day not just for women (a strategy which, a number of historians show, some male workers used to control female labour) but for all workers:[18]

> Up, workin' men, yo'r needs assert,
> No moor be tramplt into dirt,
> Bu', banded in a bawd array,
> Refuse to work ten heawrs a day. (ll. 53–6)

> ... let no odds yo'r courage da'nt,
> Bu' feight until yo'n dun away
> Wi' workin' hard ten heawrs a day. (ll. 70–72)[19]

Thus, in an inversion of dominant Victorian constructions of gender and work identities, 'Ten Heawrs a Day' makes the representative wage-earning worker a mother.

There were few women among the ranks of dialect writers, but more than a few male poets tried to assume the perspective of women in some of their work and this also tends to disturb the smooth surface of ideological domesticity in the genre. Waugh wrote a number of these songs and poems, but I want to consider four others, all by different writers. These are Thomas Brierley's 'God Bless These Poor Wimmen That's Childer', Sam Fitton's 'Th' Childer's Holiday', Laycock's 'The Courtin' Neet: Part Second', and James Standing's 'Wimmen's Wark Es Niver Done (As if bi a womman hersel)'. More than merely sympathizing with woman's lot (though they obviously do that) these poems positively counter the ideological notions that housework is not really

---

[18] See, for example, Michelle Barrett and Maureen McIntosh, 'The "Family Wage"', *Capital and Class* 11 (Summer 1980), pp. 51–72; Robert Gray, 'Factory Legislation and the Gendering of Jobs in the North of England, 1830–1860' *Gender and History*, 5 (Spring 1993), pp. 56–80; Rose, 'Gender Antagonism and Class Conflict'.

[19] Burgess, 'Ten Heawrs a Day', in Maidment, *The Poorhouse Fugitives*, pp. 91–93.

work at all but a 'natural' function like breathing and bearing children, and is too far below men's dignity to warrant them performing it.[20]

Standing structures 'Wimmen's Wark' as a line after breathless line account of an harassed mother's round of domestic chores; chores that demand superhuman faculties and with which her husband never deigns to trouble himself:

> Aw think sometimes aw should be made
> To do beawt rest or bed,
> Wi' double hands at ether side,
> An' een all round mi yed ...(ll. 78–81)   *eyes*

> An' as for him, he takes no part
> I' keepin' corners square;
> Heawiver heedless th' childer be,
> He niver seems to care;
> An' stead o' leyin' on a hand,
> An' helpin' what he con,
> He leovs all t' bits a jobs to me,
> Whol mi warks niver done. (ll. 122–9)[21]

In 'Th' Coartin' Neet: Part Second', Laycock also invokes the image of a husband who is a wastrel but uses him as a foil for his young narrator who, newly engaged to be married, rejects an ideal of masculinity that precludes housework and childcare. Lines 31–2 offer a wonderfully appealing version of working-class masculinity:

> Yo'll noan find me like some; for lo!
> As soon as th' weddin's o'er,
> There's sich a change, theyr'e nowt at o
> Like what they wur before.
> Aw'll turn mi hond to ony job,
> Keep Johhny eawt o'th' dirt,
> Or sit bi th' hob an' nurse eawr Bob,
> While Rosy mends mi shirt.

> Aw never wish to be admired
> For handlin' broom or cleawt;   *cloth*
> But when aw see th' lass getting' tired

---

[20] Waugh, *Poems and Songs*, pp. 23–4, 55–8, 62–5, 93–4, 126–8, 135–7; Brierley, 'God Bless These Poor Wimmen That's Childer!' in John Harland (ed.), *Ballads and Songs of Lancashire, Ancient and Modern* (George Routledge and Sons and L.C. Gent, London, 1875), pp. 402–3; Sam Fitton, 'Th' Childer's Holiday', *Songs of the People*, pp. 78–80; James Standing, 'Wimmen's Wark Es Niver Done (As if bi a womman hersel)', *Songs of the People*, pp. 75–8.

[21] Standing, 'Wimmen's Wark' *Songs of the People*, pp. 77, 78.

> Aw meon to help her eawt.
> Aw'll try an' save her o aw con,
> An' when hoo's noan so well,
> Aw'll poo mi coat off, like a mon,
> An' wesh an' bake misel'. [emphasis added] (ll. 17–32)[22]

Fitton's 'Th' Childer's Holiday', is patterned like 'Wimmen's Wark' but is more patently comic in intent. Still, the humour does not negate the central thrust of the poem: the assertion that a 'tidy whoam'[23] and 'rook o' childer'[24] demanded a degree and kind of hard work too rarely characteried as 'real' labour:

> Eh, dear, I'm welly off my chump!
> I scrub, an' wesh, an' darn;
> Eawr childer han a holiday,
> An th' heawse is like a barn.
>
> Yo talk abeawt a home sweet home!
> My peace is flown away;
> I have to live i' Bedlam for
> A fortnit an' a day. (ll. 1–12)
>
> I'd lock 'em up I' thi' schoo for good
> If I could ha' my will;
> I'd see they had another clause
> I thi' Education Bill.
>
> I've clouted 'em an' slapped 'em till
> My honds an' arms are sore;
> I'st fancy I'm i' Paradise
> When th' holidays are o'er. (ll. 57–64)[25]

Brierley's 'God Bless These Poor Wimmen That's Childer' comes the closest of these works to an apotheosis of women and their domestic labour and much of it is positively mawkish. Despite this, the poem both makes an important statement against the ideology of domesticity by acknowledging that caring for the home and family is hard slog, and challenges men to recognize their own failings as domestic partners:

> God bless these poor wimmen that's childer!
> Shuz whether they're rich or poor,
> Thur's nob'dy can tell whot a woman

---

[22] Laycock, 'Th' Courtin' Neet: Part Second', *Collected Writings*, pp. 145–6.
[23] Ramsbottom, 'Coaxin', *Songs of the People*, p. 51.
[24] Standing, 'Wimmen's Wark', *Songs of the People*, p. 77.
[25] Fitton, 'Th' Childer's Holiday', *Songs of the People*, pp. 78–80.

Wi' little uns has to endure;
The times that hoo's wakken i' th' neet-time,
Attendin' thur wailin and pain,
Un' smoothin' thur pillow of sickness,
Would crack ony patient mon's brain. (ll. 1–8)

God bless these poor wimmen that's childer!
Aw know that they'n mony a fort,
But chaps as no 'kashun to chuckle,
Men's blemishes are not so short;
Then have a kind word for these wimmen,
If t'maddest and vilest o' men
Wurn just made i'wimmen a fortneet,
They'd never beat wimmen agen. (ll. 25–32)[26]

Historians have indicated that the violence against women to which this poem alludes was frequently occasioned by wives' failure to provide the domestic services to which husbands felt they were entitled.[27] One of the achievements of domestic ideologies was to mask these aspects of gender and married relations, to airbrush them out and to paint family life only in the rosiest, coziest hues. This representation of working-class domesticity is perhaps best seen in much of Waugh's work but it is not difficult to find a number of powerful counters to it in poems that expose and challenge some men's thoughtlessness, selfishness and brutality. Though not part of the north-western dialect literature that is the focus here, Ellen Johnston's work should be mentioned in this regard. Johnston was a Scottish textile worker whose autobiography hints strongly that she suffered sexual abuse at the hands of her stepfather. Her autobiography and poems also illuminate the extent of to which men's power – be it as an abusive parent, false friend or lover, driving employer or foreman – governed her life. Several of Johnston's poems are vehicles for her angry protest against the male violence and power she knew at home and at work. 'The Drunkard's Wife', written in Standard English, was for the poet's aunt:

When I look on thee now, and think what thou hast been,
When thy young hopes, unclouded, flew on like a dream –
When I see thee a victim by drunkenness curs'd,
Mem'ry seems but a phantom that fancy hath nursed.

---

[26] Brierley, 'God Bless These Poor Wimmen', *Ballads and Songs of Lancashire*, p. 402.
[27] Ellen Ross, '"Fierce Questions and Taunts": Married Life in Working-Class London, 1870–1914', *Feminist Studies*, 8 (1982), p. 580; Nancy Tomes, 'A "Torrent of Abuse": Crimes of Violence Between Working-Class Men and Women in London, 1840–1875', *Journal of Social History*, 11 (1978), p. 330.

And that child which so fondly thou hold'st to thy breast,
Unconscious of woe, it now slumbers at rest;
Should it live unto manhood, like its father to turn,
Ah! far better for thee it were laid neath the urn.

For its father's a drunkard! The lone hours of the night
Beholds thee poor Hannah, sit trembling with fright,
And the weak dying embers to ashes decay,
Whilst thou wait on his coming till dawning of day.

Alas! Wretched Hannah, how I feel for thy woes,
And I long to behold thee in peaceful repose,
For thy heart wears a history of heartrending strife,
But death will soon release thee, thou poor drunkard's wife! (ll. 41–56)[28]

Among the Lancashire dialect poets, Laycock often raises his voice in condemnation of men's behaviour and to denounce the baneful impact it could have on women and family life. 'Owd Fogey', 'Eawr Jim', 'Uncle Dick's Advoice to Wed Men', and 'A Little Bit of Boath Sides', all either sympathize with the hardships wives endured in marriage or criticize the way in which husbands failed to meet the obligations marriage placed on them.[29] 'Owd Fogey', for example, was known by all as:

> ... sich a foo'.
> Last week he pawned his Sunday clooas,
> An' sowed a favourite tit;   horse
> An' neaw he hasn't a haupeny left,
> He's drunk it every bit. (ll. 4–8)
>
> Ther's nowt ov ony value left,
> Except poor Jane, his wife;
> An' hoo's so knocked abeawt i'th' world,
> Hoo's weary ov her life.
> An' nobbut th' week afore they'rn wed,
> He took her on his knee,
> An' swore he'd allus treat her weel;
> But has he done? not he! (ll. 17–24)[30]

---

[28] Johnston, 'The Drunkard's Wife', *Autobiography, Poems and Songs*, pp. 6–10, 13, 59–61. On Johnston's life and work see Dorothy McMillan, 'Selves and Others: Non-fiction Writing in the Eighteenth and Early Nineteenth Centuries', in Douglas Gifford and Dorothy McMillan (eds), *A History of Scottish Women's Writing* (Edinburgh University Press, Edinburgh, 1997), pp. 83–4.

[29] Laycock, 'Owd Fogey', 'Eawr Jim', 'Uncle Dick's Advoice to Wed Men', 'A Little Bit o' Boath Sides', *Collected Writings*, pp. 67–9, 176–7, 107–9, 170–75.

[30] Laycock, 'Owd Fogey', *Collected Writings*, p. 67.

'Eawr Jim' similarly decries a husband's drunkenness, the damage it causes his family, and the despair it brings to his wife:

> Aw hardly know what to do wi' eawr Jim,
> For he's drunk every neet of his life;
> He's crackin' a skull, or breakin' a limb,
> An' often ill-usin' his wife. (ll. 1–4)
>
> If [Jim] 'd some wit, an' would put it to use,
> He'd buy [his] lad a pair o' new clogs,
> But he'd rayther be spendin' his time at 'Th' Owd Goose,'
> Makin' matches wi' pigeons an' dogs.
>
> It pains me to look at his poor, patient wife
> 'At wur once so good-lookin' an' fair:
> Sich a harrasin', wretched, an' comfortless life
> Must drive her to hopeless despair. (ll. 13–20)[31]

Not only were poets such as Laycock not loath to represent failed domesticity and attribute it to men, they were ready to point out possible remedies. Zlotnick uses 'Toothsome Advice' by Waugh and Laycock's 'Uncle Dick's Advoice to Wed Women' to illustrate how a distinct sub-genre of dialect advice poems placed the burden of achieving domestic content on wives. However, Laycock penned a series of these, each directed to either a female or male agent whom he insisted bore equal responsibility for establishing and maintaining domesticity. 'Uncle Dick's Advoice to Sengle Men' and 'Uncle Dick's Advoice to Sengle Women' counsel on the kinds of qualities for which prospective wives and husbands should look in their betrothed. Laycock cautions young women 'To try an' foind aewt if he's fond ov his books, / Never mind what he wears, nor heaw pretty he looks.'[32] Young men should also beware of being drawn by superficialities, but must:

> ... look eawt for a lass
> wi' some brains an' good fingers, care nowt abeawt brass,
> For iv that's o tha gets ta'll repent o thi loif
> At tha' didn't get howd of a sensible wife. (ll. 37–40)[33]

In his advice to those already married, Laycock declaims against drunken, brutal husbands and slatternly, bad-tempered wives and exhorts both to try and

---

[31] Laycock, 'Eawr Jim', *Collected Writings*, p. 176.

[32] Zlotnick, 'A Thousand Times I'd Be a Factory Girl', p. 15; Laycock, 'Uncle Dick's Advoice to Sengle Men', 'Uncle Dick's Advoice to Sengle Women', 'Uncle Dick's Advoice to Wed Men', Uncle Dick's Advoice to Wed Women', 'A Little Bit o' Boath Sides' *Collected Writings*, pp. 104–6, 101–3 (quote), 107–9, 98–9, 170–75.

[33] Laycock, 'Uncle Dick's Advoice to Sengle Men', *Collected Writings*, p. 105.

'mend a bit.' 'A Little Bit o' Boath Sides' is a slightly more oblique admonitory poem in two parts that presents a marriage successfully made into an ideal of domesticity through just such a dual effort. Significantly – for Laycock was a teetotaler – the first step in the transformation is the husband's signing of a pledge of abstinence.[34]

The foregoing all suggest the complexity of working-class understandings of domesticity as represented in Lancashire dialect poetry. Some poets clearly did not shrink from acknowledging the dark side of marriage and yet could also write as fondly and lyrically of the joys of home life as any of their peers. Laycock's work especially exemplifies this duality and tension, but it is present in dialect poetry as a whole where domesticity as an ideology does not go unchallenged and domesticity as an ideal is shown as a fragile, precious thing towards which both women and men should strive. The vulnerability of the domestic world is a consistent theme. Endemic poverty, disease, death, unemployment, under-employment, hard-hearted landlords, and bullying bailiffs: in addition to the fallibility of wives and husbands, these are all identified as representing a threat to even the most modest working-class aspiration to domestic security and cheer. It is the genre's general tendency to avoid articulating a political analysis of these forces that has led scholars to see it as conservative and consolatory. And 'the invocation of the pleasures of domesticity' has been identified as one of the main strategies that dialect poets used to defuse social and political tension.[35] But it is also feasible to see domesticity as something which working women and men were struggling to construct and to read dialect poetry as an expression of that struggle. Some dialect poets articulated a vision of domesticity that did not preclude a wage-earning wife, that did not insist upon female subservience and self-sacrifice, that did recognize women's desires and acknowledge men's failings, that allowed for, indeed insisted upon, the necessity of equity and mutuality between wives and husbands. They were hardly, therefore, apolitical or quiescent, or the architects and mouthpieces of an uncontested ideology; rather, they were emotionally invested and engaged in the politics of gender, in imagining the kind of family life that might be possible under more propitious circumstances. To deny this tendency in dialect poetry and the culture of which it was a part is to come close to denying that working-class people had the capacity to aspire to such things, to deny them the right to shape their own Golden Age visions. And that is surely something historians – especially those with a relativist bent – should avoid doing.

---

[34] Laycock, 'Uncle Dick's Advoice to Wed Men', and 'A Little Bit o' Boath Sides', *Collected Writings*, pp. 109 (quote), 170–75.

[35] Maidment, *Poor House Fugitives*, p. 227.

Chapter 18

'All our Past Proclaims our Future': Popular Biography and Masculine Identity during the Golden Age, 1850–1870

Judith Rowbotham

An air of confident British masculinity is one of the most readily apparent features of the Golden Age – and it was a complacency stemming from male perceptions of linkages between the increasing levels of individual and national success in the period. But closer examination reveals that, while it may have been the most socially dominant expression of masculinity, certainly amongst middle-class men, it was not the exclusive expression of male identity during this period. It also reveals the extent to which this manifestation of masculine confidence was accompanied by fear – not in the shape of doubts about the virtues and benefits, individually and nationally, of the masculine identity thus represented. Rather, it related to concerns over what were identified as less desirable expressions of masculine identity, because in contemporary perceptions these were associated at mass level with social unrest, and at the individual level with a degree of emotionally-driven illogicality that was seen as damaging to that exercise of reason which was essential to true 'manliness'. While social turmoil was apparently under control after the upsets of the 1830s and 1840s, there was a constant fear that it could return so long as unacceptable individual manifestations of masculinity remained unreformed.[1] There was a class dimension to this perception, in that such threats were widely accepted as emanating mainly (though never exclusively) from working-class men, yet this cannot be laboured too far. Victorian respectability crossed class boundaries, and the purveyors of respectability were not exclusively middle class. Victorian respectable masculinity was essentially what Tosh has termed 'hegemonic' masculinity, particularly when practised by economically successful males, but the targets in this period were quite as much non-compliant males as women,

---

[1] For comments on the issues raised by Victorian masculinity, see John Tosh, 'What should historians do with masculinity? Reflections on nineteenth-century Britain', *History Workshop*, 38 (1994), pp. 179–202, especially p. 180.

usually seen as the traditional targets of attempts at patriarchal manipulation.[2] Bluntly, women as a social category were seen during this period as providing a less acute problem, individually and communally, than the category of males associated by contemporary respectability with recent threats to national prosperity and security.

It is in this context that a proliferation, between 1850 and 1870, of popular biography aimed at working-class men and boys needs to be comprehended, as major planks in an exercise to widen the appeal of respectable masculinity and so to remove, or diminish significantly, threats to national prosperity from that quarter by reinventing the nature of working-men's lives. Examination of this consciously didactic genre is illuminating. It aids more nuanced understandings of the complexities of masculinity, in particular highlighting the characteristics seen by contemporaries as crucial to the achievement of respectable masculinity. It also places such constructs of gender identity in the context of the evolution of the concept of the period as a Golden Age – underlining both the cultural nature of this concept and its crucial connection to respectable masculinity. It is, of course, very difficult to assess the impact of such biography on the intended audience. This is not the focus of this chapter, but it is worth pointing out that one of the undoubted bestsellers of this period, *Self-Help*, widely read by working-class men, falls within this genre. Its discourse is typical, in being essentially a text identifying useful exemplars from the recent past, whose lives were interpreted so as to demonstrate the rewards of living and striving according to a set of what were presented as essentially timeless practical conduct rules for men.[3]

Employing historical contexts for such lessons was not new: history was, stereotypically, a powerful didactic medium. But, as practised by a Macaulay or Gibbon, it was seen as having less relevance to mass audiences. To acquire popular appeal, history needed to be 'humanized' and fitted to the comprehensions of 'ordinary' people, through biography. One feature of this period was the acceptance that the individual had a crucial role to play in the continued success of Britain. An age that happily accepted the grouping of numbers of people into broad categories, stereotyping them according to a range of characteristics presumed to belong to gender in particular, as well as location, social hierarchy, age, income and occupation, was paradoxically eager to assure its denizens that they were valued primarily as individuals, and that their individual contributions were critical to the continued success, prosperity

---

[2] Tosh, 'Masculinity', p. 192.

[3] Here the term 'popular', when used to describe certain types of publication, refers to works intended by producers for a mass reading audience. Such may not always have actually achieved wide circulation, though much of it did do so. In terms of its production, however, it is important to identify this aspect, because it had profound implications for the concepts behind the material, and the messages that it was supposed to convey to its intended mass readership.

and happiness of the State: 'Hence the value of ... biography ... [its] charm for all ages ... [its] mighty power'.[4]

In commercial terms, biographies provided one of the staple assets of any publishing list in this period. The Religious Tract Society (RTS) commented on the 'appetite for biography'.[5] But besides the obvious economic considerations, a primary objective for the mid-Victorian producers of popular biography was the education of their readership in a cultural, rather than an academic, sense; with an emphasis on education as an exercise 'to fashion the character, to cultivate and regulate the affections, to instil right principles', more than adding to 'the store of knowledge' for its own sake.[6] But historical men and women featuring in such productions were credited with characters and attributes appropriate more to their gender according to mid-Victorian comprehensions than to either their original period or their social status. Of course, contemporary interpretations of popular history from any period incline towards the identification of 'heroes' as a way into the world of the past. Because of this, the personal heroic characteristics highlighted are generally revealing of the priorities and prejudices of the period in which the history is written, though this is not always a conscious process. But the biographical style of history that was widely disseminated by the middle classes during the Golden Age was very self-consciously didactic. The period saw a very deliberate and self-conscious effort to utilize the past to speak to the present and so to shape the future of the nation. Through the use of biographical narratives it was intended that invocation of Britain's past would become a central element in the exercise to spread a sense of sharing in a golden era as widely as possible; and helping, thereby, to sustain it. Comparison and contrast would, the authors of such works often stated, incontrovertibly demonstrate the origins of the current happy state and the nature of any threats to it.[7] Gilded identifications of this period are not simply creations of twentieth-century observers. They echo a contemporary consciousness; one that paid less attention to the economics of the period and more to a generic sense of national confidence.[8]

---

[4] A.J. Morris, *Glimpses of Great Men: or, Biographic Thoughts of Moral Manhood* (Ward and Co., London, 1853), p. 1.

[5] Religious Tract Society Archives, SOAS, Meeting of the Religious Tract Society Publishing Committee, 10 August 1850.

[6] Connop Thirlwall, *The Advantages of Literary and Scientific Institutions for all Classes. A Lecture, delivered at the Town Hall, Camarthen, on December 11 1849* (London, 1850), p. 7.

[7] Joseph Johnson, *Living to Purpose; or, Making the Best of Life* (Nelson, London, 1868), p. 30.

[8] Geoffrey Best, *Mid-Victorian Britain 1851–75* (Penguin, London, 1979), pp. 19–23. For contemporary comment see, for example, William Bathgate, *Essays on the Characteristics of a Superior Popular Literature* (Ward and Co., London, 1854), pp. viii–ix.

That confidence was, to a considerable extent, rooted in a popular belief in the superiority of the British (practically understood as the English) over other peoples. Many had been taught to recite as children:

> I thank the goodness and the grace
> Which on my birth have smiled
> And made me, in these Christian days,
> A happy, English child.[9]

Such expressions of national pride were based on interpretations of Britain's (or, again, predominantly England's) history, showing how the contemporary pinnacle had been built, as part of an apparently inexorably onward and upward progress, upon the achievements of the past.[10] Yet, closer examination reveals that this confidence in a clear historical progression was accompanied by a belief that lasting success was not an automatic 'given'. It was earned success, merited by past virtues displayed by numbers of Englishmen (and women).[11] Its continuance required endeavour from ever greater numbers of the population in the context of continuing threats to national prosperity posed by temptations to individual delinquency of duty. Through reformulating history as biography, history could be made to seem a very personal affair, relevant and useful at an individual and practical level. After all, 'Example is more forceful in its teaching than precept; precept appeals to faith, example to sight ... the dullest intellect can see the force of example'.[12] Thus, real examples would remind readers that 'National progress is the sum of individual industry, energy and uprightness, as national decay is of individual idleness, selfishness and vice', to quote Smiles.[13]

By the mid-century, those considered as being most in need of such education as biography could provide were, for a variety of reasons, also believed to be largely out of the reach of formal schooling, while being possessed also of dangerous amounts of leisure time. In other words, prime targets were men and boys of the working classes, precisely those categories who had so recently been implicated in the Chartist agitation. It was thus necessary to focus efforts on texts which could be construed as both appropriate and appealing for this audience in their leisure hours. It was 'on the

---

[9] Jane and Anne Taylor, *Hymns for Infant Minds* (London, 1824).
[10] See, for instance, T. Milner, *History of England*, (London, 1855) p. 1; Valerie Chancellor, *History for their Masters. Opinion in the English History Textbook: 1800–1914* (Adams and White, Bath, 1970).
[11] Samuel Smiles, *Self-Help. With Illustrations of Conduct and Perseverance* (John Murray, London, 1859), p. 2.
[12] Johnson, *Living to Purpose*, pp. 50–51.
[13] Smiles, *Self-Help*, p. 2. For further comments on individual action, see also Joseph Reed, *English Biography in the Early Nineteenth century, 1801–1838* (Yale University Press, New Haven, 1966), pp. 14–15.

way in which they are employed [that] the young man's character and prospects mainly depend'.[14] Accepting this, the mid-Victorian publishing industry believed it had an interest and a duty to promote appropriate reading (including biography) for its various categories of popular readership, since the anticipated profits would be more than financial. Then as now, of course, biography did not need to be a strictly historical exercise. But the hagiographic biography produced for this readership tended to concentrate more on past figures than contemporary ones. It was easier to present the dead, even if only recently departed, in ways that both disarmed criticism (*de mortuis nil nisi bonum*)[15] and substantiated the didactic messages that were at the heart of the genre. The only real exceptions were those heroes, such as David Livingstone, who were, for most of the time, separated from a potential audience for their heroics by geographical distance rather than the distance of history, and were therefore equally susceptible to the processes of hagiography.[16] A hagiographic approach moulded events and characters into useful exemplary lessons, with inconvenient facts glossed over or omitted, in order to emphasize the moral dimension and to identify the trends that the producers desired should guide the future. Thus, despite the amusingly quaint incidents and customs often incorporated to underline the desirability of the present, the British past was depicted as no foreign country, but as territory familiar to nineteenth-century males, especially in the depiction of national character and virtues. The past was presented in this discourse as the key to both their individual masculinity and to their claims on a share of national identity.

Identifying the models suitable for both this Golden Age and the stereotypes producers had of the working-class male was central to the exercise, since such exemplars had to make the characteristics of respectable masculinity appear both concrete and achievable. Biography was believed to teach by simultaneously informing and delighting, a process presumed to be especially appealing to masculine readers.[17] After all, 'Moral truths and qualities are best discerned and best appreciated when they are embodied in real forms. And, to men, the needed forms are human ... the only medium through which instruction and impression can come.'[18] The upheavals of the previous decades seemed to indicate that writing about a more traditional canon of 'great men', nobles, statesmen and military heroes would be unlikely to inspire an emulative spirit in the increasingly urbanized and industrialized working-class male.

---

[14] Ibid., p. 14.
[15] Roughly, 'of the dead, speak nothing but good'.
[16] During this period, Livingstone was only in Britain for a period of months in 1859, a factor which undoubtedly helped to establish his heroic status.
[17] Reed, *Biography*, p. 25.
[18] Morris, *Glimpses*, p. 1.

Evangelical experience seemed to suggest that the Bible provided the inspiration for an alternative style of popular hero:

> The idea [of biography] is, in fact, a scriptural one. What mean those four biographies, or, including the Book of Acts, those five which form the first half of the New Testament? ... That the law or ideal of human life, in order to tell, must be translated from the dead language of precept into the native dialect of example.[19]

The humble background of Jesus, the ultimate exemplar, could be emphasized; and his disciples shown as working men, labouring with their hands. Parables, and the Bible's stress on 'real' characters and deeds, demonstrated ways of teaching the most profound social as well as moral lessons: 'There is no fear that a man thus instructed will ever become a leveller, or a wild revolutionist. Such a man will be the friend of order'.[20] Using the Bible as both model and justification, Victorian biographical texts could be employed as latter-day parables to disseminate details of acceptable beliefs and behaviour by commenting and advising on ideal actions in a range of situations and dilemmas. Biographical discourse was thus intended to remind individuals and groups of individuals that it was their own responsibility to preserve their own respectability, and that this was crucial not just to their own immediate best interests and hopes of Heaven, but also to the continuation of the national golden aura. The respectable publishing industry of the period (including authors and reviewers) took it upon itself to establish such a canon.

Acceptance of a widespread ability amongst working-class men, urban and rural, to read at least simple texts was fundamental to Golden Age respectable publishing culture, with its agenda focused on improving literature for a mass audience.[21] However, there was the problem that (as the Chartist experience had so recently demonstrated) these literates were not confining their reading activities to the Bible and other worthy texts which would provide them with suitable hints on self-improvement or the rewards of conformity to the established social order.[22] What affected publishing policy here was the 'crucial nineteenth-century social and educational assumption' that, when reading, a process of identification took place between the reader and characters in the text who seemed 'sympathetic' in terms of sharing background, experience or

---

[19] *Biography of Self-Taught Men* (Nelson, London, 1869) pp. v–vi.

[20] J. Parker, 'Literature of the Working Classes' in Viscount Ingestre (ed.), *Meliora; or, Better Times to Come* (J. Parker, London, 1852), p. 181.

[21] See David Mitch, *The Rise of Popular Literacy in Victorian England. The Influence of Private Choice and Public Policy* (University of Pennsylvania Press, Philadelphia, 1992) in particular ch. 3; also comments made by Cassells in 1861. Preface, *Quiver*, 1 (1861), n.p.

[22] The independence of working-class readers in their choice of reading matter is discussed by Jonathan Rose, 'Rereading the English Common Reader: A Preface to a History of Audiences', *Journal of the History of Ideas*, 53 (1992), p. 56.

personality traits, including ambitions or hopes for the future.[23] However, it was supposed to be an emotional rather than a logical process, leaving impressionable readers, such as working-class males, vulnerable to misdirection through consuming 'diseased' reading matter. This was worrying because of the association between masculinity and the use of the mind, rather than the emotions. The antidote lay in the provision of the well-told 'story' with 'real' characters that would promote manly ideals by stimulating reasoning faculties – biography, in other words.[24]

It was not an age where commentators looked first to government, or government-inspired policy, for remedies to identified cultural problems at least, even if government was beginning to be looked to, to provide remedies in other more tangible areas. But in cultural terms individuals, and groups of individuals, sought to develop strategies and solutions appropriate to particular issues of concern to them.[25] It should not be forgotten (it certainly was not at the time) that George Reynolds, the key figure in the success of *Reynolds' Miscellany*, had been an avowed Chartist during the 1840s, using his *Miscellany* 'to expound his Chartist views'. Indeed, a distinctly Radical tone continued to flavour Reynolds's writing even after the collapse of the Chartist movement.[26] Given the high circulation figures for Reynolds's productions, it is easy to see why men such as John Parker were concerned:

> *The Mysteries of London*, and *The Mysteries of the Courts of London* ... from the pen and process of G.W.M. Reynolds are extensively circulated ... [and] are calculated to debauch and demoralise the young, the credulous, and the undisciplined. None but a man of depraved mind could thus pander to the worst passions and instincts of young men ... filling their imaginations with impure imagery, and deluding their minds with false social and political economy.[27]

It was in this context that the respectable publishing industry believed it had a significant contribution to make.[28] This is not to say that publishers formally met together and solemnly decided on joint policy. Rather that, sharing the same broad codes of masculine identity, they reacted in similar ways to the

---

[23] Kate Flint, *The Woman Reader 1837–1914* (Clarendon, Oxford, 1993) p. 36.
[24] L.E. Bather, *Footprints on the Sands of Time* (J.H. Parker, 1860) p. vi.
[25] John Feather, *A History of British Publishing* (Croom Helm, London, 1988) p. 144, comments on the extent to which 'The Victorian publishing house existed very much in the image of its owner'.
[26] Louis James, *Fiction for the Working Man 1830–1850. A study of the Literature produced for the working classes in early Victorian Urban England* (Oxford University Press, London, 1963), p. 41.
[27] Parker, *Literature*, p. 186.
[28] See, for example, Simon Nowell-Smith, *The House of Cassell 1848–1958* (Cassell, London, 1958) p. 22.

concerns of the day, and pursued similar strategies, based on similar perspectives on the nature of the writing needed to bring about 'improvement' in the British male character as a whole.[29]

The expansion of respectable publishing, accompanied by a decline in the working-class and/or Radical elements of the publishing industry that had flourished earlier, was one feature of the period. Publishing was another industry which established itself as increasingly masculine, respectable and middle class. With few exceptions, such as Reynolds, the majority of working-class printers and publishers with their wider sympathies found that their products were becoming increasingly marginalized.[30] The practical factors involved included advances in the technology of publishing.[31] Few of the working-class publishers active earlier in the century had the capital, or access to the capital, that would permit them to make the initial financial investment necessary to acquire this new technology. Nor did they have the contacts and capital to support distribution of books and periodicals to more than an essentially local market. The small printing and publishing firms which did remain in business thus had to concentrate their efforts on the essentially small and local tasks, including production of tracts, pamphlets, advertisements and books with a more limited appeal, either in term of price or content. By contrast, the respectable publishing industry was increasingly national in terms of the market distribution of its products. Firms like Cassells were able to utilize the networks linked to religion, mirroring the techniques employed by religious publishing houses like the RTS and the Society for the Propagation of Christian Knowledge (SPCK) for the dissemination of their products.[32] And even the efficiency of these established networks was improved by development of the railway network. This all had a significant impact on purchasing patterns and thus publishing profitability. To summarize, developments in technology made it possible to produce larger print-runs of higher quality which were cheaper than previous products. It even became possible to provide good illustrations without sending costs soaring. To quote one commentator, as a result 'cheap and wholesome literature found its way into the abodes where food for the mind had never before entered'.[33]

This expansion was also associated with a perceived improvement in the nation's purchasing power. The commercial aspect was a critical factor in the

---

[29] Charles Knight, *Passages of a Working Life during Half a Century*, 2 vols (London, 1864) vol. 2, pp. 329–30, for example.
[30] Though G.W.R. Reynolds was, in fact, of middle-class origins. However, he became firmly identified with both Chartism and with working-class interests. See James, *Fiction*, pp. 40–42.
[31] Feather, *British Publishing*, pp. 133–5.
[32] For some comment on tracts, see James, *Fiction*, pp. 114–19.
[33] Pamphilius, *Patient Boys; and How by Patience they became Great Men* (Ward, Lock and Taylor, London, 1865) p. 59.

policy of any publishing house. The publishing committee of the RTS, for instance, considered in 1848 the publication of 'a cheap weekly periodical to counteract the pernicious works, now largely circulated', but decided that the 'current depressed condition of the working classes' and consequent 'depressed state of the Book Trade' meant that the project would be too great a financial risk. However, the same committee, three years later, decided that matters had improved to such an extent that the project was now viable. The book trade had picked up, and the need was even more urgent to explain to the poor, using 'clear and full information' through publications of a 'useful tendency', what were 'the real causes' of their recent distress so they could ensure the perpetuation of present prosperity through the 'counteraction of popular errors'.[34] This leads to another factor involved in the expansion of respectable publishing: consciousness of the need, without jeopardizing moral content, to make what was between the covers as attractive to consumers as the frequently gilt-edged externals, but also as attractive as the products of its competitors, the working-class press, in terms of price and text.[35] In the early 1850s, the RTS was not alone in its confidence that the economic climate was now set fair for an expansion of 'useful' publishing, adapted to the 'circumstances and wants of the working classes' and to other 'needy' sectors of the population. Commercial publishing concerns shared both the economic confidence of the RTS and its belief that there was a crying need for a 'better type' of publication combining the apparently conflicting qualities of being 'wholesome', an enjoyable popular read, and affordable. As the author of an essay on literature for the working classes trumpeted: 'At the present time, books are plentiful, and books are cheap. Formerly books were written for the privileged few; now they are printed for the million. Books of every description, and at almost any price, are to be met with ... "Everything for a penny"'.[36]

Commentators such as Burns have pointed to an apparent 'avidity' on the part of the reading public – one by no means confined to the comfortably-off – for cheering and inspiring aphorisms and for tales of more moderate success and happiness for those of deserving character and hard-working tendencies from humble backgrounds.[37] It cannot be claimed that it was one type, or class, of male reader that bought *Reynold's Miscellany* and another that purchased improving literature in its varying forms. Large numbers of working-class men consumed both, and with apparent enjoyment.[38] This gives rise to a range of

---

[34] Religious Tract Society Archives, SOAS, Meeting of the Religious Tract Society Publishing Committee, 23 February 1848; 16 August 1848; 14 March 1851.
[35] See, for instance, Knight, *Passages*, vol. II, p. 330.
[36] Parker, *Literature*, p. 182.
[37] W.L. Burns, *The Age of Equipoise A Study of Mid–Victorian generation* (George Allen and Unwin, London 1964), p. 294.
[38] Rose, *Rereading*, pp. 48, 51.

essentially unquantifiable questions, such as whether both forms were comprehended by their readership merely as forms of escapism, or whether (as many contemporaries believed and feared), literature had to be taken more seriously, as something used by an unsophisticated readership to form its values and guide its behaviour. It can only be concluded that many working-class males read works that might now be presumed by historians to have opened their eyes to a wider sphere of action, but that they also continued to consume the works designed for them, and indeed to collude in their production through their support in purchasing and/or reading them.[39]

One mid-century belief was that the Chartism which had marred the previous age had been able to catch the imagination of so many impressionable members of the working classes because their imaginations had, in too many cases, been vitiated by their reading of 'unwholesome' literature.[40] It was claimed of this genre that 'The poor ... buy them by the millions and this introduces mischievous ideas into their minds'.[41] Chartism was believed to have capitalized on the discontent created by unwholesome writing through its literature. In particular, they were believed to have won hearts by elevating to heroic status their leaders, real and fictional, while maintaining their popular accessibility through the medium of this discourse.[42] Such men had served as models through which an inchoate dissatisfaction with the status quo was encouraged to develop into active protest against the wise ordinations of Providence as expressed in the class and political system of the 1830s and 1840s. Thus, the example of Chartism helped to create a belief that this would be countered most effectively by providing an alternative range of figures who reinforced the virtues of the present system in any recitation of their lives. So now popular biography would not just focus on 'great' men in the traditional sense of the word. It would be used to highlight individuals whose lives could be considered more directly relevant to the concerns of a working-class audience. This meant concentration upon figures who achieved 'greatness' through hard labour and perseverance in the face of difficulties, often over a sustained period of time, instead of upon the lives of those whose paths had been eased by birth, fortune and favour. This new style of hero was likely to be at least partly self-taught, and from a relatively, if not actually, humble background. It was the concern of the respectable publishing industry for the 'sensitive unformed minds' of the intended readership which saw a return to

---

[39] Ibid., p. 55.
[40] See, for instance, J.A. St John, *The Education of the People* (London, 1858), ch. 12 in particular.
[41] Ibid., pp. 140–41.
[42] Martha Vicinus, *The Industrial Muse. A Study of Nineteenth-century British Working Class Literature* (Croom Helm, London, 1974) especially p. 114. Vicinus points up the tradition in Chartist fiction of the honest working-class man as hero, ideally making a linkage with political activity.

what Joseph Reed terms 'an old biographical bugaboo, the exemplary principle'.[43]

The impact such life-stories were intended to have upon their readership was emphasized (and the paucity of exciting detail in the storylines obscured) by featuring these exemplars in collections. The intention of gathering together so many lives was to make it plain that it was not an occasional accident that such men should be rewarded by honours and financial security:

> What man has already done surely man may do. Nothing, it has been said, is impossible to genius and perseverance ... what is called genius, and does the work of genius, is generally the effect of culture and of self-culture. Those who have accomplished most did not at first show signs of uncommon talent; but they manfully set to work upon themselves and their circumstances; and they persevered ... Let those bent on self-improvement bear in mind that there was never yet great excellence without great effort ... that *nothing is impossible to perseverance*.[44]

It was recasting concepts of 'genius' and 'greatness' in forms suitable for incorporation into working-class masculine identity. In seeking to develop a relevant heroic canon for an industrial age, the traditional hero had to develop a more practical perspective as well as being possessed in some form of more traditional and moral 'great' qualities, such as courage and fortitude. As the comments of the publishing committee of the RTS in 1851 reveal, it was the combination which was felt to give the new icons their impact.[45] The general rule was that models for emulation had to be accessible, something achieved by perceived relevance to the contemporary age. Readers were believed to relate very directly to such exemplars in terms of their own experience and expectations, and, if the process of incorporating their qualities into popular models of desirable masculinity was to be successful, they also had to feed a reader's sense of self-importance. Thus, it could not be just a focus on hard work and constant struggle instead of easy triumph. To promote ready identification by males whose own lives were anyway ones of endurance and struggle, there was also an emphasis on figures whose occupations were directly linked to the present occupations that had brought about national prosperity. The result was a canon of heroes for an industrial age, including Crompton, Arkwright and, of course, the Stephensons. Such men were not just the most accessible, but also the most British of heroes.

The particular reasoning behind such exercises in biography, and the factors leading to the singling-out of a remarkably uniform line-up of heroes, was

---

[43] Reed, *Biography*, p. 26.
[44] *Self-Taught Men*, pp. vii–viii.
[45] Religious Tract Society Archives, SOAS, Meeting of the Religious Tract Society Publishing Committee, 14 March 1851.

clearly laid out. These texts set out, very explicitly, to demonstrate how such individuals conformed to the 'best' in age, class and gender stereotypes in their lives. In this way even the humblest contemporary man living a life of shining virtue could be constructed as 'great' or 'heroic'.[46] Particularly in view of recent discontents, it was also important to demonstrate that, as Joseph Johnson put it, 'Grumbling is caused by thoughtlessness'.[47] In his prolific collections, Johnson was always at pains to indicate that 'Poverty is not always a curse; it is ... ever an incentive, and frequently a blessing ... Without the pinching of poverty, we might never have heard the names of Samuel Drew, of Morrison, of Dr. Samuel Lee, and of Dr. Kitto'.[48] Ultimately, 'right' behaviour in all situations, and under all provocations, difficulties and temptations, from childhood on, was the only universal guarantee of worldly success, even when it only came late in life: one major reason why popular biography usually included comment on the youth of their subjects. In Famous Boys: And How They Became Great Men. Dedicated to Youths and Young Men, as a Stimulus to Earnest Living, the preface insisted that:

> Biography has no truer lesson to reach than this, that as sure as any object is pursued with diligence, with industry, with unfaltering perseverance ... the end desired is certain to be attained. There is no law so sure, there is no end so certain, as that industry meets with its just reward.[49]

Of course, despite such exhortations, and the huge success of biographically-rooted self-help tomes, there was the complication of a middle-class consciousness that not all humble adherents to right behaviour and earnest living actually would gain their just deserts (and, almost certainly, no great desire that it should be anything other than an out-of-the-ordinary occurrence). As *The Times*, for instance, commented in 1859, the odds on any major success were far from high.[50] But working-class boys and men continued to be the target for works that stressed that 'Heaven helps those who help themselves', because the self-interest and patriotism of the producers of these tomes saw it as essential to gain the cooperation of these males if masculine respectability was to be effectively extended.[51] The historian may well doubt the success of such strategies, but many contemporaries genuinely believed it possible to

---

[46] This is also something that echoes the tradition of 'death-bed' memoirs espoused by the evangelical tradition from the late eighteenth century. See, for instance, Reed, *Biography*, pp. 29–30.
[47] Johnson, *Living to Purpose*, p. 138.
[48] Ibid., pp. 86–87.
[49] *Fa.mous Boys; and How They Became Great Men. Dedicated to Youths and Young Men, as a Stimulus to Earnest Living* (Darton and Co., London, 1860), p. vi.
[50] *The Times*, 12 August 1859.
[51] This aphorism is the opening line of Smiles, *Self-Help*, p. 1.

promote social harmony through such texts. These books were bought, accepted and, when read, probably read by their working-class audiences as much to see what was expected of them as to identify moral models. At least such biographies demonstrated the rules which were accepted as qualifications for joining respectable society. After all, acceptance as a member of the respectable male brotherhood was, in one sense, a small triumph of social mobility. Where this didactic exercise was 'successful', it was almost certainly as part of a process of socialization rather than simple emulation, but its patriotic dimension means that it cannot simply be dismissed by historians considering the development of patterns of British masculinity across the social hierarchy.

Alpha Beta, the author of a collection entitled *How Young Men May Become Great Men*, used his preface to direct the thoughts of his readers to 'the country which, with pride, we call our own'. He stated, unequivocally, 'Great is her glory', because:

> *Our commercial prosperity is wonderful* ... such is our energy, that we go everywhere to promote the interests of trade ... We have ... the palm of commercial superiority ... *in science we have effected much* ... *Then again, our political privileges are great*. Our property is safe, our liberty is sure ... the rich cannot oppress the poor. While we may have some political grievances, we are, nevertheless not prepared to exchange our policy for that of any people under heaven.[52]

He then proceeded to ask what was 'the secret' behind this dazzling list of blessings 'which make us the envy of the world'. His conclusion was that it was the result of the efforts of the nation's 'great men' and the 'difficulties' they had used as 'stepping-stones' on their upward path.[53] It all helped to create an illusion, at least, of balance and consensus and it cannot be denied that there were significant numbers of men from a wide range of social backgrounds who were willing to collude in sustaining such a mirage.

This was an age where hegemonic masculinity sought to identify itself externally practical, reasonable and respectable, ignoring the claims of less complacent and conformist masculine identities to serious consideration as part of a cultural shoring-up of Britain's Golden Age. Looking at the complacency and sanctimoniousness expressed in hagiographic biography in particular, however, it is easy now to argue that the will to substitute what historians would see as verisimilitude for veracity or honest assessment of the practical pressures faced by Victorian working men undermines claims that, culturally, it was a Golden Age for a majority of Victorian men. Yet respectable

---

[52] Ibid., pp. 6–7.
[53] Alpha Beta, *How Young Men*, pp. 6–8.

contemporaries seriously sought to build 'brotherhood' through sharing their perspectives on the cultural attributes of masculinity. It must also be accepted that there was no serious attempt to challenge the desirability of the external features, at least, of this respectable masculinity. Instead, men such as Watt, Wedgwood and the Stephensons were widely accepted as genuine heroes. In 1868, the prolific popular biographer Joseph Johnson could extol the positive impact of the discourse when referring to the recent cotton famine in Lancashire, claiming that the conduct of operatives was now 'most admirable and exemplary'. It 'gave excellent evidence that they had in their days of prosperity, made industrious use of their leisure hours' by reading works such as popular biography. Such interpretations of the past 'served them and saved them from excesses which could not have improved their condition, but which must have made it materially worse' by leading them towards a broad acceptance, at least outwardly, of constructs of masculine identity more in line with those considered desirable by the economically and socially successful masculine community.[54]

---

[54] Johnson, *Living to Purpose*, pp. 90–91.

Index

Abbott, B. 93
Aberdeen 156
accidents 16, 39–42
Ackersley, J. 42
Adams, W. 12, 14
Admiralty 95–103
agency and innovation 131–8
agrarian capitalism 189–90
agricultural employment and wages 51–2, 57–60, 225–36
agriculture 46–60
Airy, G.B. 97
Albert Exhibition Club (Manchester) 158–9
Amalgamated Society of Engineers xiii
Amalgamated Weavers' Association 176
American Civil War xvi, 63, 82
Anglican ideal 195
Anglo-American Telegraph Company 82
Anti-Corn Law League 148, 151
anti-industrialism 6; *see also* industrialism
Apreece, J. 93
apprenticeship 93–4, 139, 179
Arch, J. 51
Armstrong, W. 155
Arrow War (1856) xv
Artisan 90
artizanal culture 4
artizans and tradesmen 4, 117, 129–39, 149–63
Artisans' Lecturing Society (Liverpool) 170
assize circuits 238
Atlantic cable (1858) 78–9
Atlantic Telegraph Company 77–83, 85, 115

Babbage, C. xiv, 90
Bagehot, W. 4–5, 16, 171
Baker, J.P. 38
Baker, V. Coln. 245–6

Ballot Act (1872) xviii
Baltic Wharf 160
Barnard, S. 93
Barnett, C. 108
Belgian patent system 137
Berman, M. 101
Bessemer, H. xiv, 137
Best, G. 143–4
biblical reference 267
Biggs, W. 150–51
biography 262–75
Birmingham 124–7
Birmingham and Midlands Institute 4
Black Book (1820–47) 24
Bodkin, A. 208
Bolton, 146, 154–7, 159
bondagers 52
Boole, G. xiv, xvii
Bowker, B 178
Bradford 158
Brande, W.T. 95
Bridges, A. 233
Brierley, S. 255, 257–8
Briggs, T. 147
Bristol Association of the Great Exhibition 158
British-Indian Submarine Telegraph Co. 82
British manufacturing 112
British technology system 121–39
Broadhurst, H. 13–14
Brougham, H. 3
Brunel, I.K. xv
bubble companies 204
Budd, W. xvii
Burgess, J. 254–61
Burns, W.L. 143, 270
Burnley Weavers' Association 176
Bury, Curtis and Kennedy Co. 132
business class 25–6
business crime 205–13

Butler, J. 20

cable and telegraph companies 80–85
Cable and Wireless Co. 84
Caird, J. 190, 226
Cambridge 90
Cannadine, D. 22–6
capital and investment 70–71, 77, 80–85, 99, 189, 203
capital exports 5
Cardwell, D. 106, 109–10, 113, 118
Carnot, S. 95
Cassells Co. 269
casual labour 50
Census of 1851 164
Central Agency of the Great Exhibition 159
Central Committee of Social Propaganda 149
Chadwick, E. 208
Charity Organization Society 14
Chartism 21, 59, 62, 79, 147–9, 220–21, 265–8, 271
Chamberlain, J. 4
chemical process technologies 128
Chester, G. Revd 194
children 12, 16–20
Children's Employment Commission 225
cholera 16
Church, R.A. 29
Church of England 184–98
City of Glasgow Bank 206, 210
City Philosophical Society 93
civility and civil culture 3, 21–6, 165, 180–81
Clapham, J. 108
Clark, A. 253
class conflict 9, 21–6
class unity 149
Clerkenwell Magistrates Court 242
clothiers in agriculture 54
Clyde valley 68
coal famine 35, 41
coalmining 30, 32–45
 and the navy 95
coal output statistics 33
coal prices 35
Coats, J., and P. and Co. 67
Cole, H. 113

Cole, J.W. 210
Coleridge, S.T. 5
Collier, Messrs W. and Co. 132
colliery owners 42–5
collodian 137
compound engine 69
Conley, C. 240
constitutionalism 147, 171
consumerism 147
Contagious Diseases Acts (1663–71) 20
Conversations on Chemistry 93
Cook, T. 146, 151
cooperative movement 25
copper sheeting 96
core–periphery relations 56
Corliss valve 69
corn and wheat acreage 50
Cornhill Magazine 239
Corn Laws, repeal (1846) 47–8
corporate patenting 132–3, 137
corrosion and electro-chemistry 96
corruption 24
cottage industries 253–4
cotton exports 72–3
Cotton Famine xvi, 14, 29, 61, 70–74, 176, 275
cotton firms 67–8
Cotton Supply Association (1857) 70, 72
cotton textile industry 61–74, 145, 175–83
cotton weaving 145, 176–9, 253–61
Cowper-Temple Clause (1870) xviii
crime and criminals 17–20, 153–5, 194, 208–13; *see also* white-collar crime
Crimean War xiv, 79, 98, 181, 242
cross-skilling 131–3, 137
Crotchet Castle 167
Cruikshank, G. 17
Crystal Palace xiii, 3
Culliford, R. 137
cultural constructions 220

dairy farms 55
day-labourers 191
Dance, W. 93
Danish Great Northern Telegraph Company 81
Dare, J. 151
Darwin, C. xv
Darwinism xv, 14

Davidoff, L. 238
Davy, H. 93–7
Day, P.E. 149
day-employment 232–6
death rates in mines 40–42
declinism 1–8, 90–91, 105–21, 223–4
deep-sea cables 76–7
Dent, J. 187
Department of Science and Art 110
dialect poetry 250–61
Dickens, C. 12, 16, 19, 139
Dickenson, K. 245–6, 248
diseases 15–17, 41–2
Disraeli, B. xvi–vii, 3
Dissenters xiii, 6
dividends and financing 80–83
Dock Warrant frauds 210
doctrine of recent complaint 244, 247
domesticity and domestic sphere 222, 235–6, 237, 249–61
domestic service 228–36
Dore, G. 12, 15
double standard 239
Driffield 228–9
drink and disorder 17–20, 150, 153–4, 194, 258–61
Dumas, J.B. 109

Eastern markets 64, 72–3
Eastern Telegraph Company 84
East India Company 96
East Riding of Yorkshire 184–98, 226–36
Eastwood, D. 2
Ecclesiastical Titles Act (1871) xviii
Ecole Centrale des Arts 109
Economist 83–4
Edgerton, D. 108, 139
Edinburgh 118
education (general) 150
electrical engineering industry 29, 76–85
electrical standards 78
electric generator xvi
electricity 75–86
electrification of lighthouse illumination 101–3
electro-magnetic induction 97
Elementary Education Act (1870) xviii
Elliot, G. 171
Elliot, Sir G. 30

elite patenting 136
Emancipation Proclamation (American, 1862) xvi
employers and workmen 111, 131–3
employment statistics 35, 57, 65
   Lancashire cotton industry 66
enclosure 189–92
Encyclopaedia Britannica 93
Engels, F. 15, 143
engineers and engineering 105, 129–37
English Republic 12
enterprise culture 139
entrepreneurial ideal 23–6, 143
European revolutions 153, 221
evangelicalism 187–8, 193
Evans, D.M. 201–5
Evans, J. 114
exclusion 171
excursion train fares 159
Exeter Hall 3
Exhibition Herald 146, 162
expertise and advisors 92, 96–104
Exposition of 1851 90

factories 254–61
Factory Act (1847) xii
Factory Act Extension Act (1867) xvii
Facts, Failures and Frauds 205
Fairlie's locomotives
family farms 53–8, 234–6
Faraday, M. 89, 92–104
farmworkers and servants 47–53, 59, 144, 184–98, 227–36
Farnie, D. 69, 179
Faucher, L. 15
female farm servants 186–7
female labour 65–8, 70, 179–83, 186–7, 225–36, 253–61
female membership of voluntary association 164–74
female scientific education 171
financial crime 199–215
Fitton, S. 255, 257
Flitcham Home Farm 233–4
Foden, F. 110
foreign competition 116–20
foreigners 153–60
foreign patentees 133–8
foreign trade 5, 7, 48, 62–65, 71–4

INDEX

Foreland lighthouse xv, 100, 102–3
Forester, F. 251
Fowler, J. 113
Fox, W.J. 3
France 113–14, 137
Frankland, E. 116
Fraser, J. 230
Fraternal Democrats 153
French Atlantic cable (1869) 82, 84
Fresnel lenses 100
friendly societies 25
Friend of the People 149–50

Galloway tubes 70
Galton Committee (submarine telegraphy) 78–86
gang (labour) system 230–32
Garfield's assassination 85
gender (general) 219–75
gender relations 164, 166–72, 251–61
gender segregation 186–7
gender stereotypes 237–8, 240
genius 272–3
gentlemen 213–15; patentees 131
Gilbert, D. 97
gilded age 59, 264
Giffen, R. 10
Gladstone, W.E. 31
Glass, Elliot and Co. 81
Glass, R. 82
global communication 75–86
Golden Age of Agriculture 46–60, 225–36
Gooch, D. 82, 84
good behaviour 154
government policy 6, 30, 79–80, 86, 92, 95–104
Graham, P. 114
Gray, T. 9
Great Eastern 82
Great Exhibition xiii, 3, 6, 9, 47, 89–91, 144, 146–174
gutta percha 76
Gutta Percha Company 81

Halifax 54, 137
Hamilton, K. 19
Hancock, W. 158
hand-weaving 67, 69, 175
Harney, J. 148, 153

Harris, J. 221
Harrison, T. 160
Headrick, D. 115
Heath, R. 225
Henley, J.E. 225
Henley, W.T. 81
Hensman, H. 91
Herapath, S. 160
Herculaneum Dock 71
heroic literature 264–72
Heward, C. 110
Hicks, B. 156–7
hierarchy 22–6
Higginson, A. 43
Higgs, E. 229
high art crime 145, 199–215
high farming 46, 189
hind house system 192
hiring and hiring fairs 184–98, 226–8
history and its uses 265–8
Holborn 134
Holden, R. 176
Hole, J. 22
Hollingworth, B. 250–52
Holmes, D. 176
Holmes, F.H. 102
Home Counties 166, 168
 patenting in 123–5
Hornsby, W. 54
housework 255–61
Howell, G. 25
Howkins, A. 227
Huddersfield 171
Hudson, G. 19
Hull Advertiser 228
Hume, J. 3
Huth, E. 116
Huxley, T.H. 4
Hyde Park demonstrations (1866) 21

ideal woman 237
Illustrated London News 81, 85
import duties 73
inclusion 166
India and Britain 62, 70–73, 79, 82
Indian Mutiny xv, 73, 79
indiarubber 76
individualism 263–4
industrialism 5–8, 111–16, 144, 147–9

industrial education 109
industrial employment 7
industrial knowledge 138
industrial relations 42–5
industrial revolution 1, 10, 89
industry (general) 29–86
industry–agriculture relations 49–60, 179, 191–4, 226–36
inflation 10
information diffusion 121, 127–30
Institute of Mechanical Engineers 30
Institution of Civil Engineers 113
intellectual public sphere 164–74
International Convention for the Protection of Industrial Property (1883) 123
internationalism 148
inventors and artizans 90–91
inventors' clubs 149
Ireson, J. 13
Irish Church Bill (1869) xviii

Jackson and Graham and Co. 114
James, H. 15
Japan 138–9
Jenkin, F. 78
Jevons, W.S. 2, 32, 43
Johnson, J. 273–5
Johnston, E. 251, 258–9
Journal of the Society of Arts 6, 108, 112, 115
Joyce, P. 250

Kennedy, J. 132
Kent trials 240
Kentish London 166
Kirkman 156
Knight, M. xviii
Krupps steelworks 119

laboratory work 96
labour market 193–5, 227–36
labour productivity 33, 69
labour recruitment and supply 36, 70, 227–36
labour relations 62, 111
Lambeth Police Court 243
Lancashire 54–5, 62, 65–71, 175–83, 249–61
Lancashire boiler 69

Lancashire dialect poetry 176, 249–63
Land Act (1870) xviii
law and civility 22, 237–48
Laycock, S. 176, 250–52, 255–7, 259–61
Legard, G. 231
Leicester 150–51, 162
Leicester Working Men's Association 151
Levi, L. 10
liberal intelligentsia 4, 7, 91, 138–9, 148
Liberty, Equality and Fraternity 239
lighthouses 92, 99–103
lighthouse technology 101–3
Lincoln 241
Lincoln's assassination 85
liquefaction of gases 96
literacy 267
literary and scientific institutions 164–74
little mesters 179
Liverpool 169–70
Liverpool Mechanics' Institution 170
Livingstone, D. 266
Local Government Act (1858) xv
London 3–4, 11, 15, 18–20, 53, 55, 90, 123–6, 131, 133–5, 153–7, 159–63, 166–9
  police courts 238–48
long-wall extraction 34
Lowenthal, D. 249
lung diseases 41
Lyell, C. xv

McConnell, J. 116
MacCormack, C. 137
machinery imports 119
machinofacture 4, 121–39
  defined 138
Mackworth, H.F. 41
MacLeod, R. 110
male–female wage gap 231
male hegemony 251–61
Mallett, R. 113
managerial associations 38–9
Manchester, 4, 15, 55, 82, 124–32, 151–3, 156–60, 167, 244
Manchester Free Library 128–9
Manchester Guardian 62–3, 74
Manchester Working Men's Committee 151–2
manhood suffrage 21

Mannheim, K. 7
Marcet, J. 93
masculinity 220–22, 255–61, 262–75
Mason, M. 239
Master and Servant Amendment Act (1867) 22
masters and servants 185–98, 228–36
Matrimonial Causes Act (1857) 20
Matts, J. 151
Maxwell, J. Clerk 78
Mechanics' Homes (1851) 155, 160–62
mechanics' institutes 25, 128, 148, 164–74, 181
mechanization 65–6
Mediterranean telegraphs 79
membership of intellectual and technical institutions 165–74
Michel, F. 242–3
Middlesex Assizes 242, 246
migration 52–3, 180
Mill, J.S. 2, 16
mine contractors 36
mine management 36–9
miners and agriculture 53
miners' asthma 41–2
Miners' Federation of Great Britain (1889) 45
Mines Regulation Act (1872) 39
mining engineers 38
mining technologies 34
misery 60
Moira collieries 43–4
mollies and rent boys 19
money clubs 157
Monmouth 246
Mont Cenis Railway Tunnel xviii
moral compact and moral nexus 185, 239
moral crisis and outrage 144, 193, 199–201, 213–15
Morley, H. 16
Municipal Reform Act (1835) 6
Mushet, R. xiv

Nadir (balloon) xv
National Bazaar (Covent Garden 1845) 148
national income 10
nationalism 147, 174
National Reform Association 3

natural laws 102
Neilson's steam hammer 137
Nelson (Lancashire) 177–83
Nelson, J. 177–8
Newall, R.S. 81
Newall and Co. 81
Newcastle 51, 53
Newcastle Commission xvi
Newman, J. 94
new model unions 21
newspapers xiv, 85, 152–3, 157, 180, 238–48
New York 135
New York Industrial Congress 149
Norfolk 226–36
North-East Lancashire 175–83
Northern Star 153
North of England Institute of Mining Engineers 37–8
Nottingham 167, 170
nystagmus 42

occupational diseases 41
occupations of patentees 129–38
O'Connor, F. 21, 153
Old Corruption 24
Oldham 253
Ollerenshaw, J.C. 72
optical glass 97
oral traditions 251–61
Outdoor Relief Regulation Order (1851) xiii
overland telegraph system 76
overproduction 74
Owen, R. 149
Oxford Movement 188

Palmerston, H.J.T. xvi
parish system 188
Parisian innovators 133–7
Paris World Exhibition 6, 90, 105–20
Parker, J. 160
Parker, J. 268
Parker, W. 151
Parliamentary elections xiii–xviii, 98–9
Parliamentary Reform Act (1867) 21,
partnerships in innovation 131–7
patent costs 135–6
patentee statistics 122–37

Patent Law Amendment Act (1852) 123
patent material and journals 128
Patent Office Library and Museum
patents and patenting 69, 91, 101, 114, 121–39
    foreigners 133–8
paternalism 197, 226
patrons and patronage 94
Paxton, J. xiii
Peacock, T.L. 167
Pearl, C. 19
Peking, Treaty of xv
Pelly, J.H. 100
Pender, J. 82–4
Penn, M. 55
Perkin, H. xiv
periodization 2–8, 11, 24–6, 45, 58–60, 143–5, 175, 178, 183, 219–24, 262–3
Piddington, J. 137
Playfair, L. 6, 90–91, 105–20
police and policing 18–22, 199–203, 238–48
Police Act, County and Borough xiv
Pollard, S. xix, 139
Poor Law 11–15, 24, 46
    statistics 11–12
Poor Law Amendment At (1847) xiii
popular culture 184–98, 262–75
population xiii, xv, 46–8, 178
Portsmouth Harbour tests 96
poverty 11–17
powerlooms 66, 69, 137
Preston Guardian 180
Price, D. 117
Primitive Methodists 197
printing and dyeing 65, 68–9
propaganda 149–51
prostitution 11, 17–20, 246–8
Protection of Women (law) 241
Prothero, R. 188–9
Public Health Act (1848) xii
Public Health Commission (1864) 20
public houses 157–8, 170–71, 186
public sphere 164–74, 252
public subscriptions 148–63
publishing industry 266–75
Pullinger, W.G. 19

radicalism 148, 249, 269; *see also* socialism
railways xviii, 71, 145, 147, 157–63, 203–4, 211, 213, 245–6, 269
Ramsbottom, J. 250, 252
Ramsden's steam boilers 137
rape and indecent assault 238–48
rates of return on patenting 135–6
rational recreation 150, 186, 249
raw cotton imports 62–5
reading public 262–75
real wages 10
Reed, J. 272
Redpath, L. 19
Red Sea cable 79–81
Reflexions on the Motive Power of Fire 95
reformist politics 162
regional distribution of innovation 121–38
regionalism 150–62, 166–74, 227
registration societies (farm hiring) 185–6
religious beliefs 102–3
religious census xiii
religious tests xiv
Reports of Artisans 114
reputation 241–8
resistance standard 78
respectability 199–215, 221–2, 230, 237–48, 262–3, 266–7
Reuter, J. 83, 85
Revised Education Code (1862) xvi
Reynolds' Miscellany 268, 270
Reynolds, G. 268–9
Riebau, G. 93
Roby, H.J. 111
Rochester, G. 155
Rogers, T. 4
Roget, P.M. 97
room and power 179
Rosehill seminar 171
Ross, F.G. 155
Rosser, H. 150
Rothstein, T. 148
Rothschild, L.N. xv
Rowbotham, S. 219
Royal British Bank 206–7, 212
Royal Commission on Lighthouses (1860) 100
Royal Commission on Trade Unions (1867) 25

Royal Commission on the Depression of Trade and Industry (1886) 45
Royal Institution 92–103
Royal Military Aademy Woolwich 98
Royal School of Mines 113
rural family economy 226–36
rural manufacturing 49–58
rural sub-cultures 198
rural women workers 225–36
Russell, J.S. 116

saccharine 137
safety lamps 34, 42, 95–6
Salford Royal Museum 128
Salford Working People's Association 158
Samuelson (Technical Instruction) Committee 119
Sandemanians 89, 93, 101
Sanderson, M. 106, 110, 118
Sanitary Act (1866) xvii
Saturday Review 20
science and technology 75–86, 92–104
Scientific Committee of the Admiralty (1828) 98
science provision in institutions 169–71
scientific education 105–20
Scott, J. 220
Scottish education 6
secured knowledge 138
self-actor mule 66, 69
Self-Help 263
settlement patterns 191, 193
sewing machines 128, 137
sex and sexuality 20, 194–6
sexual division of labour 179, 231–6
sexual violence 222, 237–48, 258–9
shared technical expertise 133
Sheppard, J.A. 227
skills and trades 89, 91, 117, 150–53
skill-specific patenting 131–3
Siemens, F. xiv
slums 15–17
Smiles, S. 265
Smyth, W. 113
Snow, J. 16
social class 18, 21–6, 219, 262
social harmony 21–2, 24–6, 143, 263, 274
social institutions (general) 143–215

socialism 148–62, 249
Society for the Propagation of Christian Knowledge 269
Soho Ironworks (Bolton) 156
Solvay, E. xv
Sommerville, M. 171
South Derbyshire coalfield 43
Southampton 157
Spear, J. 132
specialization 49–53
speculative mania 203, 211
Spencer, H. 5, 14
spinning and weaving capacity 67
Standing, J. 255–6
Stanley, R. 158
steam intellect 144, 164, 167–71
defined 168
Stephen, F. 239–40
stock of technical knowledge 121, 135–6
Strahan, Paul and Bates (bankers) 208–12
strikes and lockouts 21–2, 44
Stronach, R. 210
Submarine Cables' Trust 84
submarine cable telegraphs xviii, 75–86, 106, 115–16
Sunderland 155
Surdam, D. 64
Sutherland, E. 200–201
Sutton Glassworks 132
Swaffham 230

Taunton (Schools' Inquiry) Commission 108–11
Taylor, G.R. 155
technical associations 164–74
technical change and innovation 4, 33–5, 46–7, 65–71, 75–86, 89–139, 149–63, 189–90, 223, 269–70
  in Home Counties and industrial counties 123–5
  and information system 127–30, 149
technical education 3, 6, 37–9, 116–20, 149–50
technical information culture 127, 164–74
technological backwardness 90–91, 116–20, 138–9
technology (general) 89–139
technology transfer 121, 123, 133–9

telegram prices 84
Telegraph Construction and Maintenance
    Co. Ltd 80–84
telegraphs: *see* electricity
telegraph scheme guarantee 79
temperance 17–19
Ten Hours' Act (1847) 16, 150
textile engineering 71
Thackray, A. 164
Thesiger, F. 212
Thomas, J. 153–4
Thompson, E.P. 175, 178
Thompson, F. 235
Thompson, F.M.L. 48
Thomson, W. (Lord Kelvin) 78
threshing machines 51
Times 238–48, 273
Tosh, J. 220, 262
town patriotism 182
trade cycles 9–11, 73, 203, 226, 270
trade skills 89–91, 114–39
Trades Union Congress (1868) 25
trade unions 21–5, 44, 47–8, 51, 111, 157,
    176, 249
tramping 14, 50–3
Trinity House 92, 99–104
Turner, H.A. 179
Tyndall, J. 92, 111–12, 116–17

underground safety 40–41
unemployment 14
Union blockade 64
United States 62–4, 69–70, 72, 74, 133–8
unwholesome literature 271
urbanism 15–20, 178, 251
    and voluntary associations 167–74
urban ranking of British patents 124–5
urban–rural interactions 59–60

Varley, C. xv
Vaughan, R. 15
ventilation 42, 101
Vernon, T. 132

vertical integration 80–85
Vicinus, M. 271
Vickery, A. 221, 238
victims and victimhood 238, 246–8
Victorianism 17, 24–6, 178, 183, 188, 196,
    213–15, 226, 236, 239
Victoria Regia Lily House xiii
violence 21–6, 237–48, 258–9
voluntary associations 144, 164–74
Vulcan Foundry (Warrington) 132

wages (agricultural) 227–36
Wallace, A.R. xv
Walters, C. 19
Warner, F. 151
Warwickshire 127
Wapping Dock 71
waterproofing 137
Waugh, E. 250–52, 255, 258, 260
Weiner, M.J. 1, 7, 108, 138
Westinghouse, G. 137
West Riding 50, 166
Wheatstone, C. xv, 76
white-collar crime 19, 199–215
Whitworth, J. 110, 132
Wilberforce, R.I. 144, 195–6
Willis, R. 90
women 6, 225–75
    as a social category 263
    see also gender
working-class culture and politics 146–63,
    249–61, 271–5
working conditions 15–17, 39–45, 51–8
Woodcroft, B. 123
workers' travel clubs 156–60
workhouses 13–14, 48

York Crown Court 246
Yorkshire 123–7
Yorkshire Wolds 190, 227
Young, G.M. 144

Zlotnick, S. 251–2, 260